T0214171

Mathematical Modeling of Groundwater Pollution

Springer Science+Business Media, LLC

Springer Science+Business Media, LLC

Ne-Zheng Sun

Mathematical Modeling of Groundwater Pollution

With 104 Illustrations

Translation by Fan Pengfei and Shi Dehong

Originally published by Geological Publishing House,
Beijing, People's Republic of China

 Springer

Ne-Zheng Sun
Civil and Environmental Engineering Department
University of California
Los Angeles, CA 90024
USA

Originally published as 地下水污染－数学模型和数值方法 (*Groundwater Pollution—Mathematical Models and Numerical Methods*). © 1989 Geological Publishing House, Beijing, People's Republic of China. Translators: Fan Pengfei and Shi Dehong. Editor: Zhu Xiling.

Library of Congress Cataloging-in-Publication Data
Sun, Ne-Zheng.
  Mathematical modeling of groundwater pollution / Ne-Zheng Sun.
    p.   cm.
  Includes bibliographical references and index.
  ISBN 978-1-4757-2560-5   ISBN 978-1-4757-2558-2 (eBook)
  DOI 10.1007/978-1-4757-2558-2
  1. Groundwater—Pollution—Mathematical models. I. Title.
TD426.S86   1995
628.1′68′015118—dc20   94-10680

Printed on acid-free paper.

Production managed by Laura Carlson; manufacturing supervised by Jeffrey Taub.
Typeset by Asco Trade Typesetting Ltd., Hong Kong.

9 8 7 6 5 4 3 2 1

ISBN 978-1-4757-2560-5

To the memory of
my father Chi-Peng Sun
and
my mother Xin-Ru Wang

Preface

Groundwater is one of the most important resources in the world. In many areas, water supplies for industrial, domestic, and agricultural uses are dependent on groundwater. As an "open" system, groundwater may exchange mass and energy with its neighboring systems (soil, air, and surface water) through adsorption, ion-exchange, infiltration, evaporation, inflow, outflow, and other exchange forms. Consequently, both the quantity and quality of groundwater may vary with environmental changes and human activities.

Due to population growth, and industrial and agricultural development, more and more groundwater is extracted, especially in arid areas. If the groundwater management problem is not seriously considered, over-extraction may lead to groundwater mining, salt water intrusion, and land subsidence. In fact, the quality of groundwater is gradually deteriorating throughout the world. The problem of groundwater pollution has appeared, not only in developed countries, but also in developing countries. Groundwater pollution is a serious environmental problem that may damage human health, destroy the ecosystem, and cause water shortage.

In the protection and improvement of groundwater quality, two challenging problems have been presented: for uncontaminated aquifers, it is required to assess the potential dangers of pollution; for contaminated aquifers, it is required to draw up remediation projects. In both situations, we need a tool to predict the pollutant distribution in groundwater. Obviously, field experiments cannot serve this purpose. The only tool that we can use is mathematical modeling. In the past two decades, mathematical modeling techniques were extensively used in the study of mass and heat transport in groundwater and soil. Presently, we can simulate a three-dimensional multi-component transport in a multi-phase flow using a computer without any essential difficulty. Since simulation models can provide forecasts of future states of groundwater systems, the optimal protection or rehabilitation strategy may be found by incorporating a simulation model into a management model.

There are, however, several difficult problems in groundwater quality modeling that have not yet been adequately solved. The first one is called the "scale effect problem." The identified dispersivities may vary with the scale of experiment and the size of element of numerical discretization. The second one is called the "numerical dispersion problem." Sharp concentration fronts are difficult to simulate accurately using a numerical method. The third difficulty is caused by the "uncertainty enlargement problem." The uncertainties associated with hydraulic conductivities, porosities and head distributions may be enlarged and propagated to the calculation of velocity fields through Darcy's Law. Incorrect velocity distributions may cause large computational error in the determination of both advection and dispersion components of contaminant transport and fate. The fourth difficulty is caused by the "data insufficient problem." Tracer tests can only be carried out on a small region and it is difficult to observe concentration plumes in three-dimensional space. Generally, we do not have enough data for calibrating the mass transport model of regional problems. These difficulties make the modeling of groundwater quality more challenging than the modeling of groundwater flow. Although the importance of mathematical modeling in the study of groundwater quality problems is significant, the accuracy of a mass transport model and, thus, the reliability of management decisions derived from the model, are often questionable.

This book introduces all primary aspects of groundwater quality modeling. The emphasis, however, is on numerical techniques. Besides introducing basic concepts, theories, methods and applications, special attentions are paid to three-dimensional models, model selection criteria, tracer test design, dispersion parameter identification, and reliability analysis. The overall purpose is to develop an applicable methodology for groundwater quality modeling. This book is designed to provide a course text at the graduate level. The materials are presented in such a manner that the book can also be used as a reference for hydrogeologists, geochemists and environmental engineers.

Chapter 1 is an introduction, in which the problem of modeling groundwater pollution is depicted, and the relationships between simulation, parameter identification, and groundwater quality management are explained.

Advection-Dispersion equations (ADE) that can simulate multi-component transport in multi-phase flow are derived in Chapter 2. Hydrodynamic dispersion coefficients and other parameters in the ADE are defined. Various combinations of sink/source terms, initial conditions, and boundary conditions are listed.

Chapter 3 gives analytical solutions for some one-, two-, and three-dimensional advection-dispersion problems. Several often-used techniques for finding analytical solutions are introduced.

In Chapter 4, conditions of convergency and stability of finite difference methods for solving the ADE are presented. The phenomena of overshoot and numerical dispersion associated with finite difference solutions for problems with large Peclet numbers are then analyzed. The method of characteristics and the random walk method are also discussed.

Chapter 5 introduces the family of finite element methods and their variations. In the solution of three-dimensional problems, the Galerkin finite element method, the mixed finite difference and finite element method, and the multiple cell balance method are discussed and compared.

Chapter 6 is devoted to advection dominated problems. After a Fourier analysis of numerical dispersion, various Eulerian, Lagrangian, and Eulerian-Lagrangian methods are given in detail.

Chapter 7 considers how to select and build a model for a practical groundwater quality problem. Models of simulating tracer tests in different scales are presented for the purpose of parameter identification. A coupled inverse problem of groundwater flow and mass transport is then presented and solved by a conjugate gradient method or a modified Gauss-Newton method. The final section of this Chapter is a short introduction to the statistic theory of mass transport in porous media which is a rapidly developing field. Methods of evaluating the reliability of groundwater quality models are also included.

Chapter 8 discusses the main application fields of groundwater quality models. Prediction of groundwater pollution in saturated and unsaturated zones, simulation of mass transport in fractured aquifers, sea water intrusion, optimal design for aquifer remediation, and groundwater resources management are also discussed.

At the end of the text, there is a short conclusion which lists some open problems in this field. A FORTRAN program is given in Appendix B, which can be used to simulate contaminant transport either in steady-state or transient-flow fields.

The original edition of this book was written in Chinese and published by the Geological Publishing House of China in 1989. It was translated into English in 1991 by Mr. Fan Pengfei and Mr. Shi Dehong at the Hydrogeology Institute, Ministry of Geology and Mineral Resources of China. The English translation was corrected by Ms. Zhu Xiling, who was the director of the international exchange division, the Geological Publishing House of China, and modified by the author for teaching the course "Mathematical Modeling of Contaminant Transport in Groundwater," in the Civil Engineering Department, University of California, Los Angeles. At that time, exercises were added to each chapter. In 1994, the English manuscript was again revised to include some late developments in this field.

The author wishes to express heartfelt thanks to Dr. Susan D. Pelmulder, who read the entire manuscript and helped the author revise it for publication. The manuscript was also read, in part, by Dr. Marshall W. Davert, Dr. William A. Moseley, Dr. Suresh Lingineni, Dr. Ming-Chin Jeng, and Dr. Ming-Jame Horng. They suggested many improvements. The author also wishes to thank Mrs. Cathy Jeng, secretary of the Civil Engineering Department, UCLA, who typed the first six chapters of the manuscript. My son, Yi-Shan Sun, typed Chapters 7 and 8, and checked all of the equations in the book. The author also acknowledges the work of Mr. Bi Lijun, editor of the Geological Publishing House of China, and Mr. Bernd Grossmann, the Vice President of Springer-Verlag, and the contributions of Ms. Elizabeth Sheehan, Editor, in the production of this book.

Los Angeles, California   NE-ZHENG SUN
August 1995

Contents

1
Introduction

1.1 Groundwater Quality

The chemical and biological constituents contained in groundwater depend
on two factors: the natural environment of groundwater storage and movement, and human activities. Precipitation infiltration and surface water percolation are the natural sources of groundwater. The total dissolved solids
(TDS) of precipitation is generally very low, but its chemical components will
be changed when infiltrated through soil beds by a series of actions, such as
solution, oxidation, reduction, ion exchange, and so on. The infiltration and
percolation water will be involved in groundwater movement in both the
vertical and the lateral directions in the aquifer. During this process, the TDS
of groundwater will continually increase as rocks and minerals are dissolved
into the water. Human activities may change the natural process and cause
groundwater to contain organisms, hydrocarbons, heavy metals and other
harmful matter. Groundwater, therefore, should be looked upon as a multicomponent fluid. The content of each component in groundwater can be
expressed by its concentration, i.e., a mass of certain component contained in
unit volume of water (M/L^3). If the concentration of component a is written
as C_α, then the *standard of water quality* for a certain use can be written in the
following common form:

$$C_{\alpha,\min} < C_\alpha < C_{\alpha,\max},$$
$$(\alpha = 1, 2, \ldots, n) \tag{1.1.1}$$

where $C_{\alpha,\min}$ and $C_{\alpha,\max}$ are the given lower and upper limits, respectively, of
the concentration of component α, and n is the total number of components
considered. According to the actual situation, the components may refer to
either single ions or multi-ion compositions.

If the groundwater quality (including its physical, chemical, and biological
properties) has been changed so that it is no longer suited to the previous
uses, then the groundwater is said to be polluted.

Since groundwater is buried under the land surface, it is not as easily
polluted as the surface water. Because of its low flow rate, the pollutants

1

spread so slowly that the groundwater contamination may remain unde-
tected for a long time. Hence, the potential danger of groundwater contamination is often ignored. Actually, groundwater is being polluted on a large
scale in some cities and districts due to human activities. Groundwater is
generally hard to clean up once polluted. The predictive calculations for
some aquifers have shown that even if the pollution source has been removed, self-purification will take several decades, or even hundreds of years.
It is more difficult to clear away groundwater pollution in fine porous media.
High costs and a long period of time are needed for remediating groundwater
pollution compared to what is required for surface water. Therefore, prefer-
ence should be given to prediction and prevention, as well as regular moni-
toring and analysis, of groundwater quality.

The *analysis of water quality* involves the measurements of physical properties, chemical and/or biological components of a water sample. The result
of water quality analysis gives the concentration value C_α of any designated
component α in a sample. In order to describe the degree of contamination in
an aquifer and to monitor the changes in groundwater quality, it is necessary
to sample and analyze the water quality at different locations and times.

The distribution of concentration C_α in space and time is called the *concentration field* and is written as $C_\alpha(x, y, z, t)$. The results obtained from moni-
toring and analyzing are discrete observation values of the concentration
field:

$$\{C_\alpha(x_i, y_i, z_i, t_j)\},$$

$$i = 1, 2, \ldots, I,$$ (1.1.2)

$$j = 1, 2, \ldots, J,$$

where I is the number of observation points,
J is the number of observation times,
$I \cdot J$ is the total number of samples.

From the mathematical point of view, any distributed scalar field, such as
a concentration field $C_\alpha(x, y, z, t)$, can be displayed approximately by a set
of discrete values in space and time, such as the observed concentrations
given in Eq. (2.1.2), provided that the observation locations and times are
appropriately arranged.

1.2 Groundwater Quality Management

The groundwater resource is limited. The best use of the limited resource
with maximum economic benefit can only be obtained through scientific
management and rational utilization. An incorrect management decision
not only causes economic waste and resources destruction, but also brings

serious consequences, such as the depletion of water, the deterioration of water quality, and the subsidence of land.

When the groundwater system is utilized in accordance with some special purposes and demands, it is necessary to make some decisions. For example, if the aquifer is regarded as a supply source, then the location, yield, the number of the pumping wells, as well as the water quantity and quality of artificial recharge, must be determined. There are many ways to reach the same objective, so the factors mentioned above should be regarded as variables, which are called *decision variables*.

However, not all the decisions are feasible. The decision variables must obey certain restrictions which are called *constraints*. Suppose that an aquifer management plan is beneficial in one aspect, but causes the water table to decline on a large scale. If either the pumping wells are unable to draw water or polluted water comes from a neighboring region, the plan is not feasible. In mathematical terms, we say that the relevant decision variables do not satisfy the restrictive conditions or *constraints*.

In general cases, there are a number of feasible decisions which satisfy all of the restricted conditions. Thus, an important problem is how to select the *optimal decision*. Before solving this problem, we must set up some criteria for judging "bad" or "good" decisions. For example, the cost of water per ton may be taken as a criterion. Since different decisions correspond with different costs, the optimal decision can be selected as the one with the lowest cost. The selected criteria, which are functions of decision variables, are called *objective functions*. The minimum (or maximum) of an objective function is a goal that we endeavor to achieve. The *management of groundwater* then becomes the selection of the optimal feasible decision or decisions for attaining one or several goals. Mathematically, it involves solving the following single- or multiple-objective *optimization problem* with a series of constraints:

$$\min \mathbf{J}(\mathbf{q}), \qquad \mathbf{q} \in \mathbf{Q} \tag{1.2.1}$$

subject to

$$h(\mathbf{q}; \mathbf{x}_i, t) \geq \underline{h}_i, \quad (i = 1, 2, \ldots, I), \tag{1.2.2}$$

$$C_\alpha(\mathbf{q}; \mathbf{x}_j, t) \leq \overline{C}_{\alpha, j}, \quad (j = 1, 2, \ldots, J). \tag{1.2.3}$$

In Eq. (1.2.1), $\mathbf{J}(\mathbf{q})$ is a vector objective function; $\mathbf{q}$ is the vector of decision variables; and $\mathbf{Q}$ is the admissible set of decision variables. Eqs. (1.2.2) and (1.2.3) are all constraints, where $h(\mathbf{q}; \mathbf{x}_i, t)$ is the water head in assigned location $\mathbf{x}_i$, not to be smaller than the given head $\underline{h}_i$, $i = 1, 2, \ldots, I$; and $C_\alpha(\mathbf{q}; \mathbf{x}_j, t)$ is the concentration of component α in assigned location $\mathbf{x}_j$, not to be greater than the given concentration $\overline{C}_{\alpha, j}$, $j = 1, 2, \ldots, J$. The objective function is generally proposed by the manager, while the water head distribution $h(\mathbf{q}; \mathbf{x}, t)$ and the concentration distribution $C(\mathbf{q}; \mathbf{x}, t)$ are determined by a simulation model. All of them are functions of the decision variables $\mathbf{q}$.

The distributions of water head, concentration, temperature, and land subsidence are variables representing the states of a groundwater system, which are called *state variables*. To find the distributions of state variables in accordance with given decision variables is called the *prediction problem*.

The management problem involves the prediction problem. To answer whether or not a decision is feasible, we should first predict the response of the system to the decision. For example, let the number, locations and rates of pumping wells be decision variables of a management problem. For each decision, the distribution of water level, especially in the wells, can be obtained by solving the prediction problem. The prediction results can tell us whether the lowering of water table exceeds the given value, and allow us to calculate the power consumption for drawing water. Another example occurs when the locations of the recharge wells and the quality of the recharged water are considered as decision variables. We can solve the prediction problem first in order to compute the concentration of the contaminants in the supply wells. Then, we can judge whether the quality of the water supply meets the given standard, and calculate the treatment cost of the recharged water.

How is the prediction problem solved? The most reliable way, of course, is to conduct a field test and directly observe the state of the aquifer. Unfortunately, this is unrealistic because the field test cannot tell us what will happen in the future. It is impossible to conduct field tests for all feasible decisions and then compare them. Although theoretical analysis is important for solving practical problems, it cannot give prediction results for an individual problem. The modeling method, therefore, becomes the only way for solving the prediction problem. A physical or mathematical model is built based on the internal structures and external conditions of a real system. It combines common physical rules (mass balance, Darcy's law, for example) with particular conditions (values of parameters, initial and boundary conditions, for example) of the system. The excitation-response relation of the real system can be described by the input-output relation of the model. Thus, the model is helpful for better understanding the system and may provides invaluable tools for prediction purposes. All feasible decisions can be inputted to the model, and the resulting states can be observed after the model is run. With a model, we can search for the optimal decisions based on the given objective functions.

1.3 Groundwater Modeling

Many types of models have been used for simulating groundwater flow, such as sand trough, vertical or horizontal Hele-Shaw, membrane, *R-R* and *R-C* electric analogue, as well as various *mathematical models*.

TABLE 1.1. Applications of groundwater modeling.

Model	Groundwater flow	Mass transport	Heat transport	Media deformation
Application	Water supply	Groundwater contamination	Geothermal utilization	Land subsidence
	Regional aquifer analysis	Sea water intrusion	Heat and cold storage under the ground	Engineering geological problem
	Relationship between ground-water and surface water	Migration of pesticides and chemical fertilizers in plant root zone	Heat pollution in groundwater	
	Precipitation, in-filtration and evaporation	Soil reformation		
	Drainage and dewatering engineering	Radionuclide waste repositories		
	Artificial recharge	Impact of river pollution on groundwater		

Mathematical models have many advantages compared with the others. They can reflect complex physical structures and irregular geometric shapes of an aquifer system; they are convenient and flexible to use and easy to calibrate; and they can describe not only the phenomena of water flow in porous media but also the mass and energy transports and other complex physical-chemical-biological phenomena in porous media. Therefore, mathematical modeling has become the major tool of basin studies. Table 1.1 shows its major applications.

Each one of the mathematical models in Table 1.1 consists of a set of partial differential equations (PDE) with appropriate initial and boundary conditions. The equations describe the characteristics of the system and the physical-chemical processes going on in it. The initial and boundary conditions represent the system's initial state and its relationships with the surrounding systems. Mass and heat transport models couple the flow model with the solute mass conservation equation, and energy conservation equation, respectively. The model of media deformation couples the flow model with the equation expressing the skeletal strain. Generally, these mathematical models are very complex, and computers must be used to solve them.

There are two kinds of distributed parameter models in the simulation of groundwater quality: pure advection and advection-dispersion. The pure

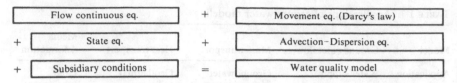

FIGURE 1.1. The constitution of the water quality model on advection-dispersion.

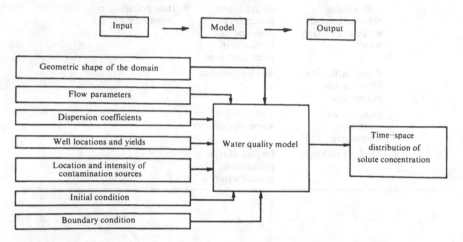

FIGURE 1.2. Input and output of the water quality model.

advection model is a simple one, in which the solute transport is assumed to be completely determined by groundwater flow. On the other hand, as shown in Figure 1.1, the advection-dispersion model is based on the theory of hydrodynamic dispersion in porous media. Thus, the dispersion phenomena must be studied and some new parameters must be introduced for deriving the advection-dispersion equation. Since the advection-dispersion model of groundwater quality consists of several partial differential equations, numerical solution techniques must be used.

In order to build a water quality model for an aquifer system, besides the selection of model structure, various hydrogeological parameters (especially the dispersion-related ones) appearing in the governing equations and initial and boundary conditions must be determined. The input and output of a groundwater quality model is shown in Figure 1.2. Model parameters, as well as boundary conditions, are generally determined by a parameter identification procedure, i.e., using tracer test data and other prior information to calibrate the model. In a certain sense, the parameter identification problem is the *inverse problem* of the prediction problem. As shown in Figure 1.2, the input and output of the two problems are exchanged.

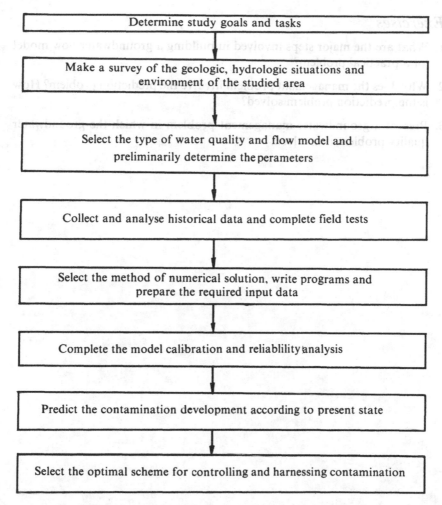

FIGURE 1.3. The major steps in the application of water quality model to the research of groundwater contamination.

Because of the uncertainties in model structure and model parameters, the uncertainties in model predictions, and thus in the optimal management decisions, can never be avoided. An important problem in groundwater modeling is the estimation of these uncertainties either in the deterministic framework or in the statistical framework.

Thus, to build a groundwater quality model, we have to solve both prediction and inverse problems and provide a reliability analysis. Figure 1.3 shows the major steps of using the modeling approach to solve a real problem of groundwater quality. This flow chart can also be seen as an outline of this book.

Exercises

1. What are the major steps involved in building a groundwater flow model for a practical problem?

2. Why does the management problem involve the prediction problem? How is the prediction problem solved?

3. Present a groundwater management problem in which the groundwater quality problem is involved.

2
Hydrodynamic Dispersion in Porous Media

2.1 Physical Parameters

2.1.1 Spatial Average Method

When fluid flows through the interconnected voids and passages of a porous medium, the walls of these voids and passages form many small tunnels, and the fluid flows inside them. The study of physical phenomena in a porous medium on such a scale (pore scale) is called the *microscopic method*. Due to the complexity of the micro-geometry of porous media, it is unrealistic to study the details of the flow on this scale. Therefore, one has to describe the flow phenomena in porous media on a macroscopic rather than microscopic basis. The *spatial average method* is a way to transfer properties of porous media from the *microscopic level* to the *macroscopic level*.

Consider a mathematical point x in a flow region, with coordinates (x_1, x_2, x_3) in a three-dimensional coordinate system. A small volume, $[U_0(\mathbf{x})]$, which can be either spheric or cubic with its center at x, is defined as a particle of the porous medium. On the one hand, the volume $[U_0(\mathbf{x})]$ must be large enough to cover a sufficient number of solid particles and pores so that the stable mean values of certain physical properties associated with $[U_0(\mathbf{x})]$ can be obtained over the volume. For example, if the pore space of $[U_0(\mathbf{x})]$ is marked as $[U_{0,v}(\mathbf{x})]$, and $[U_0(\mathbf{x})]$ varies within a certain range, then the ratio of volume

$$n(\mathbf{x}) = \frac{U_{0,v}(\mathbf{x})}{U_0(\mathbf{x})} \tag{2.1.1}$$

may be nearly a constant. As a result, it can be defined as the *porosity* at point x. On the other hand, the volume of $[U_0(\mathbf{x})]$ must also be small enough in comparison with the whole region so that it can be treated as a point. The particle thus defined is also called a *Representative Elementary Volume* (REV) of porous media (Bear, 1972).

If every mathematical point in a porous medium is associated with a particle, then the porous medium, which is constructed with solid matrices

9

and pores, can be considered as a continuum fully filled with those particles. All relevant parameters and properties, such as the water head, concentration, porosity, and permeability may be regarded as continuous and even differentiable functions. The difficulty in getting details of the microstructure of porous media is thus avoided. The study of flow phenomena in porous media based on the scale of porous medium REV is called the *macroscopic approach*.

Let us now consider saturated flow in a porous medium. This occurs when the pores are filled with fluid. Suppose that $a(\mathbf{x}')$ is a microscopic property (scalar or vector) of the fluid, where $\mathbf{x}'$ represents a point in the microscopic level. The *spatial average* of $a(\mathbf{x}')$ within the pore volume $[U_{0,v}(\mathbf{x})]$ of REV $[U_0(\mathbf{x})]$ is defined by

$$\bar{a}(\mathbf{x}) = \frac{1}{U_{0,v}(\mathbf{x})} \int_{[U_{0,v}(\mathbf{x})]} a(\mathbf{x}') \, dU_{0,v}. \tag{2.1.2}$$

The so defined $\bar{a}$ is directly related to the particle of porous media, and thus, is a macroscopic property. From this point of view, every microscopic property of the system can be transferred to its macroscopic counterpart by means of spatial averaging, as in Eq. (2.1.2). If a property a can be assumed to be zero within the solid matrices, then Eq. (2.1.2) can be rewritten as

$$\bar{a}(\mathbf{x}) = \frac{1}{nU_0(\mathbf{x})} \int_{[U_0(\mathbf{x})]} a(\mathbf{x}') \, dU_0, \tag{2.1.3}$$

where n is the porosity determined by Eq. (2.1.1). Two examples are given below to show the utilization of the spatial average method.

Example 1

Let a be the density, ρ_α, of component α in a multicomponent fluid. It is a microscopic property and is defined as a function of coordinates of every point in the fluid space. The spatial average method gives

$$\bar{\rho}_\alpha(\mathbf{x}, t) = \frac{1}{U_{0,v}(\mathbf{x})} \int_{[U_{0,v}(\mathbf{x})]} \rho_\alpha(\mathbf{x}', t) \, dU_{0,v}. \tag{2.1.4}$$

After spatial averaging, the mean density, $\bar{\rho}_\alpha(\mathbf{x}, t)$, becomes a function defined on the region of porous medium and is now a macroscopic property. It represents the density distribution of component α in a porous medium. If there is no component α in the solid matrix, Eq. (2.1.4) can be modified to

$$\bar{\rho}_\alpha(\mathbf{x}, t) = \frac{1}{nU_0(\mathbf{x})} \int_{[U_0(\mathbf{x})]} \rho_\alpha(\mathbf{x}', t) \, dU_0. \tag{2.1.5}$$

Example 2

Assume that a is the microscopic velocity $\mathbf{V}$ of fluid particles inside the pores. The spatial averaging gives

$$\bar{\mathbf{V}}(\mathbf{x}, t) = \frac{1}{U_{0,v}(\mathbf{x})} \int_{[U_{0,v}(\mathbf{x})]} \mathbf{V}(\mathbf{x}', t) \, dU_{0,v}. \tag{2.1.6}$$

where $\overline{V}$ is called the mean pore velocity in the region of porous media. If the velocity within solid matrices is defined to be zero, Eq. (2.1.6) can be changed to

$$\overline{V}(\mathbf{x},t) = \frac{1}{nU_0(\mathbf{x})} \int_{[U_0(\mathbf{x})]} V(\mathbf{x}',t)\,dU_0. \qquad (2.1.7)$$

A few simple rules of the spatial averaging are given below. Let $\overset{\circ}{a}$ be the difference between a and its spatial average $\overline{a}$, i.e., let

$$\overset{\circ}{a} = a - \overline{a}, \qquad (2.1.8)$$

then we have

(1) $\overline{\overset{\circ}{a}} = 0$.

In fact, it can be directly obtained from the definition of spatial averaging in Eq. (2.1.2) that

$$\overline{\overset{\circ}{a}} = \frac{1}{U_{0,v}(\mathbf{x})} \int_{[U_{0,v}(\mathbf{x})]} (a - \overline{a})\,dU_{0,v}$$

$$= \frac{1}{U_{0,v}(\mathbf{x})} \int_{[U_{0,v}(\mathbf{x})]} a\,dU_{0,v} - \overline{a} = \overline{a} - \overline{\overline{a}} = 0.$$

(2) For the average of a product of two properties a_1 and a_2, we have

$$\overline{a_1 a_2} = \overline{a_1}\,\overline{a_2} + \overline{\overset{\circ}{a_1}\overset{\circ}{a_2}}. \qquad (2.1.9)$$

Since $\overset{\circ}{a_1} = a_1 - \overline{a_1}$ and $\overset{\circ}{a_2} = a_2 - \overline{a_2}$, the product of the two gives

$$\overset{\circ}{a_1}\overset{\circ}{a_2} = a_1 a_2 - a_1 \overline{a_2} - a_2 \overline{a_1} + \overline{a_1}\,\overline{a_2}.$$

Taking spatial average over both sides, and noting that

$$\overline{a_1 \overline{a_2}} = \overline{a_2 \overline{a_1}} = \overline{\overline{a_1}\,\overline{a_2}} = \overline{a_1}\,\overline{a_2},$$

we then have

$$\overline{\overset{\circ}{a_1}\overset{\circ}{a_2}} = \overline{a_1 a_2} - \overline{a_1}\,\overline{a}.$$

For an unsaturated system, or a more general multi-phase one, the definition of spatial averaging in Eq. (2.1.2) needs some modification. For a microscopic property a which is relevant to phase γ, we define

$$\overline{a}(\mathbf{x}) = \frac{1}{U_{0,\gamma}(\mathbf{x})} \int_{[U_{0,\gamma}(\mathbf{x})]} a(\mathbf{x}')\,dU_{0,\gamma}. \qquad (2.1.10)$$

where $[U_{0,\gamma}(\mathbf{x})]$ is the volume occupied by phase γ within REV $[U_0(\mathbf{x})]$.

Note that there are uncertainties associated with the definition of REV. First, the so defined REV may not exist; Second, even the REV exists, we do not know how to determine its size. In the statistical theory of porous media, the microscopic property $a(\mathbf{x}')$ in Eq. (2.1.2) is regarded as a *random function* and the averaging volume (REV) is replaced by a volume V with a certain scale. The *spatial average* of $a(\mathbf{x}')$ is then defined by

$$\bar{a}(\mathbf{x}) = \frac{1}{V} \int_V a(\mathbf{x}')\, d\mathbf{x}', \tag{2.1.11}$$

or more general, by

$$\bar{a}(\mathbf{x}) = \frac{1}{V} \int a(\mathbf{x}')U(\mathbf{x} - \mathbf{x}')\, d\mathbf{x}', \tag{2.1.12}$$

where integration is carried out over the entire space and the weight function $U(\mathbf{x} - \mathbf{x}')$ is equal to unity when $\mathbf{x}' \in [V]$, and equal to zero outside it (Cushman, 1983). Spatial average $\bar{a}(\mathbf{x})$ is also a random function and can be characterized by its mean $\langle \bar{a} \rangle$ and variance $\sigma_{\bar{a}}^2$. When $\sigma_{\bar{a}}^2$ is close to zero, $\bar{a}$ becomes deterministic. Mathematical expectation $\langle \bar{a} \rangle$ is often referred to as the *ensemble averaging*.

To study flow and mass transport phenomena in a porous medium, we must first define the "scale" of the problem of interest. Dagan (1986) suggested three different macroscopic scales as follows:

- the *laboratory scale* $(10^{-1} \sim 10^0 m)$,
- the *local scale* $(10^1 \sim 10^2 m)$, and
- the *regional scale* $(10^3 \sim 10^5 m)$.

The statistical theory may provide a unified approach for studying all of these scales. In this book, however, we will mainly deal with deterministic models, because this kind of model is simple in concept and any numerical technique designed for solving deterministic models can also be used to solve statistical models. In Chapter 7, we will return to the statistical approach for discussing the "scale effect" problem and estimating the uncertainties associated with model predictions.

2.1.2 Fluid, Medium and State Parameters

Fluid Density

Consider a multicomponent fluid consisting of N components. From the viewpoint of continuum, it can be considered as a sum of N independent continua, i.e., at a certain mathematical point, particles of various components can be stacked. Hence, we can define the particle density for every component. Let ρ_α be the *density* of component α, and ρ the density of the whole fluid system. It is then apparent that

$$\rho = \sum_{\alpha=1}^{N} \rho_\alpha. \tag{2.1.13}$$

The dimensionless parameter

$$\omega_\alpha = \frac{\rho_\alpha}{\rho} \tag{2.1.14}$$

is called the *mass fraction* of component α, and represents the mass ratio of component α to the fluid mixture per unit mass. Similarly, we can define the *volumetric fraction*, v_α, of component α as the volume ratio of component α to the fluid mixture per unit volume. By means of spatial averaging, both ω_α and v_α can be transferred from the microscopic level to the macroscopic level. The mean density of liquid phase β in a porous medium is given by:

$$\bar{\rho}_\beta(\mathbf{x}) = \frac{1}{U_{0,\beta}(\mathbf{x})} \int_{[U_{0,\beta}(\mathbf{x})]} \rho_\beta(\mathbf{x}') \, dU_{0,\beta}, \qquad (2.1.15)$$

where ρ_β is the microscopic density of the liquid phase β and $[U_{0,\beta}(\mathbf{x})]$ is the volume of phase β in the REV of porous media.

The *specific gravity*, γ, of a fluid is defined as

$$\gamma = \rho g, \qquad (2.1.16)$$

where g is the gravitational acceleration. Then the mean specific gravity of phase β in porous media is

$$\bar{\gamma}_\beta = \bar{\rho}_\beta g. \qquad (2.1.17)$$

Solute Concentration

The *concentration* of component α in a multicomponent fluid is actually its density ρ_α. It is conventionally expressed as C_α. For a binary system with only one solute in a solvent, a single letter C can be used to express the solute concentration. If the density of a solution is independent of the solute concentration, it is more convenient to use *dimensionless concentration*,

$$c = \frac{C}{\rho}, \qquad (2.1.18)$$

which gives the mass ratio of solute to solution. From Eq. (2.1.14), we know that c is, in fact, the mass fraction of the solute.

As to a porous medium, component α can be either in the liquid phase, or in the solid phase. We will use $C_{\gamma,\alpha}$ to denote the concentration of solute α in phase γ. Its mean value is given by the following equation:

$$\bar{C}_{\gamma,\alpha}(\mathbf{x}) = \frac{1}{U_{0,\gamma}(\mathbf{x})} \int_{[U_{0,\gamma}(\mathbf{x})]} C_{\gamma,\alpha}(\mathbf{x}') \, dU_{0,\gamma}, \qquad (2.1.19)$$

where $[U_{0,\gamma}(\mathbf{x})]$ is the volume of phase γ within the REV. If there will be no misunderstanding, the subscripts γ and α may be omitted.

Fluid Viscosity

A continuous deformation of the fluid, i.e., a flow, will occur if a shear force acts on it. The resistance of the fluid to such deformation is called its *viscosity*.

All Newtonian fluids obey the following law:

$$\tau = \mu \frac{\partial u}{\partial n}, \qquad (2.1.20)$$

where τ is the shear stress; $\partial u/\partial n$, the velocity gradient normal to liquid surface; and μ, a coefficient which is called the *dynamic viscosity*. The ratio of dynamic viscosity to density is called the *kinematic viscosity*, $v = \mu/\rho$. All of these properties can be transferred to their macroscopic values by the spatial average method.

Note that the density and viscosity of a solution will change with pressure p, solute concentration C and temperature T. At a constant temperature, their relationship can be expressed as

$$\rho = \rho(C, p), \qquad \mu = \mu(C, p), \qquad (2.1.21)$$

which are called *state equations*. As a first approximation, density and viscosity can be regarded as linear functions of C and p.

Flow Velocity of Multicomponent Fluids

Let vector $\mathbf{V}_\alpha$ be the flow velocity of component α on a microscopic level, then two macroscopic mean velocities can be defined for this fluid system, i.e., *the mean mass velocity*

$$\mathbf{V} = \sum_{\alpha=1}^{N} \omega_\alpha \mathbf{V}_\alpha, \qquad (2.1.22)$$

and the *mean volume velocity*

$$\mathbf{V}' = \sum_{\alpha=1}^{N} v_\alpha \mathbf{V}_\alpha. \qquad (2.1.23)$$

The former uses mass fractions as weighting coefficients, and the latter uses volume fractions. We will mainly use the mean mass velocity throughout this book. The mean volume velocity is used only when the volume of fluid mixture is dependent on the densities of the components.

The relevant macroscopic averages of Eq. (2.1.22) and Eq. (2.1.23) can be obtained by spatial averaging. The averages of $\mathbf{V}$ and $\mathbf{V}'$ in a REV of porous media are defined as

$$\overline{\mathbf{V}} = \frac{1}{U_{0,\beta}(\mathbf{x})} \int_{[U_{0,\beta}(\mathbf{x})]} \mathbf{V} \, dU_{0,\beta}, \qquad (2.1.24)$$

$$\overline{\mathbf{V}'} = \frac{1}{U_{0,\beta}(\mathbf{x})} \int_{[U_{0,\beta}(\mathbf{x})]} \mathbf{V}' \, dU_{0,\beta}, \qquad (2.1.25)$$

where $[U_{0,\beta}(\mathbf{x})]$ is the volume occupied by a multicomponent fluid in the REV. If a porous medium is saturated with an incompressible homogeneous

FIGURE 2.1. Solute transport in a
dead-end pore.

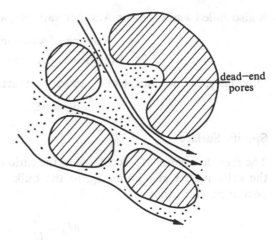

dead–end
pores

fluid, we have

$$\overline{V} = \overline{V}' = q/n,\qquad(2.1.26)$$

where q is the *Darcy velocity* and *n* the porosity.

Volumetric Fraction

We have defined porosity *n* based on the REV of porous media in Equation
(2.1.1), which is, in fact, the volume fraction of the pore space in the REV.
Hence, the fraction of solid matrix can be written as $(1 - n)$. Those pores,
which are not connected to other pores, will not take part in flow or mass
transport in groundwater, and are normally not treated as part of the pore
space. The porosity defined in this way is called the *effective porosity*, which
will be used throughout this book unless otherwise specified.

Some pores, though connected to others, are dead-end pores as shown in
Figure 2.1. The flow velocity inside them is extremely low and the fluid is
almost stagnant. In the study of solute transport, the effect of dead-end pores
must be taken into account. The pollutants inside these pores can only be
transported by molecular diffusion, and not by advection. Since molecular
diffusion is a slow process, removal of pollutants from these pores may take
a long time.

The saturated zone and unsaturated zone need to be treated separately
when we study the solute transport. In a saturated zone, the volume fraction
of fluid equals porosity *n* of the porous media, while in an unsaturated zone,
there exists a gas phase, as well as solid and liquid phases. The volumetric
fraction of water

$$\theta_w = \frac{\text{Volume of water in REV}}{\text{Volume of REV}}$$

is also called *water content*. Another variable, *water saturation*, is defined as

$$S_w = \frac{\text{Volume of water in REV}}{\text{Volume of pore space in REV}}.$$

Obviously, the two variables θ_w and S_w are interrelated by the equation

$$\theta_w = nS_w. \tag{2.1.27}$$

Specific Surface

The *specific surface*, M_0, is defined as the ratio of the interfacial area between the solid matrix and pores, S_0, to the bulk volume, U_0, of the REV in a porous medium, that is,

$$M_0 = \frac{S_0}{U_0}. \tag{2.1.28}$$

Obviously, the smaller the solid particle size, the larger the specific surface area. The magnitude of M_0 is dependent on the porosity, particle-size distribution, and the shape and configuration of the particles. Since adsorption, desorption, and ion exchange all occur at the solid-liquid boundary interface, the specific surface area plays an important role in the study of solute transport.

If there are multiple phases in the REV of a porous medium, the specific surface of phase γ can be defined as

$$M_\gamma = \frac{S_{0,\gamma}}{U_0}, \tag{2.1.29}$$

where $S_{0,\gamma}$ is the total area of interfaces between phase γ and other phases.

Tortuosity

At the microscopic level, the fluid flow within a porous medium is actually a movement along the tortuous three-dimensional passages in voids, as shown in Figure 2.2. The local velocities in the passages are different from their macroscopic average values, both in magnitude and in direction. In a physical model, the void space of a porous medium may be regarded as a network of curved channels. One of the channels is shown schematically in Figure 2.3. It is so placed that the axis of the channel is on the same plane as the mean flow direction (x). Its length is L_e and its projection on axis x has a length of L.

If the mean flow velocity along the channel axis is $\bar{V}$, and its projection on axis x is $\bar{u}$, then from $L_e/\bar{V} = L/\bar{u}$, we have

$$\bar{u} = \bar{V}(L/L_e). \tag{2.1.30}$$

Let Δh be the water head difference between the two ends of the channel. The mean velocity along the channel, $\bar{V}$, is proportional to the mean hydraulic gradient $\Delta h/L_e$, i.e.,

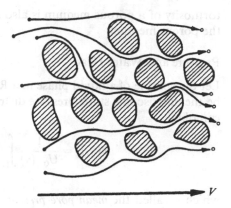

FIGURE 2.2. Microscopic movement of fluid particles in pores. • = initial position; o = terminal position; → = movement path.

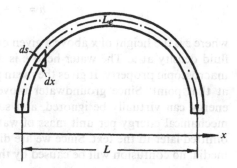

FIGURE 2.3. Geometric illustration of the tortuosity of a porous medium.

$$\bar{V} = -K\Delta h/L_e \qquad (2.1.31)$$

where K is a coefficient of proportionality. By inserting Eq. (2.1.31) into Eq. (2.1.30), we have

$$\bar{u} = -K\left(\frac{L}{L_e}\right)^2\frac{\Delta h}{L} = -KT\frac{\Delta h}{L}, \qquad (2.1.32)$$

where $\bar{u}$ and $\Delta h/L$ are the mean flow velocity and hydraulic gradient along the x axis, respectively. The proportional coefficient

$$T = (L/L_e)^2 \qquad (2.1.33)$$

is called the *tortuosity* of the channel. It is obvious that $0 < T < 1$. The value of T depends on the shape of the channel. The larger L_e is, the smaller T will be. From Eq. (2.1.32), it can be seen that a decrease of T is equivalent to an increase in the flow resistance along the x direction. As we have seen before, if the microscopic tortuosity function within the REV is known, the mean tortuosity of the porous media can be obtained by the spatial average method. The tortuosity of an isotropic medium is a scalar, while the tortuosity of an anisotropic one is a second rank symmetric tensor (Bear, 1972). The tortuosity of a porous medium is also a basic parameter to the depiction of the porous medium.

Pressure and Water Head

Let the volume of a liquid phase in a REV of porous media be $[U_{0,\gamma}]$, and p be the microscopic static pressure distribution, then by spatial averaging we can obtain

$$\bar{p}(\mathbf{x}, t) = \frac{1}{U_{0,\gamma}(\mathbf{x})} \int_{[U_{0,\gamma}(\mathbf{x})]} p(\mathbf{x}', t) \, dU_{0,\gamma}, \qquad (2.1.34)$$

which is called the *mean pore pressure* of the porous medium. We can also define *water head* at point $\mathbf{x}$ as

$$h = z + \frac{\bar{p}}{\bar{\rho} g}, \qquad (2.1.35)$$

where z is the height of $\mathbf{x}$ above a given datum and $\bar{\rho}$ is the macroscopic mean fluid density at $\mathbf{x}$. The water head h is dependent on $\bar{p}$ and $\bar{\rho}$, and is also a macroscopic property. It gives the mean potential energy of unit mass of fluid at that point. Since groundwater movement is relatively slow, its kinetic energy can virtually be ignored, and so h can also be treated as the total mechanical energy per unit mass of water. The bars above $\bar{p}$ and $\bar{\rho}$ will be omitted later in the text. Since we are discussing flow phenomena in porous media, no confusion will be caused by this omission.

Coefficient of Permeability

Darcy's Law is a basic law in seepage theory. For an anisotropic porous medium, Darcy's Law is expressed as

$$\begin{cases} q_1 = -K_{11}\dfrac{\partial h}{\partial x_1} - K_{12}\dfrac{\partial h}{\partial x_2} - K_{13}\dfrac{\partial h}{\partial x_3}; \\[2mm] q_2 = -K_{21}\dfrac{\partial h}{\partial x_1} - K_{22}\dfrac{\partial h}{\partial x_2} - K_{23}\dfrac{\partial h}{\partial x_3}; \\[2mm] q_3 = -K_{31}\dfrac{\partial h}{\partial x_1} - K_{32}\dfrac{\partial h}{\partial x_2} - K_{33}\dfrac{\partial h}{\partial x_3}, \end{cases} \qquad (2.1.36)$$

where (q_1, q_2, q_3) are the components of Darcy velocity $\mathbf{q}$. The second rank symmetric tensor

$$\mathbf{K} = \begin{bmatrix} K_{11} & K_{12} & K_{13} \\ K_{21} & K_{22} & K_{23} \\ K_{31} & K_{32} & K_{33} \end{bmatrix} \qquad (2.1.37)$$

is called the *hydraulic conductivity* of porous media. Using Einstein's summation convention, Darcy's Law, Eq. (2.1.36), can be written in a compact form:

$$q_i = -K_{ij} \frac{\partial h}{\partial x_j} \qquad (i = 1, 2, 3) \qquad (2.1.38)$$

which will be used repeatedly. For an isotropic medium, the hydraulic conductivity reduces to a scalar, and the relevant Darcy's Law becomes

$$q_i = -K \frac{\partial h}{\partial x_i} \qquad (i = 1, 2, 3). \qquad (2.1.39)$$

Hydraulic conductivity K is a macroscopic parameter which depends on properties of both the porous matrix and the liquid, and may be expressed as

$$K = \frac{k \rho g}{\mu}, \qquad (2.1.40)$$

where ρ and μ are the macroscopic averages of density and viscosity of the fluid, i.e., $\bar{\rho}$ and $\bar{\mu}$, g is the gravitational acceleration, and k is called the *permeability* which is independent of the properties of the fluid and has the dimensions of $[L^2]$. For an anisotropic porous medium, k is a second rank symmetric tensor.

With Eqs. (2.1.26), (2.1.35), and (2.1.40), Darcy's Law, Eq. (2.1.36), can be rewritten as

$$V_i = \frac{k_{ij}}{n\mu} \left(\frac{\partial p}{\partial x_j} + \rho g \frac{\partial z}{\partial x_j} \right) \qquad (2.1.41)$$

$$(i = 1, 2, 3).$$

Storage Coefficients

Specific storativity, S_s, is defined as the volume fraction of water released from a REV when the water head declines by one unit, i.e.,

$$S_s = \frac{\Delta U_w}{U_0 \cdot \Delta h}, \qquad (2.1.42)$$

where ΔU_w denotes the volume of water released from $[U_0]$ as a result of solid matrix deformation or water expansion, and Δh is the decline of water head. The dimension of S_s is $[1/L]$ and we have the following relationship (Bear, 1972)

$$S_s = \rho g [\alpha(1 - n) + n\beta], \qquad (2.1.43)$$

where α and β are the compressibilities of the solid matrix and water, respectively.

Specific yield S_y is a property describing the water release ability of an aquifer. It is defined as the volume of water released from a unit area of aquifer per unit decline of water table, i.e.,

$$S_y = \frac{\Delta U_w}{A_0 \cdot \Delta H}, \qquad (2.1.44)$$

where A_0 is the area element at the point considered, ΔH is the drawdown of water table, and ΔU_w is the volume of water releasing from the aquifer with the bulk of A_0 as its base and ΔH as its height. Generally speaking, S_y is smaller than porosity n, since capillary water is still retained in the volume even when the water table is lowered. If this amount of water is negligible, S_y may be approximated by the effective porosity of the medium.

Some other properties relevant to mass transport in a porous medium, such as the hydrodynamic dispersion coefficient and molecular diffusion coefficient, will be introduced later. Appendix A gives a list of symbols of the parameters mentioned above as well as others, along with their dimensions.

2.2 Phenomena and Mechanism of Hydrodynamic Dispersion

2.2.1 Hydrodynamic Dispersion Phenomena

Mass transport in a porous medium is carried out in the moving fluid and the movement of the fluid takes place in a very complicated intersticial system. These conditions give rise to a very special phenomenon, i.e., *hydrodynamic dispersion*. Two examples are given below to show the existence of the hydrodynamic dispersion phenomenon on a macroscopic scale.

Example 1

Consider continuous injection of a tracer into a well which is situated in a uniform one-dimensional flow field in the x direction. We can observe a gradual spreading of the tracer around the well. The extent that the tracer spreads is larger than the region expected from the average flow alone. It not only spreads longitudinally along the mean flow direction, but also expands transversely. There exists a transition zone rather than an abrupt interface between the tracer-containing solution and the original fluid around the well as shown in Figure 2.4. This is called the *hydrodynamic dispersion phenomenon*. Without dispersion, the tracer would move in accordance with the mean flow velocity. There would be no transversal spreading and an abrupt concentration change would be seen.

Example 2

Consider a steady flow in a homogeneous sand column saturated with water. At a certain time, the water with a tracer concentration C_0 starts to replace the original water. The concentration at the other end of the column, $C(t)$, is monitored, and a curve of relative concentration $C(t)/C_0$ versus time t is drawn. The curve is called the *breakthrough curve* which is shown in Figure 2.5.

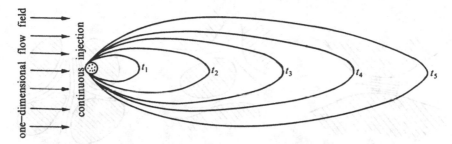

FIGURE 2.4. Tracer distribution in an one-dimensional flow field with a continuously injected point source.

FIGURE 2.5. Schematic breakthrough curve in one dimension.

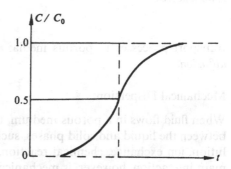

If there is no dispersion, the breakthrough curve should be of the shape shown by the dashed line in the figure, that is, there should be an abrupt concentration interface moving at the mean flow velocity. The observed curve, however, is the one shown in the solid line in Figure 2.5. The mixing of the tracer-containing water and the original water results in a transition zone where the tracer concentration changes gradually. This example also shows the existence of hydrodynamic dispersion.

According to Fried (1975), *hydrodynamic dispersion* is defined as the process of occurrence and development of a transition zone observed in a porous medium, when two miscible moving fluids, each having its own components, are mixed up. Such dispersion is a nonsteady, irreversible process that occurs over time. The original distribution of tracer cannot be recovered by inverting the flow direction.

2.2.2 Mechanisms of Hydrodynamic Dispersion

Hydrodynamic dispersion is a macroscopic phenomenon, but the actual causes are the complex microstructures of the porous media and the non-uniform microscopic movement of the fluid. The explanation for the hydrodynamic dispersion can only be given by a microscopic analysis of the system. In fact, hydrodynamic dispersion is the combined result of two mass

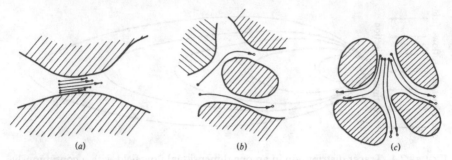

FIGURE 2.6. Non-uniformity of the microscopic flow velocity due to the existence of solid matrices.

transport processes in porous media: *mechanical dispersion* and *molecular diffusion*.

Mechanical Dispersion

When fluid flows in a porous medium, there exist very complex interactions between the liquid and solid phases, such as adsorption, precipitation, dissolution, ion exchange, chemical reaction, and even biological processes. The main interaction, however, is mechanical in nature. As a result of the microstructure of the medium, the fluid velocity varies both in magnitude and in direction inside the pore system. The velocity distribution can roughly be divided into three categories. First, as a result of the fluid viscosity, the velocity in a small channel is at its maximum along the axis, and its minimum near the walls of the channel, as shown in Figure 2.6a. Second, because of the variation in size of the pore space, the maximum velocities along the axis also vary, as shown in Figure 2.6b. Third, the actual movement of fluid particles is on a zigzag path, because of the resistance of the solid matrices and the microscopic stream line fluctuations along the mean flow direction. This is shown in Figure 2.6c.

Because of the stochastic nature of the pore space in porous media and the nonhomogeneity of the microscopic velocity distribution, the tracer particle groups are being separated continuously during the flow process, flowing into finer and more closely woven channels, and occupying ever-increasing space. The result is that the tracer spreads out more than what is expected from just the mean flow velocity. This mass transport phenomenon, resulting from the heterogeneity of the microscopic velocity distribution, is called the *mechanical dispersion*.

Molecular Diffusion

Molecular diffusion is caused by the nonhomogeneous distribution of tracer in a fluid. The tracer molecules in high concentration will move to the low

FIGURE 2.7. The longitudinal and transversal dispersion in hydrodynamic dispersion phenomenon.

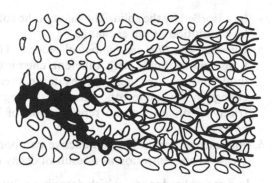

concentration areas to form a uniform concentration distribution. Even in a stationary fluid, the concentration gradient will still cause the tracer to spread out to a larger area.

When a fluid flows in a porous medium, both mechanical dispersion and molecular diffusion are in action. The total effect of them forms the hydrodynamic dispersion. The mechanical dispersion makes the tracer particles move along the channels, while the molecular diffusion causes the tracer concentration to become homogeneous, not only within a channel, but also among channels. Mechanical dispersion plays a major role in the total dispersion. However, when the flow velocity is extremely low, molecular diffusion may become more prominent. Obviously, both mechanical dispersion and molecular diffusion will force the tracer to move longitudinally and transversally with respect to the mean flow direction. The former is called *longitudinal dispersion* and the latter *transversal dispersion*. Figure 2.7 gives an illustration of hydrodynamic dispersion.

2.2.3 Mass Transport in Porous Media

There are many factors which cause changes of solute concentration in porous media. The primary factors are listed below:

1. *Macroscopic advection.* The solute flows on a macroscopic scale with its carrier, the solvent, from one place to another.
2. *Hydrodynamic dispersion.* As mentioned above, the combined effects of mechanical dispersion and molecular diffusion make the solute spread to an even larger area than simple advection.
3. *Sources and sinks of the solute.* The solute may enter or leave the porous area through sources or sinks. Discrete and distributed pollutant sources, wells, discharge ditches and recharge wells are examples of solute sources and sinks.
4. *Adsorption and ion exchange.* Adsorption and ion exchange occur at the interface between the solid and liquid phases. The solute in the liquid may

be adsorbed by the solid. The mass in the solid may also get into the liquid by dissolution or ion exchange.

5. *Chemical reaction and biological process*. There may exist chemical reactions among fluids with different chemical compositions and between fluids and solid particles. For example, precipitation may cause a certain change in the solute concentration. Biological processes such as the putridity of organisms and reproduction of bacteria will also change the concentration.

6. *Radioactive decay*. The radioactive components within the fluid will decrease in concentration as a result of decay over time.

In a generalized model which describes solute concentration behaviors in a porous medium, all of the factors mentioned above should be taken into account. The importance of each factor, however, may differ from case to case.

To study the mass transport in porous media, we can follow two major steps. First, we must clarify the mechanism of solute transport from a microscopic viewpoint and derive the relevant mass conservation and advection-diffusion equations. Then, we can transfer these equations to the macroscopic level by means of spatial averaging.

2.3 Mass Conservation and Convection-Diffusion Equations in a Fluid Continuum

2.3.1 Diffusive Velocities and Fluxes

In a multicomponent fluid, the transport of a component can be resolved into two parts: one is the transport along with the bulk fluid at its mean flow velocity, which is called advection; the other is the molecular diffusion, which is caused by the concentration gradient of the component in the fluid.

The difference between the velocity V_α of component α, and the mean flow velocity V of the fluid, written as $(V_\alpha - V)$, is called the *mass diffusion velocity* of component α. The product of the mass diffusion velocity and the density of component α, is called the *mass diffusion flux*, written as J_α

$$J_\alpha = \rho_\alpha(V_\alpha - V). \qquad (2.3.1)$$

According to the famous *Fick's Law*, mass diffusion obeys the following equation:

$$J_\alpha = -\rho D_d \operatorname{grad}(\rho_\alpha/\rho), \qquad (2.3.2)$$

where ρ is the density of the fluid. For a binary system containing only a solvent and a solute, Eq. (2.3.2) can be written as

$$J = -\rho D_d \operatorname{grad}(C/\rho), \qquad (2.3.3)$$

where J is the mass diffusion flux of the solute and C is the solute concentration. For a dilute solution, the fluid density ρ can be approximated as a constant and the above equation can be simplified to

$$J = -D_d \operatorname{grad} C. \tag{2.3.4}$$

With these assumptions, Fick's Law may be expressed in the following way: the diffusion flux is proportional to the concentration gradient of the solute. The negative sign on the right side indicates that the diffusion goes towards the lower concentration fluid. Coefficient D_d is called the *molecular diffusion coefficient* in the solution. It may be considered as a constant if the solute concentration is relatively low. D_d has the dimensions of $[L^2/T]$, and is a function of the composition and temperature of the solution (Robinson and Stokes, 1965).

2.3.2 Mass Conservation Equation of a Component

Assume that (U) is a cubic elementary volume in a fluid domain centered at $\mathbf{x}$, with the dimensions of $dx_1 \times dx_2 \times dx_3$ as shown in Figure 2.8. Now let us consider the mass balance of a component α in the volume. From the mass conservation principle, the mass change of α in the volume over time dt is equal to the sum of the net mass out flow (or inflow) of component α and the production (or elimination) of α due to chemical reactions or other reasons.

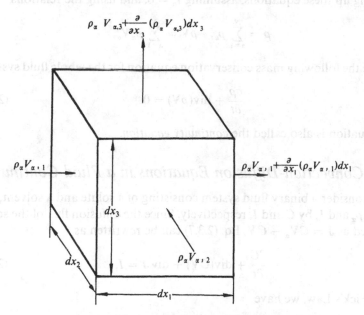

FIGURE 2.8. Mass balance of a component in an elementary volume.

Hence,

$$\left[\frac{\partial}{\partial x_1}(\rho_\alpha V_{\alpha,1}) dx_1 \right] dx_2\, dx_3\, dt + \left[\frac{\partial}{\partial x_2}(\rho_\alpha V_{\alpha,2}) dx_2 \right] dx_1\, dx_3\, dt$$

$$+ \left[\frac{\partial}{\partial x_3}(\rho_\alpha V_{\alpha,3}) dx_3 \right] dx_1\, dx_2\, dt - I_\alpha\, dx_1\, dx_2\, dx_3\, dt$$

$$= -\left(\frac{\partial \rho_\alpha}{\partial t} dt \right) dx_1\, dx_2\, dx_3, \tag{2.3.5}$$

where $V_{\alpha,1}$, $V_{\alpha,2}$, $V_{\alpha,3}$ are components of V_α, the velocity of component α; x_1, x_2 and x_3 are coordinates, and I_α is the rate of producing α per unit volume. Rearranging Eq. (2.3.5), we have

$$\frac{\partial \rho_\alpha}{\partial t} + \frac{\partial(\rho_\alpha V_{\alpha,1})}{\partial x_1} + \frac{\partial(\rho_\alpha V_{\alpha,2})}{\partial x_2} + \frac{\partial(\rho_\alpha V_{\alpha,3})}{\partial x_3} = I_\alpha \tag{2.3.6}$$

or

$$\frac{\partial \rho_\alpha}{\partial t} + \mathrm{div}(\rho_\alpha V_\alpha) = I_\alpha. \tag{2.3.7}$$

This is the mass conservation equation for a component in the solution. The same rule may be applied to all the other components in the fluid system. Summing up these equations, assuming $I_\alpha = 0$, and using the relations

$$\rho = \sum_{\alpha=1}^{N} \rho_\alpha, \quad \rho V = \sum_{\alpha=1}^{N} \rho_\alpha V_\alpha,$$

we have the following mass conservation equation for the whole fluid system:

$$\frac{\partial \rho}{\partial t} + \mathrm{div}(\rho V) = 0. \tag{2.3.8}$$

This equation is also called the *continuity equation*.

2.3.3 Convection-Diffusion Equations in a Fluid Continuum

Let us consider a binary fluid system consisting of a solute and a solvent, and replace ρ_α and I_α by C and I, respectively. Since the diffusion flux of the solute is defined as $J = CV_\alpha - CV$, Eq. (2.3.7) can be rewritten as

$$\frac{\partial C}{\partial t} + \mathrm{div}(CV) + \mathrm{div}\, J = I, \tag{2.3.9}$$

Using Fick's Law, we have

$$\frac{\partial C}{\partial t} + \mathrm{div}\left[CV - \rho D_d \mathrm{grad}\left(\frac{C}{\rho} \right) \right] = I. \tag{2.3.10}$$

This equation is called the *convection-diffusion equation* of the solute. It is a second-order parabolic partial differential equation. With certain initial and boundary conditions, a unique solution, the concentration distribution C, can be obtained.

Equation (2.3.10) is derived for a continuous fluid system. If the solute transport problem is discussed at the microscopic level, boundary conditions have to be given along the solid matrices of the porous medium. As mentioned above, however, these boundary conditions are impossible to define due to the complexity of microscopic constructures of porous media. We have to study the problem macroscopically by means of the spatial average method.

2.4 Hydrodynamic Dispersion Equations

2.4.1 The Average of Time Derivatives

In order to obtain the spatial average for Eq. (2.3.9), the following integration must be calculated:

$$\left(\overline{\frac{\partial C}{\partial t}}\right) = \frac{1}{U_{0,\gamma}} \int_{[U_{0,\gamma}]} \frac{\partial C}{\partial t} \, dU_{0,\gamma}, \tag{2.4.1}$$

where $[U_{0,\gamma}]$ is the volume occupied by phase γ (solution) in the REV $[U_0]$. If θ_γ is the volumetric fraction of phase γ, then we have $U_{0,\gamma} = \theta_\gamma U_0$. Consider a generalized case which is applicable for both unsaturated zones and deformable solid matrices, that is, $[U_{0,\gamma}]$ or θ_γ is time dependent. Write $[U_{0,\gamma}]$ at times t and $t + \Delta t$ respectively as follows:

$$[U_{0,\gamma}]_t = [U_1] + [U_2],$$

and

$$[U_{0,\gamma}]_{t+\Delta t} = [U_2] + [U_3].$$

As shown in Figure 2.9, $[U_2]$ is their common part. From the definition of derivative, we have

$$\frac{\partial}{\partial t} \int_{[U_{0,\gamma}]} C \, dU_{0,\gamma} = \lim_{\Delta t \to 0} \frac{1}{\Delta t} \left\{ \int_{[U_2]+[U_3]} C(t + \Delta t) \, dU_{0,\gamma} - \int_{[U_1]+[U_2]} C(t) \, dU_{0,\gamma} \right\}$$

$$= \lim_{\Delta t \to 0} \frac{1}{\Delta t} \left\{ \int_{[U_2]} [C(t + \Delta t) - C(t)] \, dU_{0,\gamma} \right.$$

$$\left. + \int_{[U_3]} C(t + \Delta t) \, dU_{0,\gamma} - \int_{[U_1]} C(t) \, dU_{0,\gamma} \right\}. \tag{2.4.2}$$

Furthermore, with the moving velocity $\mathbf{u}$ and the normal direction $\mathbf{n}$ of the boundary (ABC) for $[U_{0,\gamma}]$ in Figure 2.9, the infinitesimal volume element of

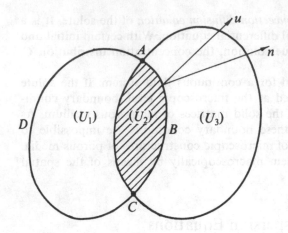

FIGURE 2.9. Schematic explanation of spatial average for the time derivative.

$[U_3]$ may be expressed as

$$dU_{0,\gamma} = \mathbf{u} \cdot \mathbf{n} dS_{0,\gamma} \Delta t = u_n dS_{0,\gamma} \Delta t,$$

where $[S_{0,\gamma}]$ is the interface between phase γ and other phases. Thus, the second integral on the right side of Eq. (2.4.2) can be simplified to

$$\int_{[U_3]} C(t + \Delta t) dU_{0,\gamma} = \Delta t \int_{(ABC)} C u_n dS_{0,\gamma}, \qquad (2.4.3)$$

and similarly,

$$\int_{[U_1]} C(t) dU_{0,\gamma} = -\Delta t \int_{(ADC)} C u_n dS_{0,\gamma}. \qquad (2.4.4)$$

By substituting Eqs. (2.4.3) and (2.4.4) into Eq. (2.4.2), and letting $\Delta t \to 0$, we then have

$$\frac{\partial}{\partial t} \int_{[U_{0,\gamma}]} C \, dU_{0,\gamma} = \int_{[U_{0,\gamma}]} \frac{\partial C}{\partial t} dU_{0,\gamma} + \int_{[S_{0,\gamma}]} C u_n dS_{0,\gamma}. \qquad (2.4.5)$$

By obtaining $\int_{[U_{0,\gamma}]} \partial C/\partial t \, dU_{0,\gamma}$ from the above equation, and inserting it into Eq. (2.4.1), we have

$$\left(\overline{\frac{\partial C}{\partial t}} \right) = \frac{1}{\theta_\gamma} \left\{ \frac{\partial}{\partial t} \left(\frac{1}{U_0} \int_{[U_{0,\gamma}]} C \, dU_{0,\gamma} \right) - \frac{1}{U_0} \int_{[S_{0,\gamma}]} C u_n dS_{0,\gamma} \right\}$$

$$= \frac{1}{\theta_\gamma} \left\{ \frac{\partial}{\partial t} (\theta_\gamma \overline{C}) - \langle C u_n \rangle M_\gamma \right\}, \qquad (2.4.6)$$

where $\langle C u_n \rangle$ is the average value over the area defined by

$$\langle C u_n \rangle = \frac{1}{S_{0,\gamma}} \int_{[S_{0,\gamma}]} C u_n dS_{0,\gamma}, \qquad (2.4.7)$$

and $M_\gamma = S_{0,\gamma}/U_0$ is the specific surface of phase γ.

If $u = 0$, that is, there is no deformation of $[U_{0,\gamma}]$, Eq. (2.4.6) is simplified to

$$\left(\overline{\frac{\partial C}{\partial t}}\right) = \frac{\partial \overline{C}}{\partial t}. \tag{2.4.8}$$

The left-hand side of Eq. (2.4.8) is the spatial average of the derivative of concentration with respect to time, while the right side is the derivative of spatial averaged concentration with respect to time. The former indicates the macroscopic mean of concentration change rate while the latter indicates the macroscopic change rate of the mean concentration. Only when $[U_{0,\gamma}]$ does not change with time can they be equal to each other.

2.4.2 The Average of Spatial Derivatives

We can now prove that for a differentiable function, G_α, defined at the microscopic level for phase γ, the following equation is always true:

$$\left(\overline{\frac{\partial G_\gamma}{\partial x_i}}\right) = \frac{1}{\theta_\gamma}\left\{\frac{\partial}{\partial x_i}(\theta_\gamma \overline{G_\gamma}) + \langle G_\gamma n_i \rangle M_\gamma\right\}, \tag{2.4.9}$$

where x_i ($i = 1, 2, 3$) is the spatial coordinate, n_i the component of a unit outer vector normal to $[S_{0,\gamma}]$ in x_i direction, M_γ the specific surface of phase γ, and $\langle G_\gamma n_i \rangle$ is the average of $G_\gamma n_i$ over the surface $[S_{0,\gamma}]$.

Let the volumes of $[U_{0,\gamma}]$ at point $\mathbf{x}$ and $\mathbf{x} + \Delta x_i e_i$ be, respectively:

$$[U_{0,\gamma}(\mathbf{x})] = [U_1] + [U_2],$$

and

$$[U_{0,\gamma}(\mathbf{x} + \Delta x_i e_i)] = [U_2] + [U_3],$$

where e_i is the unit vector in the i-direction. Similar to the last section, we have:

$$\frac{\partial}{\partial x_i}\int_{[U_{0,\gamma}]} G_\gamma \, dU_{0,\gamma} = \int_{[U_{0,\gamma}]} \frac{\partial G_\gamma}{\partial x_i} dU_{0,\gamma} + \lim_{\Delta x_i \to 0} \frac{1}{\Delta x_i}\int_{[U_3]} G_\gamma \, dU_{0,\gamma}$$

$$- \lim_{\Delta x_i \to 0} \frac{1}{\Delta x_i}\int_{[U_1]} G_\gamma \, dU_{0,\gamma}. \tag{2.4.10}$$

The infinitesimal volume element of $[U_3]$ can be expressed as $dU_{0,\gamma} = -n_i \Delta x_i \, dS_{0,\gamma}$, and $[U_1]$ as $dU_{0,\gamma} = n_i \Delta x_i \, dS_{0,\gamma}$, so Eq. (2.4.10) is changed to

$$\frac{\partial}{\partial x_i}\int_{[U_{0,\gamma}]} G_\gamma \, dU_{0,\gamma} = \int_{[U_{0,\gamma}]} \frac{\partial G_\gamma}{\partial x_i} dU_{0,\gamma} - \int_{[S_{0,\gamma}]} G_\gamma n_i \, dS_{0,\gamma} \tag{2.4.11}$$

and, finally, we have:

$$\left(\overline{\frac{\partial G_\gamma}{\partial x_i}}\right) = \frac{1}{\theta_\gamma U_0}\int_{[U_{0,\gamma}]} \frac{\partial G_\gamma}{\partial x_i} dU_{0,\gamma}$$

$$= \frac{1}{\theta_\gamma}\left\{\frac{\partial}{\partial x_i}\frac{1}{U_0}\int_{[U_{0,\gamma}]} G_\gamma \, dU_{0,\gamma} + \frac{1}{U_0}\int_{[S_{0,\gamma}]} G_\gamma n_i \, dS_{0,\gamma}\right\}$$

$$= \frac{1}{\theta_\gamma}\left\{\frac{\partial}{\partial x_i}(\theta_\gamma \overline{G_\gamma}) + \langle G_\gamma n_i \rangle M_\gamma\right\},$$

which is Eq. (2.4.9). From this equation the average divergence of an arbitrary vector $\mathbf{A}$, $(\overline{\text{div}\,\mathbf{A}})$, can be determined. By applying Eq. (2.4.9) to the three components of vector $\mathbf{A}$ and summing them up, we then have:

$$(\overline{\text{div}\,\mathbf{A}}) = \frac{1}{\theta_\gamma}[\text{div}(\theta_\gamma\overline{\mathbf{A}}) + \langle A_n\rangle M_\gamma], \tag{2.4.12}$$

where A_n is the projection of vector $\mathbf{A}$ onto the outer normal of $[S_{0,\gamma}]$. From the above equation we find that only when $A_n = 0$ and θ_γ is constant can we have

$$(\overline{\text{div}\,\mathbf{A}}) = \text{div}\,\overline{\mathbf{A}}. \tag{2.4.13}$$

2.4.3 Advection-Dispersion Equations in Porous Media

By adding a source and sink term to Eq. (2.3.9) and applying spatial averaging, we then have

$$\left(\overline{\frac{\partial C}{\partial t}}\right) + \overline{\text{div}(C\mathbf{V})} + \overline{\text{div}\,\mathbf{J}} = \bar{I}. \tag{2.4.14}$$

Substituting Eqs. (2.4.6) and (2.4.12) for the relevant terms into Eq. (2.4.14) yields:

$$\frac{\partial(\theta_\gamma\overline{C})}{\partial t} + \text{div}(\theta_\gamma\overline{C\mathbf{V}}) + \text{div}(\theta_\gamma\overline{\mathbf{J}}) + \langle J_n + C(V_n - u_n)\rangle M_\gamma = \theta_\gamma\bar{I}. \tag{2.4.15}$$

From Eq. (2.1.9) we know

$$\overline{C\mathbf{V}} = \overline{C}\,\overline{\mathbf{V}} + \overset{\circ}{C}\overset{\circ}{\mathbf{V}}, \tag{2.4.16}$$

Substituting Eq. (2.4.16) into Eq. (2.4.15) and rearranging the equation, we have

$$\frac{\partial(\theta_\gamma\overline{C})}{\partial t} = -\text{div}(\theta_\gamma\overline{C}\,\overline{\mathbf{V}}) - \text{div}(\theta_\gamma\overset{\circ}{C}\overset{\circ}{\mathbf{V}}) - \text{div}(\theta_\gamma\overline{\mathbf{J}})$$
$$- \langle J_n + C(V_n - u_n)\rangle M_\gamma + \theta_\gamma\bar{I}. \tag{2.4.17}$$

The is a mass conservation equation. With reference to the microscopic mass conservation Eq. (2.3.9), the meaning of each term in Eq. (2.4.17) is explained as follows. The left-side term denotes the change rate of solute with time in the REV. The first term on the right-hand side shows the macroscopic advection effect; the second term is an extra one generated during the spatial averaging procedure. This extra term is related to the fluctuation of microscopic velocity with respect to the mean flow velocity. Hence, it is the result of mechanical dispersion. The third term is the macroscopic molecular diffusion term. The fourth term is also an extra one which shows the solute exchanges through interfaces between phase γ and other phases by means of diffusion dissolution, adsorption, ion exchange and so forth. The fifth term is the source and sink term, which gives the increase, or decrease, of solute in the system due to recharge, discharge, chemical reactions, radioactive decay, and so forth.

All possible factors relevant to mass transport in porous media have been taken into consideration in Eq. (2.4.17). Thus, the equation is also applicable for the cases of multicomponent flow and deformable solid matrix.

The classical hydrodynamic dispersion theory assumes that both macroscopic molecular diffusion flux $\bar{\mathbf{J}}$ and mechanical dispersion flux $\overset{\circ\circ}{C\mathbf{V}}$ can be expressed in the form of Fick's Law:

$$\overset{\circ\circ}{C\mathbf{V}} = -\bar{\rho}\mathbf{D}'\,\mathrm{grad}(\bar{C}/\bar{\rho}), \qquad (2.4.18)$$

$$\bar{\mathbf{J}} = -\bar{\rho}\mathbf{D}''\,\mathrm{grad}(\bar{C}/\bar{\rho}), \qquad (2.4.19)$$

where coefficient $\mathbf{D}'$ is called the *mechanical dispersion coefficient* and $\mathbf{D}''$ the *molecular diffusion coefficient* in porous media, both of which are second rank symmetric tensors. Their expressions and relationships with other physical parameters will be discussed in detail in Section 2.6.

Adding Eq. (2.4.18) and Eq. (2.4.19) together, we have

$$\mathbf{J}^* = -\bar{\rho}\mathbf{D}\,\mathrm{grad}(\bar{C}/\bar{\rho}), \qquad (2.4.20)$$

where $\mathbf{J}^* = \bar{\mathbf{J}} + \overset{\circ\circ}{C\mathbf{V}}$ is called the *hydrodynamic dispersion flux*. $\mathbf{J}^*$ is a summation of the mechanical dispersion flux and the macroscopic molecular diffusion flux. The coefficient

$$\mathbf{D} = \mathbf{D}' + \mathbf{D}'' \qquad (2.4.21)$$

is called the *hydrodynamic dispersion coefficient*.

Let us temporarily put aside the last two terms in Eq. (2.4.17). In order to concentrate on the main terms of the equation, the effects of adsorption, ion exchange, chemical reaction, and radioactive decay will be discussed later. By inserting Eq. (2.4.20) into Eq. (2.4.17) and omitting the subscript of θ_y, which indicates the volumetric fraction of the liquid phase, we then have

$$\frac{\partial(\theta\bar{C})}{\partial t} = \mathrm{div}\left[\theta\mathbf{D}\bar{\rho}\,\mathrm{grad}\left(\frac{\bar{C}}{\bar{\rho}}\right)\right] - \mathrm{div}(\theta\bar{C}\overline{\mathbf{V}}). \qquad (2.4.22)$$

This is the *hydrodynamic dispersion equation* of a solute in a porous medium, which is also called the *advection-dispersion equation*. The first term on the right side is the *dispersion term* and the second the *advection term*. The derivation procedure shows that this equation is a combination of the mass conservation equation Eq. (2.4.17) and the linear dispersion law Eq. (2.4.20). It gives a quantitative description of the hydrodynamic dispersion mechanism given in Section 2.2.

For a solute with low concentration, $\bar{\rho}$ is approximately constant, and Eq. (2.4.22) is reduced to

$$\frac{\partial(\theta\bar{C})}{\partial t} = \mathrm{div}(\theta\mathbf{D}\,\mathrm{grad}\bar{C}) - \mathrm{div}(\theta\bar{C}\mathbf{V}). \tag{2.4.23}$$

Furthermore, assume that θ is a constant and let $\bar{V}_i$ ($i = 1, 2, 3$) be the three components of the mean flow velocity $\bar{\mathbf{V}}$, and $D_{11}, D_{12}, \ldots, D_{33}$ be the nine components of the second rank symmetric tensor $\mathbf{D}$. Using Einstein's summation convention, Eq. (2.4.23) can be expressed in the following form:

$$\frac{\partial\bar{C}}{\partial t} = \frac{\partial}{\partial x_i}\left(D_{ij}\frac{\partial\bar{C}}{\partial x_j}\right) - \frac{\partial}{\partial x_i}(\bar{V}_i\bar{C}). \tag{2.4.24}$$

Readers of this book should be familiar with this compact form.

The above derivation of the hydrodynamic dispersion equation is based on the concept of REV and the assumption that the Fick's law is valid. To derive the hydrodynamic dispersion equation from molecular physical principles needs more rigorous argument (Sposito et al., 1979, 1986).

2.4.4 The Integral Form of Hydrodynamic Dispersion Equations

The integral form of the advection-dispersion equation for a porous medium is useful, not only in describing the characteristics of the equation, but also in deriving its numerical solutions.

Consider an arbitrary mass balance volume (R) in a porous medium. Let the boundary of (R) be surface (S) with outer normal vector $\mathbf{n}$, see Figure 2.10. For simplification, assume that the fluid density, ρ, is constant. The factors that cause the solute mass to change inside (R) are listed below:

1. Due to advection, M_1 units of the solute mass enter into (R) through (S) per unit time.
2. Due to hydrodynamic dispersion, M_2 units of the solute mass enter into (R) through (S) per unit time.
3. Due to injection, chemical reaction and ion exchange between the phases, M_3 units of the solute mass are generated in (R) per unit time.

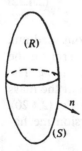

FIGURE 2.10. Mass balance volume in a porous medium.

According to mass conservation, the sum of the above three factors should be equal to the solute increase ΔM in liquid phase of (R), i.e.,

$$\Delta M = M_1 + M_2 + M_3. \tag{2.4.25}$$

From the physical meanings of the terms mentioned above, we have

$$M_1 = -\int_{(S)} \theta \overline{C}\mathbf{V} \cdot \mathbf{n}\,dS; \tag{2.4.26}$$

$$M_2 = -\int_{(S)} \theta \mathbf{J}^* \cdot \mathbf{n}\,dS; \tag{2.4.27}$$

$$M_3 = \int_{(R)} \theta \overline{I}\,dR, \tag{2.4.28}$$

where $\overline{C}\mathbf{V}$ is the advection flux, $\mathbf{J}^*$ the hydrodynamic dispersion flux, θ the volumetric fraction of the liquid phase, and $\theta\overline{I}$ gives the solute production rate in (R), that is, the mass produced in (R) per unit volume and per unit time. Note that the mass of solute per unit volume in a porous medium is $\theta\overline{C}$, therefore

$$\Delta M = \int_{(R)} \frac{\partial(\theta\overline{C})}{\partial t}\,dR. \tag{2.4.29}$$

Substituting Eqs. (2.4.26) through (2.4.29) into Eq. (2.4.25), and using the linear dispersion law

$$\mathbf{J}^* = -\mathbf{D}\,\text{grad}\overline{C}, \tag{2.4.30}$$

we then have

$$\int_{(R)} \frac{\partial(\theta\overline{C})}{\partial t}\,dR = \int_{(S)} \theta\mathbf{D}\,\text{grad}\overline{C}\cdot\mathbf{n}\,dS - \int_{(S)} \theta\overline{C}\mathbf{V}\cdot\mathbf{n}\,dS + \int_{(R)} \theta\overline{I}\,dR. \tag{2.4.31}$$

This is the *integral form of the hydrodynamic dispersion equation* with $\overline{C}$ unknown. It is evident from the derivation above that the equation is also a combination of the mass conservation equation (2.4.25) and the linear dispersion law equation (2.4.30). Similar to the groundwater flow equations, the transfer between integral and differential forms of the hydrodynamic dispersion equations can be done by using Green's formula.

2.5 Coefficients of Hydrodynamic Dispersion

2.5.1 Coefficients of Longitudinal Dispersion and Transverse Dispersion

The coefficient of hydrodynamic dispersion is a tensor, but its tensor characteristics will not be easily recognized if the study is restricted to one-dimensional dispersion phenomena. That is why in the early studies the dispersion coefficient was treated as a scalar, as the relationship between dispersion and flow direction was ignored. In fact, even in an isotropic porous medium the dispersions in the flow and the cross sectional directions are different. In an anisotropic medium, it becomes even more complicated.

A vast number of experiments have been done to determine the relationships between the dispersion coefficient and velocity distribution, as well as the molecular diffusion coefficient. Early experiments were mainly concerned with the longitudinal dispersion coefficient, D_L. In one experiment, a fluid with a constant tracer concentration is introduced at one end of a sand column. The concentration of the effluent is measured and compared with the analytic solution of a dispersion model. The longitudinal dispersion coefficient, D_L, is then obtained from this comparison. It can be proved by dimensional analysis that the dimensionless number D_L/D_d is the function of another dimensionless number

$$Pe = \frac{\bar{V}d}{D_d},\tag{2.5.1}$$

where Pe is called the *Peclet number*. $\bar{V}$ is the mean pore velocity in the porous medium and d is a characteristic length of the medium, such as particle diameter. D_d is the diffusion coefficient in solution. A large number of experiments resulted in curves such as those shown in Figure 2.11. The curve may be roughly divided into five zones.

Zone I. In this zone, molecular diffusion plays the predominant role. D_L/D_d is nearly a constant, that is, $D_L = D_d T$ (because the flow velocity is very low, the mechanical dispersion can be ignored). Proportionality coefficient T is actually the tortuosity of the medium, and thus is always less than unity, because the porous medium retards the progress of molecular diffusion in the liquid phase. With the zero mean flow velocity experiment ($q = 0$), both D_L

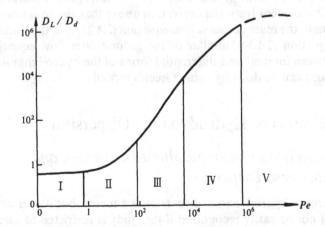

FIGURE 2.11. Relationship between longitudinal dispersion and Peclet number.

and the tortuosity T can be determined. For loose sand or similar porous media, D_L can be approximated by $2/3\ D_d$.

Zone II. The Peclet number is approximately 0.4 to 5 and the curve has an upward tendency. In this zone, mechanical dispersion and molecular diffusion become comparable, so that both effects need to be taken into consideration.

Zone III. Mass transport is mainly the result of mechanical dispersion and transverse molecular diffusion. Transverse diffusion spreads the mass and actually decreases the longitudinal mass transport. Experiments give the following result:

$$D_L/D_d \approx \alpha(Pe)^m; \ \alpha \approx 0.5, \quad 1 < m < 1.2, \qquad (2.5.2)$$

which means that the longitudinal dispersion coefficient is proportional to the mean flow velocity with power ranging from 1 to 1.2.

Zone IV. In this region, the effect of molecular diffusion can be negligible, but the flow velocity still obeys Darcy's Law. The straight line segment in Figure 2.11 gives

$$D_L/D_d \approx \beta Pe, \quad \beta \approx 1.8, \qquad (2.5.3)$$

which means the coefficient of longitudinal dispersion is proportional to the mean flow velocity.

Zone V. This is also a zone of pure mechanical dispersion with negligible molecular diffusion. Due to the high mean flow velocity, the effects of inertia and turbulence cannot be ignored. They retard the longitudinal mass transport and, as a result, the slope of this part of the curve is reduced.

As to the transversal dispersion, similar results have been obtained by experiments. For zone III, for example, we have

$$D_T/D_d \approx \alpha(Pe)^m, \quad \alpha \approx 0.025, \quad m = 1.1, \qquad (2.5.4)$$

and, in zone IV, m in Eq. (2.5.4) changes to unity. This means that D_T is also proportional to the mean flow velocity in this region when the mechanical dispersion is predominant.

Klotz et al. (1980) conducted a large number of laboratory and field experiments which dealt with the relationship between longitudinal and transversal dispersion coefficients in loose soil and the mean flow velocity. They suggested the following form for the longitudinal dispersion coefficient

$$D_L = \alpha_L \overline{V}^m, \qquad (2.5.5)$$

where α_L is called the *longitudinal dispersivity*, $\overline{V}$ is the mean flow velocity and m is the power of the velocity to be determined. Experimental results show that α_L is mainly dependent on the mean particle size and the uniformity coefficient, $U = d_{60}/d_{10}$. Laboratory test results are reproduced in Table 2.1. The range of m given in the table is

$$1.07 \le m \le 1.1. \qquad (2.5.6)$$

TABLE 2.1. Logitudinal dispersivity in laboratory scale.

Range of particle size (mm)	Mean particle size d_{50} (mm)	Uniformity coefficient	Exponential m	Logitudinal dispersivity α_L	Minimum mean flow velocity (m/s)
0.4–0.7	0.61	1.55	1.09	3.96×10^{-3}	10^{-5}
0.5–1.5	0.75	1.85	1.10	5.78×10^{-3}	8×10^{-5}
1–2	1.6	1.6	1.10	8.80×10^{-3}	1.5×10^{-4}
2–3	2.7	1.3	1.09	1.30×10^{-2}	2×10^{-4}
5–7	6.3	1.3	1.09	1.67×10^{-2}	3×10^{-4}
0.5–2	1.0	2	1.08	3.11×10^{-3}	5×10^{-3}
0.2–5	1.0	5	1.08	8.30×10^{-3}	5×10^{-3}
0.1–10	1.0	10	1.07	1.63×10^{-2}	5×10^{-3}
0.05–20	1.0	20	1.07	7.07×10^{-2}	5×10^{-3}

The longitudinal dispersivity, α_L, increases with the uniformity coefficient, U. According to Klotz et al. (1980), field experiments of single-well pumping with multi-well observations gave an approximate value of m as 1.05 and $\alpha_L = 5m$. The explanation of much larger α_L in a field test is that the uniformity coefficients of the media in the field are larger than that in the laboratory. Other reports, such as Fried (1975), Anderson (1979), and Sudicky et al. (1983) also show that α_L values determined by field tests are several orders of magnitude higher than the laboratory ones. It was found that the magnitude of α_L is dependent on the scale of experiments. Recently, Gelhar et al. (1992) reviewed 59 different field sites and denoted that the longitudinal dispersivities range from 10^{-2} to 10^4 (m) for scales ranging from 10^{-1} to 10^5 (m), but the largest value for higher reliability data was only 250 (m).

For transversal dispersion coefficient, D_T, a similar expression can be used:

$$D_T = \alpha_T \bar{V}^m, \tag{2.5.7}$$

where α_T is called the *transversal dispersivity*. Klotz et al. (1980) obtained the same m value as in Eq. (2.5.6), but α_T was 6 to 20 times smaller than α_L. For more experimental results and explanations about the transversal dispersivity, the reader may refer to Gelhar et al. (1992).

In Chapter 7, we shall introduce more field experiments designed for different scales, give the methods of identifying dispersivities based on observations, and further discuss the "scale effect" problem in the statistical framework.

2.5.2 Coefficients of Mechanical Dispersion and Molecular Diffusion

The experiments mentioned above are still unable to give a generalized explanation for the tensor characteristics of the dispersion coefficient. Spatial averaging can be used to derive the hydrodynamic equation, but to obtain expressions of dispersion coefficients, theoretical models have to be used. Bear and Bachmet (1967) and Bear (1972) recommended the following expression for mechanical dispersion coefficients using their statistical geometrical model (Bear, 1972):

$$D'_{ij} = \alpha_{ijmn} \frac{\overline{V}_m \overline{V}_n}{\overline{V}} f(Pe, \delta), \tag{2.5.8}$$

where Einstein's summation convention is implied. α_{ijmn} is a component of a fourth rank tensor, called the *dispersivity of porous media*, and has the dimension of length. $\overline{V}_m$ and $\overline{V}_n$ $(m, n = 1, 2, 3)$ are components of mean pore velocity. The function

$$f(Pe, \delta) = \frac{Pe}{Pe + 4\delta^2 + 2} \tag{2.5.9}$$

where $Pe = \overline{L}\overline{V}/D_d$, $\delta = \overline{L}/\bar{a}$ is a parameter describing the shape of channels in the porous medium with $\overline{L}$ as their mean length and $\bar{a}$ the characteristic length of their cross sections. Since Pe depends on the molecular diffusion coefficient D_d, from Eqs. (2.5.1) and (2.5.8), we can find that the mechanical dispersion coefficient, D'_{ij}, also depends on D_d. This means that mechanical dispersion coefficient and molecular diffusion coefficient are actually interrelated. In a hydrodynamic dispersion process, molecular diffusion has two effects. One is the mean macroscopic effect of microscopic diffusion, which exists even when the mean flow velocity is zero. The other is that diffusion will cause the solute to transfer between flowlines, and hence, affect the progress of mechanical dispersion. Normally, the second effect is relatively weak and $f(Pe, \delta)$, as in Eq. (2.5.8), can be approximated as unity.

In a three-dimensional flow, the dispersivity tensor generally has 81 components. However, for an isotropic medium, only 36 of them are non-zero. All of them are related to two constants, α_L and α_T. Scheidegger (1961) suggested the following expression:

$$\alpha_{ijmn} = \alpha_T \delta_{ij} \delta_{mn} + \frac{\alpha_L - \alpha_T}{2} (\delta_{im} \delta_{jn} + \delta_{in} \delta_{jm}), \tag{2.5.10}$$

where δ_{ij} is the Kronecker delta, which is defined as

$$\delta_{ij} = \begin{cases} 0 & \text{when } i \neq j \\ 1 & \text{when } i = j, \end{cases} \tag{2.5.11}$$

and δ_{mn}, δ_{im} and so forth, have the same meanings as δ_{ij}. In Eq. (2.5.10), α_L is called the *longitudinal dispersivity of the isotropic medium* and α_T the *transversal dispersivity of the isotropic medium*.

Substituting Eq. (2.5.10) into Eq. (2.5.8) and letting $f(Pe, \delta) = 1$, we then have

$$D'_{ij} = \alpha_T \overline{V} \delta_{ij} + (\alpha_L - \alpha_T) \frac{\overline{V}_i \overline{V}_i}{\overline{V}}. \tag{2.5.12}$$

In the case of three-dimensional flow, Eq. (2.5.12) determines the nine components of the mechanical dispersion coefficient tensor,

$$\mathbf{D}' = \begin{bmatrix} D'_{11} & D'_{12} & D'_{13} \\ D'_{21} & D'_{22} & D'_{23} \\ D'_{31} & D'_{32} & D'_{33} \end{bmatrix}, \tag{2.5.13}$$

they are

$$D'_{11} = [\alpha_T(\overline{V}_2^2 + \overline{V}_3^2) + \alpha_L \overline{V}_1^2]/\overline{V}$$

$$D'_{12} = D'_{21} = (\alpha_L - \alpha_T)\overline{V}_1 \overline{V}_2/\overline{V}$$

$$D'_{13} = D'_{31} = (\alpha_L - \alpha_T)\overline{V}_1 \overline{V}_3/\overline{V}$$

$$D'_{22} = [\alpha_T(\overline{V}_1^2 + \overline{V}_3^2) + \alpha_L \overline{V}_2^2]/\overline{V} \tag{2.5.14}$$

$$D'_{23} = D'_{32} = (\alpha_L - \alpha_T)\overline{V}_2 \overline{V}_3/\overline{V}$$

$$D'_{33} = [\alpha_T(\overline{V}_1^2 + \overline{V}_2^2) + \alpha_L \overline{V}_3^2]/\overline{V}$$

If axis x_1 of the Cartesian coordinates coincides with the mean flow direction, and both axes x_2 and x_3 are perpendicular to it, Eq. (2.5.14) can be simplified to

$$D'_{11} = \alpha_L \overline{V}; \quad D'_{22} = D'_{33} = \alpha_T \overline{V}; \quad D'_{ij} = 0 \ (i \neq j), \tag{2.5.15}$$

and thus, we have

$$\mathbf{D}' = \begin{bmatrix} \alpha_L \overline{V} & 0 & 0 \\ 0 & \alpha_T \overline{V} & 0 \\ 0 & 0 & \alpha_T \overline{V} \end{bmatrix}. \tag{2.5.16}$$

The axes defined in this way are called the principal axes of dispersion. If a flow field is not unidirectional, the principal axes of dispersion will change with the velocity field. The dispersion coefficient is structurally different from that of permeability, because even in an isotropic medium it is still a tensor, and its principal axes are determined by the mean flow direction, rather than the medium.

The coefficient of total dispersion is the sum of the mechanical dispersion coefficient, $\mathbf{D}'$, and the *molecular diffusion coefficient in porous media*, $\mathbf{D}'' = D_d\mathbf{T}$, where D_d is the molecular diffusion coefficient in solution, and $\mathbf{T}$ is the tortuosity of the porous medium. For an isotropic porous medium, it reduces to a scalar T, and $0 < T < 1$. Thus, the coefficient of total dispersion can be expressed as

$$D_{ij} = D'_{ij} + D''_{ij}$$

$$= \alpha_T \overline{V}\delta_{ij} + (\alpha_L - \alpha_T)\frac{\overline{V}_i \overline{V}_j}{\overline{V}} + D_d T\delta_{ij}. \tag{2.5.17}$$

For two-dimensional flow, we have

$$D_{11} = (\alpha_L \overline{V}_1^2 + \alpha_T \overline{V}_2^2)/\overline{V} + D_d T$$

$$D_{12} = D_{21} = (\alpha_L - \alpha_T)\overline{V}_1 \overline{V}_2/\overline{V} \qquad (2.5.18)$$

$$D_{22} = (\alpha_T \overline{V}_1^2 + \alpha_L \overline{V}_2^2)/\overline{V} + D_d T.$$

2.6 Extensions and Subsidiary Conditions of the Hydrodynamic Dispersion Equation

2.6.1 Hydrodynamic Dispersion Equations in Orthogonal Curvilinear Coordinate Systems

In this section we will derive expressions for the hydrodynamic dispersion equation in different coordinate systems, the source and sink terms, and the relevant initial and boundary conditions, in order to obtain a complete mathematical statement of a hydrodynamic dispersion problem. Since we are concentrating only on the problems relevant to porous media, all of the parameters under consideration are their macroscopic averages. Hence, the bars above the symbols will be omitted. For example, unless otherwise specified, parameters V, c, ρ and μ, etc., all denote their macroscopic means.

Let (x_1, x_2, x_3) be the coordinates of point P in the Cartesian coordinate system and (u_1, u_2, u_3) be the coordinates of the same point in a curvilinear coordinate system. The relationship between the two systems is

$$x_i = g_i(u_1, u_2, u_3); \quad i = 1, 2, 3. \qquad (2.6.1)$$

We require that the Jacobian

$$J = \frac{D(x_1, x_2, x_3)}{D(u_1, u_2, u_3)} = \begin{vmatrix} \dfrac{\partial x_1}{\partial u_1} & \dfrac{\partial x_2}{\partial u_1} & \dfrac{\partial x_3}{\partial u_1} \\[2mm] \dfrac{\partial x_1}{\partial u_2} & \dfrac{\partial x_2}{\partial u_2} & \dfrac{\partial x_3}{\partial u_2} \\[2mm] \dfrac{\partial x_1}{\partial u_3} & \dfrac{\partial x_2}{\partial u_3} & \dfrac{\partial x_3}{\partial u_3} \end{vmatrix} \qquad (2.6.2)$$

is non-zero, so that the two coordinate systems have a one-to-one correspondence.

If the three coordinate directions of the curvilinear coordinate system are perpendicular to each other, the system is called an *orthogonal curvilinear coordinate system*. Let

$$h_i = \left| \frac{\partial \mathbf{r}}{\partial u_i} \right|; \quad i = 1, 2, 3, \qquad (2.6.3)$$

where $\mathbf{r}$ is the radius vector and h_i is called the *size factor*. It can be proved that in an orthogonal curvilinear coordinate system the divergence of vector A can be expressed as

$$\text{div } \mathbf{A} = \frac{1}{h_1 h_2 h_3}\left[\frac{\partial}{\partial u_1}(h_2 h_3 A_1) + \frac{\partial}{\partial u_2}(h_1 h_3 A_2) + \frac{\partial}{\partial u_3}(h_1 h_2 A_3)\right], \quad (2.6.4)$$

where A_1, A_2, and A_3 are the components of $\mathbf{A}$ in the curvilinear coordinates. Furthermore, for any function ϕ having continuous second-order partial derivatives, the equation given below is true:

$$\text{div}(\text{grad } \phi) = \frac{1}{h_1 h_2 h_3}\left[\frac{\partial}{\partial u_1}\left(\frac{h_2 h_3}{h_1}\frac{\partial \phi}{\partial u_1}\right) + \frac{\partial}{\partial u_2}\left(\frac{h_1 h_3}{h_2}\frac{\partial \phi}{\partial u_2}\right) + \frac{\partial}{\partial u_3}\left(\frac{h_1 h_2}{h_3}\frac{\partial \phi}{\partial u_3}\right)\right].$$
$$(2.6.5)$$

The general concepts of curvilinear coordinates and the derivations of Eqs. (2.6.4) and (2.6.5) can be found in the field theory of advanced calculus.

Now, assume that all the three axes in the orthogonal coordinate system coincide with the principal axes of the dispersion coefficient tensor, e.g., axis u_1 is of the same direction as the mean flow velocity, axes u_2 and u_3 intersect orthogonally with u_1. For an isotropic porous medium, from Eq. (2.5.17) we have

$$\mathbf{D} = \begin{bmatrix} \alpha_L V + D_d T & 0 & 0 \\ 0 & \alpha_T V + D_d T & 0 \\ 0 & 0 & \alpha_T V + D_d T \end{bmatrix}, \quad (2.6.6)$$

where $V_1 = V$ and $V_2 = V_3 = 0$. Combining Eqs. (2.6.4) and (2.6.5) with the dispersion equation given in Eq. (2.4.24), we then have

$$\frac{\partial C}{\partial t} = \frac{1}{h_1 h_2 h_3}\left\{\frac{\partial}{\partial u_1}\left[\frac{h_2 h_3}{h_1}(\alpha_L V + D_d T)\frac{\partial C}{\partial u_1}\right] + \frac{\partial}{\partial u_2}\left[\frac{h_1 h_3}{h_2}(\alpha_T V + D_d T)\frac{\partial C}{\partial u_2}\right]\right.$$
$$\left. + \frac{\partial}{\partial u_3}\left[\frac{h_1 h_2}{h_3}(\alpha_T V + D_d T)\frac{\partial C}{\partial u_3}\right] - \frac{\partial}{\partial u_1}(h_2 h_3 CV)\right\}. \quad (2.6.7)$$

Some examples are given below.

Cylindrical Coordinates (r, θ, z)

The coordinate transformation is given by $x_1 = r\cos\theta$, $x_2 = r\sin\theta$, $x_3 = z$. We then have the size factors $h_1 = 1$, $h_2 = r$, $h_3 = 1$. By assuming that the mean flow velocity has the same direction as radius r, the advection-dispersion equation (2.6.7) becomes

$$\frac{\partial C}{\partial t} = \frac{1}{2}\frac{\partial}{\partial r}\left[r(\alpha_L V + D_d T)\frac{\partial C}{\partial r}\right] + \frac{1}{r^2}\frac{\partial}{\partial \theta}\left[(\alpha_T V + D_d T)\frac{\partial C}{\partial \theta}\right]$$
$$+ \frac{\partial}{\partial z}\left[(\alpha_T V + D_d T)\frac{\partial C}{\partial z}\right] - \frac{1}{r}\frac{\partial}{\partial r}(rCV). \quad (2.6.8)$$

Spherical Coordinates (r, θ, ϕ)

The coordinate transformation is given by: $x_1 = r \sin \theta \cos \phi$, $x_2 = r \sin \theta \sin \phi$, $x_3 = r \cos \theta$, and we have $h_1 = 1$, $h_2 = r$, $h_3 = r \sin \theta$. If the flow direction coincides with r, the relevant advection-dispersion equation is of the following form:

$$\frac{\partial C}{\partial t} = \frac{1}{r^2} \frac{\partial}{\partial r} \left[r^2 (\alpha_L V + D_d T) \frac{\partial C}{\partial r} \right] + \frac{1}{r^2 \sin \theta} \frac{\partial}{\partial \theta} \left[\sin \theta (\alpha_T V + D_d T) \frac{\partial C}{\partial \theta} \right]$$
$$+ \frac{1}{r^2 \sin^2 \theta} \frac{\partial}{\partial \phi} \left[(\alpha_T V + D_d T) \frac{\partial C}{\partial \phi} \right] - \frac{1}{r^2} \frac{\partial}{\partial r} (r^2 C V). \tag{2.6.9}$$

Polar Coordinates (r, θ)

Polar coordinates are frequently used in two-dimensional advection-dispersion problems. The coordinate transformation is given by $x = r \cos \theta$, $y = r \sin \theta$ and the advection-dispersion equation can be obtained directly from Eq. (2.6.8) by setting $\partial C / \partial z = 0$, i.e.,

$$\frac{\partial C}{\partial t} = \frac{1}{r} \frac{\partial}{\partial r} \left[r (\alpha_L V + D_d T) \frac{\partial C}{\partial r} \right] + \frac{1}{r^2} \frac{\partial}{\partial \theta} \left[(\alpha_T V + D_d T) \frac{\partial C}{\partial \theta} \right] - \frac{1}{r} \frac{\partial}{\partial r} (r C V). \tag{2.6.10}$$

If the dispersion is also axially symmetric, i.e., $\partial C / \partial \theta = 0$, the above equation is further simplified to

$$\frac{\partial C}{\partial t} = \frac{1}{r} \frac{\partial}{\partial r} \left[r (\alpha_L V + D_d T) \frac{\partial C}{\partial r} \right] - \frac{1}{r} \frac{\partial}{\partial r} (r C V). \tag{2.6.11}$$

In a radial flow we often have $rV = $ constant and normally $D_d T \ll \alpha_L V$. In this case, equation (2.6.11) can be further simplified to

$$\frac{\partial C}{\partial t} = \alpha_L V \frac{\partial^2 C}{\partial r^2} - V \frac{\partial C}{\partial r}. \tag{2.6.12}$$

This is the *radial advection-dispersion equation* most commonly seen. Eqs. (2.6.8) through (2.6.12) are especially useful in the study of point source pollution and the hydrodynamic dispersion around injection and extraction wells.

2.6.2 Extensions of Hydrodynamic Dispersion Equations

We have discussed some different forms of the hydrodynamic dispersion equation without considering the exchange term $\langle J_n + C(V_n - u_n) > M_y$ between phase γ and others, and the source and sink term $\theta_\gamma I$ in Eq. (2.4.17). In this section, some specific expressions will be given for these two terms with regard to practical cases. To simplify the text, only the case of $\theta_\gamma = $ constant is considered. With a macroscopic source or sink term, Eq. (2.4.24) is extended to

$$\frac{\partial C}{\partial t} = \frac{\partial}{\partial x_i}\left(D_{ij}\frac{\partial C}{\partial x_j}\right) - \frac{\partial}{\partial x_i}(V_i C) + I. \tag{2.6.13}$$

Injection and Extraction

Assume that water with tracer concentration C_0 is injected into an aquifer and the water injected per unit time per unit aquifer volume is W (dimensions [1/T]), we then have

$$I = \frac{W}{\theta}C_0, \tag{2.6.14}$$

where θ is the porosity of the porous medium.

In the case of extraction, the sink term becomes

$$I = -\frac{Q}{\theta}C, \tag{2.6.15}$$

where Q is the volume of water extracted away from per unit aquifer volume per unit time, C is the solute concentration at the pumping point, and is undetermined. Therefore, when both extraction and injection are taken into consideration, (2.6.13) can be written as

$$\frac{\partial C}{\partial t} = \frac{\partial}{\partial x_i}\left(D_{ij}\frac{\partial C}{\partial x_j}\right) - \frac{\partial}{\partial x_i}(V_i C) + \frac{W}{\theta}C_0 - \frac{Q}{\theta}C. \tag{2.6.16}$$

This is the advection-dispersion equation with injection and extraction conditions. Note that W and Q are both functions of time and position. At locations where no extraction or injection is performed, W and Q are both equal to zero.

When the flow field reaches a steady state, we have div $V = W/\theta$ at the injection location and div $V = -Q/\theta$ at the extraction location, and elsewhere, div $V = 0$. Thus, Eq. (2.6.16) can be modified to

$$\frac{\partial C}{\partial t} = \frac{\partial}{\partial x_i}\left(D_{ij}\frac{\partial C}{\partial x_j}\right) - V_i\frac{\partial C}{\partial x_i} + \frac{W}{\theta}(C_0 - C). \tag{2.6.17}$$

Radioactive Decay and Chemical Reactions

Radioactive decay occurs when radioactive tracers transport through a porous medium. The decay rate of the tracer is proportional to its mass, i.e.,

$$\frac{dC}{dt} = -\lambda C, \tag{2.6.18}$$

where λ is the decay constant of the tracer. dC/dt represents the mass decreasing rate of the tracer due to decay in a unit volume of the porous medium. Thus

$$I = -\lambda C. \tag{2.6.19}$$

Substituting I into (2.6.13), we have

$$\frac{\partial C}{\partial t} = \frac{\partial}{\partial x_i}\left(D_{ij}\frac{\partial C}{\partial x_j}\right) - \frac{\partial}{\partial x_i}(V_iC) - \lambda C. \tag{2.6.20}$$

This is the hydrodynamic dispersion equation with radioactive decay taken into consideration.

The increase or decrease of components due to chemical reactions can also be incorporated into the source and sink term. Assume that there are two solutes in the water, with concentrations C_1 and C_2, respectively. The hydrodynamic dispersion-chemical reaction equations can be obtained by coupling their source and sink terms as follows:

$$\frac{\partial C_1}{\partial t} = \frac{\partial}{\partial x_i}\left(D_{ij}\frac{\partial C_1}{\partial x_j}\right) - \frac{\partial}{\partial x_i}(V_iC_1) - f_1(C_1, C_2)$$

$$\frac{\partial C_2}{\partial t} = \frac{\partial}{\partial x_i}\left(D_{ij}\frac{\partial C_2}{\partial x_j}\right) - \frac{\partial}{\partial x_i}(V_iC_2) - f_2(C_1, C_2), \tag{2.6.21}$$

where f_1 and f_2 are functions of C_1 and C_2.

Adsorption and Ion Exchange

Adsorption and ion exchange phenomena occur at the interfaces between solid and liquid phases. The negative charges on a solid surface may attract cations. If the liquid phase contains some tracer ions, some of them may be adsorbed onto the solid surface and, hence, reduce the tracer concentration in the liquid. On the other hand, the tracer ions in the solid may enter the liquid through the solid surface and increase the tracer concentration in the liquid. Freeze and Cherry (1979) gave a comprehensive review on the mechanisms of adsorption and ion exchange.

The effects of adsorption and ion exchange can also be attributed to a source/sink term. To get its expression, we must simultaneously consider the mass balance within the solid and liquid phases. Assume that F is the tracer concentration in the solid, i.e., the tracer mass in a unit volume of the solid, and let $f(C, F)$ be the mass of tracer transferred from solid to liquid, per unit time and per unit volume of porous media. We have

$$\frac{\partial C}{\partial t} = \frac{\partial}{\partial x_i}\left(D_{ij}\frac{\partial C}{\partial x_j}\right) - \frac{\partial}{\partial x_i}(V_iC) + \frac{f(C, F)}{\theta}, \tag{2.6.22}$$

for the liquid phase, and

$$\frac{\partial F}{\partial t} = -\frac{f(C, F)}{1 - \theta}, \tag{2.6.23}$$

for the solid phase. Note that Eq. (2.6.23) describes the tracer mass conservation in the solid phase.

Several expressions, known as *adsorption isotherm relationships*, have been suggested for $f(C, F)$ for different adsorption cases. One example is given by

$$\frac{\partial F}{\partial t} = kC, \tag{2.6.24}$$

where k is a constant. The equation treats the solid phase as a sink of the solute and assumes that the change rate of solute concentration in the solid is proportional to the solute concentration in the liquid. Eliminating $f(C, F)$ in Eq. (2.6.22) and Eq. (2.6.23), we have

$$\frac{\partial C}{\partial t} = \frac{\partial}{\partial x_i}\left(D_{ij}\frac{\partial C}{\partial x_j}\right) - \frac{\partial}{\partial x_i}(V_i C) - \frac{1-\theta}{\theta}\frac{\partial F}{\partial t}. \tag{2.6.25}$$

Inserting Eq. (2.6.24) into the above equation, we then have the following equation, which is of the same form as Eq. (2.6.20):

$$\frac{\partial C}{\partial t} = \frac{\partial}{\partial x_i}\left(D_{ij}\frac{\partial C}{\partial x_j}\right) - \frac{\partial}{\partial x_i}(V_i C) - \left(\frac{1-\theta}{\theta}k\right)C. \tag{2.6.26}$$

Another expression for adsorption is

$$F = \beta C, \tag{2.6.27}$$

where β is a constant. This expression is known as the *linear equilibrium isotherm relationship*, which shows that the solute concentrations in the solid phase are directly proportional to those in the liquid phase. Inserting Eq. (2.6.27) into Eq. (2.6.25) and assuming

$$R_d = 1 + \frac{1+\theta}{\theta}\beta, \tag{2.6.28}$$

we then have

$$\frac{\partial C}{\partial t} = \frac{\partial}{\partial x_i}\left(\frac{D_{ij}}{R_d}\frac{\partial C}{\partial x_j}\right) - \frac{\partial}{\partial x_i}\left(\frac{V_i}{R_d}C\right). \tag{2.6.29}$$

This equation includes an implicit source/sink term. Since $R_d > 1$, both the hydrodynamic dispersion coefficient and the mean flow velocity are decreased by a certain factor and, hence, the dispersion process is weakened. The coefficient R_d is often called the *retardation factor*, which describes the retardation effect caused by adsorption.

The retardation factor may be directly measured. Developments and articles on this topic were reviewed by Faust and Mercer (1980) and Valocchi (1984). As a special case of adsorption and ion exchange, the transport of colloid and bacteria in groundwater has been extensively studied in recent years (Elimelech et al., 1995).

2.6.3 Initial and Boundary Conditions

The solution of the hydrodynamic dispersion equation (2.6.13) is the concentration distribution $C(x, t)$, which is dependent on position $\mathbf{x} = (x_1, x_2, x_3)$

and time t. To obtain a particular solution, we not only need the dispersion equation, but also some other specifications, as described below:

1. The space domain (R) and time interval $(0, T)$ of the considered problem must be provided.
2. The flow field in the considered flow domain (R) and the distributions of relevant parameters in the equation, such as mean flow velocity $V(x, t)$, $0 < t < T$, and α_L, α_T, $D_d T$, R_d, λ, W, and so forth, must be provided.
3. The initial condition for (R) and boundary conditions for its boundary (B) must be provided. The initial condition is the concentration distribution at the instant $t = 0$:

$$C(x, 0) = C_0(x); \quad x \in (R), \tag{2.6.30}$$

where $t = 0$ is an arbitrary initial time. $C_0(x)$ is a known function of position x. For example, if at $t = 0$, a solution with a tracer is injected into (R), but no tracer had existed in (R) before, then $C_0(x) = 0$.

Similar to groundwater flow simulation, there are three types of boundary condition.

The first type of boundary condition specifies concentration distribution along the boundary, i.e., it is given by

$$C(x_B, t) = g_1(x_B, t),$$

$$0 < t < T, \quad x_B \in (B_1), \tag{2.6.31}$$

where (B_1) is part of the boundary of (R), x_B a point on the boundary, and $g_1(x_B, t)$ a known function. This type of boundary condition is often called the *Dirichlet boundary condition*.

The second type of boundary condition gives a known dispersion flux along the boundary. It is often called the *Neumann boundary condition* and given by

$$-D_{ij} \frac{\partial C}{\partial x_j} n_i \bigg|_{x_B} = g_2(x_B, t)$$

$$0 < t < T, \quad x_B \in (B_2), \tag{2.6.32}$$

where (B_2) is also part of the boundary of (R), n_i $(i = 1, 2, 3)$ are components of the unit normal vector of (B_2), and $g_2(x_B, t)$ is a known function.

The third type of boundary condition defines the solute transport flux at the boundary surface. It is also called *Cauchy boundary condition* and given by

$$\left(C V_i - D_{ij} \frac{\partial C}{\partial x_j} \right) n_i \bigg|_{x_B} = g_3(x_B, t)$$

$$0 < t < T, \quad x_B \in (B_3), \tag{2.6.33}$$

where (B_3) is part of the boundary of (R), and $g_3(x_B, t)$ a known function.

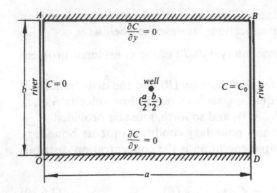

FIGURE 2.12. Diagram for Example 1.

It should be noted that Einstein summation convention has been used in Eqs. (2.6.32) and (2.6.33).

Two examples are given below which show the mathematical statement of a hydrodynamic dispersion problem when the flow field is known.

Example 1

Consider a horizontal two-dimensional groundwater pollution problem. In Figure 2.12, $OADB$ is a rectangular confined aquifer bounded by two rivers $\overline{OA}$ and $\overline{BD}$, in which $\overline{BD}$ is a polluted river with solute concentration C_0, river $\overline{OA}$ is a clean one. $\overline{OD}$ and $\overline{AB}$ are no-flow boundaries. The aquifer is not polluted until the discharge well in the center of the region begins to pump with the flow rate Q. Assuming that the water density is constant and the flow velocity field has no change. Let us try to form the mathematical model for this problem.

The flow model for this problem is quite easy to build. Assume that the water head distribution has already been obtained from a flow model. Hence, according to Darcy's Law, the velocity distribution $V(x, y, t)$ in the rectangular area $OABD$ can be calculated. Let its components be V_x and V_y. With the molecular diffusion effect neglected, the hydrodynamic dispersion coefficients will be as follows:

$$\begin{cases} D_{xx} = (\alpha_L V_x^2 + \alpha_T V_y^2)/V \\ D_{xy} = D_{yx} = (\alpha_L - \alpha_T)V_x V_y/V \\ D_{yy} = (\alpha_T V_x^2 + \alpha_L V_y^2)/V \end{cases} \quad (2.6.34)$$

and the hydrodynamic dispersion equation reads as

$$\frac{\partial C}{\partial t} = \frac{\partial}{\partial x}\left(D_{xx}\frac{\partial C}{\partial x} + D_{xy}\frac{\partial C}{\partial y}\right) + \frac{\partial}{\partial y}\left(D_{yx}\frac{\partial C}{\partial x} + D_{yy}\frac{\partial C}{\partial y}\right) - \frac{\partial}{\partial x}(CV_x)$$

$$- \frac{\partial}{\partial y}(CV_y) - \frac{QC}{\theta}\delta\left(x - \frac{a}{2}\right)\delta\left(y - \frac{b}{2}\right), \quad (2.6.35)$$

FIGURE 2.13. Diagram for Example 2.

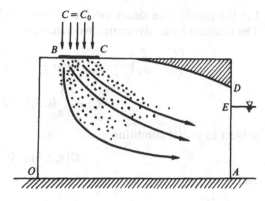

where θ is the porosity, a and b are the length and width of the rectangular aquifer, respectively, and δ is the Dirac-δ function.

The initial condition is

$$C(x, y, 0) = 0, \qquad (x, y) \in \text{rectangle } OABD \tag{2.6.36}$$

and the boundary conditions are

$$\begin{cases} C(a, y, t) = C_0, & \text{along } \overline{BD} \\[2mm] C(0, y, t) = 0, & \text{along } \overline{OA} \\[2mm] \dfrac{\partial C}{\partial y}\bigg|_{(x, 0, t)} = 0, & \text{along } \overline{OD} \\[2mm] \dfrac{\partial C}{\partial y}\bigg|_{(x, b, t)} = 0, & \text{along } \overline{AB} \end{cases} \tag{2.6.37}$$

Example 2

Consider a two-dimensional groundwater pollution problem on a vertical profile. Figure 2.13 shows an unconfined aquifer, where water is recharged from $\overline{BC}$ and flows into a river via the aquifer. Assume that a steady flow has been reached. Starting from an instant $t = 0$, the recharge source changes from clean water to polluted water with a radioactive tracer concentration of C_0. Now let us build up a mathematical model for this hydrodynamic dispersion problem.

Assume that the water head distribution is known from the steady unconfined aquifer model, and, by means of Darcy's Law, the velocity components V_x and V_z are also known. Ignoring molecular diffusion, the coefficients of hydrodynamic dispersion can be calculated by the following equations:

$$\begin{cases} D_{xx} = (\alpha_L V_x^2 + \alpha_T V_z^2)/V \\ D_{xz} = D_{zx} = (\alpha_L - \alpha_T) V_x V_z/V \\ D_{zz} = (\alpha_T V_x^2 + \alpha_L V_z^2)/V \end{cases} \tag{2.6.38}$$

Let the radioactive decay coefficient be λ and the retardation factor be R_d. The relevant hydrodynamic dispersion equation can then be written as

$$R_d \frac{\partial C}{\partial t} = \frac{\partial}{\partial x}\left(D_{xx}\frac{\partial C}{\partial x} + D_{xz}\frac{\partial C}{\partial z}\right) + \frac{\partial}{\partial z}\left(D_{zx}\frac{\partial C}{\partial x} + D_{zz}\frac{\partial C}{\partial z}\right)$$

$$- \frac{\partial}{\partial x}(CV_x) - \frac{\partial}{\partial z}(CV_z) - \lambda R_d C, \qquad (2.6.39)$$

subject to initial condition

$$C(x, z, 0) = 0 \qquad (2.6.40)$$

and boundary conditions:

$$\frac{\partial C}{\partial x} = 0, \quad \text{along } \overline{OB} \text{ (no flow boundary)}$$

$$C = C_0, \quad \text{along } \overline{BC} \text{ (constant concentration)}$$

$$\frac{\partial C}{\partial n} = 0, \quad \text{along } \overline{CD} \text{ (no solute flux)}$$

$$\frac{\partial C}{\partial x} = 0, \quad \text{along } \overline{DE} \text{ (seepage boundary)} \qquad (2.6.41)$$

$$C - \alpha_L \frac{\partial C}{\partial x} = 0, \quad \text{along } \overline{EA} \text{ (solute concentration in river is assumed to}$$

$$\text{be zero, and molecular diffusion neglected}$$

$$\frac{\partial C}{\partial z} = 0, \quad \text{along } \overline{AO} \text{ (impervious boundary).}$$

It must be pointed out that the hydrodynamic dispersion equation is a parabolic partial differential equation. In mathematics, the *well-posedness* of this kind of equation has been proved, i.e., with appropriate initial and boundary conditions the solution of a parabolic equation must be in existence, unique, and continuously dependent upon the input data. As a result, all hydrodynamic dispersion problems presented in this chapter are well-posed. In most cases, we can only find their approximate solutions by means of numerical methods. However, if the shape of the aquifer boundary is regular, the porous medium is homogeneous and isotropic, and the initial and boundary conditions are constant, it is possible to find accurate solutions by means of analytical methods.

Exercises

2.1. Prove that $\overline{a_1} \overset{\circ}{a} = 0$.

2.2. Define the hydraulic head of a confined aquifer at the macroscopic level.

2.3. List the dimensions of all parameters mentioned in Section 2.1.2.

2.4. Write the convection-diffusion equation (2.3.10) in Cartesian coordinates (1-D, 2-D, 3-D).

2.5. Explain why the convection-diffusion equation cannot be used directly to study the contaminant transport in porous media.

2.6. Prove that $\partial\theta_\gamma/\partial t = \langle u_n \rangle M_\gamma$.

2.7. Explain the physical meaning of each term in Eq. (2.4.17).

2.8. Derive the advection-dispersion equation (2.4.23) from equation (2.4.31) using Green's formula.

2.9. What is the scale dependence problem of the dispersivity coefficient?

2.10. Under what conditions can we use Eq. (2.5.14) to calculate the dispersion coefficients?

2.11. Derive Eq. (2.6.17).

2.12. What is the governing equation when radioactive decay and adsorption both exist?

2.13. Present the mathematical model for the two-dimensional experiment described by Example 1 in Section 2.2.1.

2.14. Present the mathematical model for the one-dimensional experiment described by Example 2 in Section 2.2.1.

2.15. Write the steady state flow model for Examples 1 and 2 in Section 2.6.3.

2.16. Assume that clean water $C = 0$ is injected into a homogeneous confined aquifer at rate Q from a single well and a steady radial flow is formed. Then, the concentration of injected water is changed to $C_0 > 0$. Present an advection-dispersion model to describe this experiment.

3
Analytic Solutions of Hydrodynamic Dispersion Equations

3.1 Superposition of Fundamental Solutions

3.1.1 The Fundamental Solution of a Point Source

Consider an infinite, homogeneous, and isotropic region, where it is assumed that the flow field is stationary and the solute mass M is injected instantaneously at the origin. In this case, the general hydrodynamic dispersion equation may be simplified as

$$\frac{\partial C}{\partial t} = D\left(\frac{\partial^2 C}{\partial x^2} + \frac{\partial^2 C}{\partial y^2} + \frac{\partial^2 C}{\partial z^2}\right) + \frac{M}{\theta}\delta(x)\delta(y)\delta(z)\delta(t), \tag{3.1.1}$$

where $\delta(\cdot)$ is the Dirac-δ function, D is the molecular diffusion coefficient in porous media. In this case, the use of a spheric coordinate system is convenient. In Eq. (2.6.9), let $\frac{\partial C}{\partial \theta} = 0$, $\frac{\partial C}{\partial \phi} = 0$, $V = 0$, and $D_d T = D$, we than have

$$\frac{\partial C}{\partial t} = \frac{D}{r^2}\frac{\partial}{\partial r}\left(r^2\frac{\partial C}{\partial r}\right) + \frac{M}{\theta}\delta(r)\delta(t). \tag{3.1.2}$$

The subsidiary conditions are:

$$C(r,0) = 0, \quad r > 0, \tag{3.1.3}$$

$$\lim_{r \to \infty} C(r,t) = 0, \quad t > 0. \tag{3.1.4}$$

The source term in Eq. (3.1.2) can be removed, when the instantaneous injection is represented by the following condition:

$$4\pi\theta \int_0^\infty Cr^2\, dr = M, \quad t \geq 0, \tag{3.1.5}$$

where θ is the volumetric fraction of the fluid, or the effective porosity n for a saturated flow.

By introducing the following two dimensionless variables:

$$\xi = \frac{r^2}{Dt}, \quad f(\xi) = \frac{C(Dt)^{3/2}}{M/\theta}. \tag{3.1.6}$$

Eq. (3.1.2) without its source term can be transformed into an ordinary differential equation. In fact, after the variable substitution, we have

$$\frac{\partial C}{\partial t} = \frac{M/\theta}{D^{3/2}}\left[f \frac{d}{dt}\left(\frac{1}{t^{3/2}}\right) + \frac{1}{t^{3/2}} \frac{df}{d\xi} \frac{\partial \xi}{\partial t}\right]$$

$$= -\frac{M/\theta}{D^{3/2} t^{5/2}}\left(\frac{3}{2}f + \frac{r^2}{Dt}\frac{df}{d\xi}\right), \tag{3.1.7}$$

and

$$\frac{D}{r^2}\frac{\partial}{\partial r}\left(r^2 \frac{\partial C}{\partial r}\right) = \frac{M/\theta}{D^{3/2} t^{5/2}}\left(6\frac{df}{d\xi} + \frac{4r^2}{Dt}\frac{d^2f}{d\xi^2}\right). \tag{3.1.8}$$

Thus, Eq. (3.1.2) turns into

$$\frac{d^2f}{d\xi^2} + \left(\frac{1}{4} + \frac{3}{2\xi}\right)\frac{df}{d\xi} + \frac{3}{8\xi}f = 0. \tag{3.1.9}$$

The subsidiary conditions, Eq. (3.1.3) and Eq. (3.1.4) may be expressed by the same boundary condition: $f(\infty) = 0$. Thus, condition (3.1.5) becomes

$$2\pi \int_0^\infty f\xi^{1/2}\, d\xi = 1. \tag{3.1.10}$$

Letting

$$\phi(\xi) = \frac{df}{d\xi} + \frac{1}{4}f, \tag{3.1.11}$$

Eq. (3.1.9) can be rewritten as

$$\frac{d\phi}{d\xi} + \frac{3}{2\xi}\phi = 0.$$

The solution of the above equation is $\phi = C_1 \xi^{-3/2}$, where C_1 is an undetermined coefficient. From Eq. (3.1.11) we obtain

$$\frac{df}{d\xi} + \frac{1}{4}f = \frac{C_1}{\xi^{3/2}}. \tag{3.1.12}$$

When $\xi = 0$, both f and $df/d\xi$ are finite. Therefore, C_1 must be equal to zero. As a result, the solution of Eq. (3.1.12) is

$$f(\xi) = C_2 e^{-\xi/4}. \tag{3.1.13}$$

In order to determine coefficient C_2, substitute Eq. (3.1.13) into Eq. (3.1.10) to obtain:

$$2\pi C_2 \int_0^\infty e^{-\xi/4}\xi^{1/2}\, d\xi = 1.$$

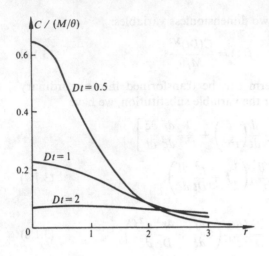

FIGURE 3.1. Fundamental solutions of the diffusion equation.

Because

$$\int_0^\infty e^{-\xi/4} \xi^{1/2} d\xi = 8\Gamma\left(\frac{3}{2}\right) = 4\sqrt{\pi},$$

we have $C_2 = \dfrac{1}{8\pi^{3/2}}$. Thus, the final solution is

$$C(r,t) = \frac{M/\theta}{(Dt)^{3/2}} f(\xi)$$

$$= \frac{M/\theta}{8(\pi Dt)^{3/2}} \exp\left(-\frac{r^2}{4Dt}\right). \tag{3.1.14}$$

Figure 3.1 shows the relationship between r and $C/(M/\theta)$ for three values of Dt. It also shows that, as time increases, the solute spreads out gradually and the maximum concentration at the origin becomes smaller and smaller. Solution (3.1.14) is called the *fundamental solution* of a point source problem.

3.1.2 Superposition Principle and Image Method

Since Eq. (3.1.1) is a linear equation, we can obtain the solutions for more problems based on the fundamental solution given in Eq. (3.1.14) by means of superposition. The simplest examples of using superposition are the solutions of transient line source and plane source problems. Assume that a line source is injected instantaneously along axis z, and the mass of solute contained in per unit length is m. Dividing the line source into many small segments with an infinitesimal length, each can be looked upon as a point source (see Figure 3.2). In terms of the fundamental solution, Eq. (3.1.14), the differential concentration at any point (x, y, z) created by the point source,

FIGURE 3.2. A representation of the line source problem.

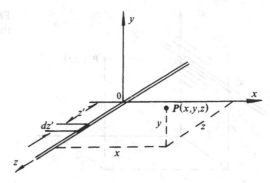

which is located at z' with mass $m\,dz'$, is

$$dC = \frac{(m/\theta)\,dz'}{8(\pi Dt)^{3/2}} \exp\left[-\frac{x^2 + y^2 + (z - z')^2}{4Dt}\right].$$

Integrating along the z axis, the solution of the *line source problem* is obtained, i.e.,

$$C(x, y, z) = \frac{m/\theta}{8(\pi Dt)^{3/2}} \int_{-\infty}^{\infty} \exp\left[-\frac{x^2 + y^2 + (z - z')^2}{4Dt}\right] dz'$$

$$= -\frac{m/\theta}{8\pi Dt} \exp\left(-\frac{x^2 + y^2}{4Dt}\right) \mathrm{erf}\left(\frac{z - z'}{2\sqrt{Dt}}\right)\Big|_{-\infty}^{\infty}$$

$$= \frac{m/\theta}{4\pi Dt} \exp\left(-\frac{x^2 + y^2}{4Dt}\right). \tag{3.1.15}$$

Using a similar method, the solution of the plane source problem can also be obtained. Assume that the solute mass, injected instantaneously into the unit area along the yz plane, is μ. The plane source may be divided into many small strips which are parallel to axis z and have differential width dy'. Each of them may be approximated by a line source (see Figure 3.3). From Eq. (3.1.15), the differential concentration at point (x, y, z) caused by a line source located in y' is

$$dC = \frac{(\mu/\theta)\,dy'}{4\pi Dt} \exp\left[-\frac{x^2 + (y - y')^2}{4Dt}\right]. \tag{3.1.16}$$

The solution of the *plane source problem* is obtained by integrating along the y axis, i.e.,

$$C(x, t) = \frac{\mu/\theta}{4\pi Dt} \int_{-\infty}^{\infty} \exp\left[-\frac{x^2 + (y - y')^2}{4Dt}\right] dy'$$

$$= \frac{\mu/\theta}{2\sqrt{\pi Dt}} \exp\left[-\frac{x^2}{4Dt}\right]. \tag{3.1.17}$$

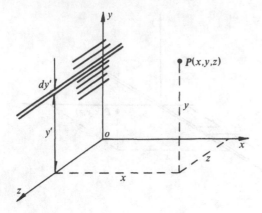

FIGURE 3.3. A representation of the plane sources problem.

The line-source fundamental solution, Eq. (3.1.15), can be used for solving the two-dimensional point-source problem perpendicular to the line. Similarly, the plane-source fundamental solution, Eq. (3.1.17), can be used for the one-dimensional point-source problem perpendicular to a plane.

It is possible to solve some diffusion problems with simple boundaries through the *super-position of imaginary sources or sinks*, as is done in ground-water flow problems. Let us consider a two-dimensional diffusion problem with a straight line boundary through which no flow passes. Assume that a solute mass m is injected instantaneously into the porous medium over a unit thickness at the point $(0, y_0)$. To obtain the concentration distribution caused by the injection, we suppose that there is another source symmetrically located on the other side of the no-flow boundary with the same injection rate. The actual concentration distribution can be obtained by superposing the solutions of the two (line) sources, that is

$$C(x, y, t) = \frac{m/\theta}{4\pi Dt} \left\{ \exp\left[-\frac{x^2 + (y - y_0)^2}{4Dt} \right] + \exp\left[-\frac{x^2 + (y + y_0)^2}{4Dt} \right] \right\}.$$

$$(3.1.18)$$

3.1.3 Continuous Injection in a Uniform Flow Field

It is assumed that there is a one-dimensional flow in a homogeneous and isotropic porous medium with a velocity u along the x-direction. If this direction is taken as the principal direction of dispersion, the advection-dispersion equation will be

$$\frac{\partial C}{\partial t} + u \frac{\partial C}{\partial x} = D_{11} \frac{\partial^2 C}{\partial x^2} + D_{22} \frac{\partial^2 C}{\partial y^2} + D_{33} \frac{\partial^2 C}{\partial z^2}. \qquad (3.1.19)$$

Letting

$$X = x - ut; \quad Y = y\sqrt{D_{11}/D_{22}}; \quad Z = z\sqrt{D_{11}/D_{33}}, \qquad (3.1.20)$$

Eq. (3.1.19) can be transformed into

$$\frac{\partial C}{\partial t} = D_{11}\left(\frac{\partial^2 C}{\partial X^2} + \frac{\partial^2 C}{\partial Y^2} + \frac{\partial^2 C}{\partial Z^2}\right). \tag{3.1.21}$$

The variable substitution in Eq. (3.1.20) means that the dispersion process is being observed in a moving coordinate system, which moves with a velocity u. Meanwhile, the scales in y and z directions are changed. Forms of Eq. (3.1.21) and Eq. (3.1.1) are exactly the same, so the solution of Eq. (3.1.19) for a *transient point source* is

$$C(x, y, z, t) = \frac{M/\theta}{8(\pi t)^{3/2}\sqrt{D_{11}D_{22}D_{33}}}$$

$$\cdot \exp\left\{-\frac{D_{22}D_{33}(x - ut)^2 + D_{11}D_{33}y^2 + D_{11}D_{22}z^2}{4D_{11}D_{22}D_{33}t}\right\}. \tag{3.1.22}$$

The corresponding solutions of Eq. (3.1.19) for the *line source* and the *plane source* are, respectively:

$$C(x, y, t) = \frac{m/\theta}{4\pi t\sqrt{D_{11}D_{22}}}\exp\left\{-\frac{D_{22}(x - ut)^2 + D_{11}y^2}{4D_{11}D_{22}t}\right\} \tag{3.1.23}$$

and

$$C(x, t) = \frac{\mu/\theta}{2\sqrt{\pi D_{11}t}}\exp\left\{-\frac{(x - ut)^2}{4D_{11}t}\right\}. \tag{3.1.24}$$

Now, let us consider the case of continuous injection for Eq. (3.1.19). Assume $D_{11} = D_{22} = D_{33} = D$, and that there is a stable point source at the origin, where solute mass $\dot{M}$ is injected in per unit time. The solution of this problem can be obtained by superposing a series of point sources. From Eq. (3.1.22), the concentration differential at point (x, y, z) created by the transient point source with mass $\dot{M} dt'$ at time t' is

$$dC = \frac{(\dot{M}/\theta) dt'}{8[\pi D(t - t')]^{3/2}}\exp\left\{-\frac{[x - u(t - t')]^2 + y^2 + z^2}{4D(t - t')}\right\}. \tag{3.1.25}$$

Integrating t' over $0 \rightarrow t$ yields

$$C(x, y, z, t) = \frac{\dot{M}/\theta}{8(\pi D)^{3/2}}\exp\left(\frac{ux}{2D}\right)\int_0^t \exp\left[-\frac{r^2}{4D(t - t')} - \frac{u^2(t - t')}{4D}\right]\frac{dt'}{(t - t')^{3/2}}, \tag{3.1.26}$$

where $r^2 = x^2 + y^2 + z^2$. After making the variable substitution $\tau = r/\sqrt{D(t - t')}$, Eq. (3.1.26) is translated into

$$C(x, y, z, t) = \frac{\dot{M}/\theta}{2D\pi^{3/2}r}\exp\left(\frac{ux}{2D}\right)\int_{r/2\sqrt{Dt}}^{\infty}\exp\left[-\tau^2 - \left(\frac{ur}{2D}\right)^2\frac{1}{4\tau^2}\right]d\tau. \tag{3.1.27}$$

Using the following integral

$$\int \exp\left(-\tau^2 - \frac{a^2}{4\tau^2}\right) d\tau = \frac{1}{2} \int \left[\left(1 - \frac{a}{2\tau^2}\right) + \left(1 + \frac{a}{2\tau^2}\right)\right] \exp\left(\tau^2 - \frac{a^2}{4\tau^2}\right) d\tau$$

$$= \frac{\sqrt{\pi}}{4}\left[e^a \operatorname{erf}\left(\frac{a}{2\tau} + \tau\right) - e^{-a}\operatorname{erf}\left(\frac{a}{2\tau} - \tau\right)\right], \qquad (3.1.28)$$

and substituting the result into Eq. (3.1.27), we obtain

$$C(x, y, z, t) = \frac{\dot{M}/\theta}{8\pi Dr}\exp\left(\frac{ux}{2D}\right)\left[\exp\left(\frac{ur}{2D}\right)\operatorname{erfc}\left(\frac{r + ut}{2\sqrt{Dt}}\right)\right.$$

$$\left. + \exp\left(-\frac{ru}{2D}\right)\operatorname{erfc}\left(\frac{r - ut}{2\sqrt{Dt}}\right)\right]. \qquad (3.1.29)$$

This is the solution of the *continuous injection problem* in a uniform flow field. When $t \to \infty$, a stable solution of continuous injection can be obtained as follows:

$$C(x, y, z) = \frac{\dot{M}/\theta}{4\pi Dr}\exp\left[-\frac{u}{2D}(r - x)\right]. \qquad (3.1.30)$$

The solution for the common case, where D_{11}, D_{22} and D_{33} are unequal, can be obtained similarly. The result is,

$$C(x, y, z, t) = \frac{\dot{M}/\theta}{8\pi \tilde{D}\tilde{r}}\exp\left(\frac{ux}{2D_{11}}\right)\left[\exp\left(\frac{u\tilde{r}}{2D_{11}}\right)\operatorname{erfc}\left(\frac{\tilde{r} + ut}{2\sqrt{D_{11}t}}\right)\right.$$

$$\left. + \exp\left(-\frac{u\tilde{r}}{2D_{11}}\right)\operatorname{erfc}\left(\frac{\tilde{r} - tu}{2\sqrt{D_{11}t}}\right)\right], \qquad (3.1.31)$$

where

$$\tilde{D} = \sqrt{D_{22}D_{33}};$$

$$\tilde{r} = \sqrt{D_{22}D_{33}x^2 + D_{11}D_{33}y^2 + D_{11}D_{22}z^2}/\tilde{D}.$$

3.2 Some Canonical Problems Having Analytic Solutions

3.2.1 One-Dimensional Dispersion Problems

Problem 1. The Transport of a Radioactive Tracer in a Semi-infinite Horizontal Sand Column $(0 \leq x \leq \infty)$

It is assumed that there is a saturated uniform flow with $q = nV$ in the sand column, where q is Darcy's velocity, n the porosity, and V the mean pore velocity. There is no tracer in the sand column at $t = 0$, while for $t \geq 0$, the concentration at the end of the sand column $(x = 0)$ always remains constant $(C = C_0)$. This case is often seen in the laboratory. For example, one-dimen-

FIGURE 3.4. One-dimensional dispersion in the field, where one end is a constant concentration boundary.

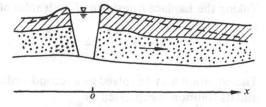

sional dispersion can be studied in a long sand column by injecting water with tracer concentration C_0 at one end of the sand column to displace the original water not containing the tracer. In the field, similar cases may occur when a polluted river cuts an aquifer and there is a stable, uniform, one-dimensional flow in the aquifer. This case is shown in Figure 3.4.

The mathematical model of this problem is

$$\frac{\partial C}{\partial t} = D_L \frac{\partial^2 C}{\partial x^2} - V \frac{\partial C}{\partial x} - \lambda C, \tag{3.2.1}$$

$$C(x, 0) = 0, \quad x > 0, \tag{3.2.2}$$

$$C(0, t) = C_0, \quad t \geq 0, \tag{3.2.3}$$

$$C(\infty, t) = 0, \quad t \geq 0, \tag{3.2.4}$$

where D_L is the longitudinal dispersion coefficient, λ the decay constant of radioactive tracer, and C_0 a given concentration.

We can find the solution of the problem using the Laplace transformation. The *Laplace transform* of unknown function C is C^*, where

$$C^*(x, p) = \int_0^\infty C e^{-pt} \, dt. \tag{3.2.5}$$

and is denoted in the form

$$C^* = L[C].$$

From Eq. (3.2.5), we can directly obtain

$$L\left[\frac{dC}{dt}\right] = pC^* - Ce^{-pt}|_{t=0}. \tag{3.2.6}$$

Due to the initial condition given in Eq. (3.2.2), the second term on the right-hand side of this equation should be equal to 0. Therefore, we have

$$L\left[\frac{dC}{dt}\right] = pC^*. \tag{3.2.7}$$

In addition, from the definition of Laplace transform (3.2.5) we have

$$L\left[\frac{\partial C}{\partial x}\right] = \frac{\partial C^*}{\partial x}, \quad L\left[\frac{\partial^2 C}{\partial x^2}\right] = \frac{\partial^2 C^*}{\partial x^2}. \tag{3.2.8}$$

Taking the Laplace transform of both sides of Eq. (3.2.1), we obtain

$$D_L \frac{\partial^2 C^*}{\partial x^2} - V \frac{\partial C^*}{\partial x} - (\lambda - p)C^* = 0. \tag{3.2.9}$$

This equation may be solved as a second-order ordinary differential equation and its solution is expressed as

$$C^* = A(p)\exp\left(\frac{Vx}{2D_L} - \frac{x}{\sqrt{D_L}}\sqrt{\left(\frac{V^2}{4D_L} + \lambda\right) + p}\right)$$
$$+ B(p)\exp\left(\frac{Vx}{2D_L} + \frac{x}{\sqrt{D_L}}\sqrt{\left(\frac{V^2}{4D_L} + \lambda\right) + p}\right). \tag{3.2.10}$$

After imposing the boundary condition (3.2.4), we must have $C^*(\infty, p) = 0$ when $x = \infty$, $C = 0$. Consequently, $B(p) = 0$, and in terms of the boundary condition (3.2.3), when $x = 0$, $C = C_0$, we have

$$C^*(0, p) = \int_0^\infty C_0 e^{-pt}\, dt = \frac{C_0}{p}.$$

Substituting this equation into Eq. (3.2.10) yields $A(p) = C_0/p$. Thus, Eq. (3.2.10) becomes

$$C^* = \frac{C_0}{p}\exp\left(\frac{Vx}{2D_L} - \frac{x}{\sqrt{D_L}}\sqrt{\left(\frac{V^2}{4D_L} + \lambda\right) + p}\right). \tag{3.2.11}$$

After C^* is obtained, the original solution, C, can also be obtained by using the inverse of Laplace transformation. Letting L^{-1} indicate the inverse transformation, we then have

$$C = L^{-1}[C^*] = C_0\exp\left(\frac{Vx}{2D_L}\right)L^{-1}\left[\frac{1}{p}\exp(-a\sqrt{b^2 + p})\right], \tag{3.2.12}$$

where

$$a = \frac{x}{\sqrt{D_L}}, \quad b^2 = \frac{x^2}{4D_L} + \lambda. \tag{3.2.13}$$

From the Table of Laplace Transforms, we know that

$$L^{-1}\left[\frac{1}{p}F(p)\right] = \int_0^t f(\tau)\, d\tau, \quad \text{where} \quad F = L[f],$$

and

$$L^{-1}[\exp(-a\sqrt{b^2 + p})] = \frac{a}{2\sqrt{\pi t^3}}\exp\left[-\left(\frac{a^2}{4t} + b^2 t\right)\right].$$

Therefore,

$$L^{-1}\left[\frac{1}{p}\exp(-a\sqrt{b^2+p})\right]$$

$$=\int_0^t \frac{a}{2\sqrt{\pi\tau^3}}\exp\left[-\left(\frac{a^2}{4\tau}+b^2\tau\right)\right]d\tau$$

$$=e^{-ab}\int_0^t \frac{a}{2\sqrt{\pi\tau^3}}\exp\left[-\frac{(a-2b\tau)^2}{4\tau}\right]d\tau$$

$$=e^{-ab}\int_0^t\left(\frac{a+2b\tau}{4\sqrt{\pi\tau^3}}+\frac{a-2b\tau}{4\sqrt{\pi\tau^3}}\right)\exp\left[-\frac{(a-2b\tau)^2}{4\tau}\right]d\tau.$$

Dividing the right-hand side of the above equation into two integrations, and making the following variable substitutions into each integral, respectively:

$$\xi=\frac{a-2b\tau}{\sqrt{4\tau}}, \quad \zeta=\frac{a+2b\tau}{\sqrt{4\tau}},$$

we obtain,

$$L^{-1}\left[\frac{1}{p}\exp(-a\sqrt{b^2+p})\right]$$

$$=-e^{-ab}\frac{1}{\sqrt{\pi}}\int_\infty^{(a-2bt)/\sqrt{4t}}e^{-\xi^2}d\xi-\frac{e^{ab}}{\sqrt{\pi}}\int_\infty^{(a+2bt)/\sqrt{4t}}e^{-\zeta^2}d\zeta$$

$$=\frac{e^{-ab}}{2}\operatorname{erfc}\left(\frac{a-2bt}{2\sqrt{t}}\right)+\frac{e^{ab}}{2}\operatorname{erfc}\left(\frac{a+2bt}{2\sqrt{t}}\right).$$

Substituting this equation into Eq. (3.2.12) and using Eq. (3.2.13), we finally get

$$C(x,t)=\frac{C_0}{2}\exp\left(\frac{Vx}{2D_L}\right)\left\{\exp\left[-\frac{x}{2D_L}\sqrt{V^2+4\lambda D_L}\right]\operatorname{erfc}\left[\frac{x-\sqrt{V^2+4\lambda D_L}\,t}{2\sqrt{D_L t}}\right]\right.$$

$$\left.+\exp\left[\frac{x}{2D_L}\sqrt{V^2+4\lambda D_L}\right]\operatorname{erfc}\left[\frac{x+\sqrt{V^2+4\lambda D_L}\,t}{2\sqrt{D_L t}}\right]\right\}. \qquad (3.2.14)$$

This is the solution of Problem 1. In case of $\lambda=0$, i.e., there is no radioactive decay, the solution is reduced to

$$C(x,t)=\frac{C_0}{2}\left\{\operatorname{erfc}\left[\frac{x-Vt}{2\sqrt{D_L t}}\right]+\exp\left(\frac{Vx}{D_L}\right)\operatorname{erfc}\left[\frac{x+Vt}{2\sqrt{D_L t}}\right]\right\}. \qquad (3.2.15)$$

Problem 2. Continuous Injection of a Radioactive Tracer into an Infinite Horizontal Sand Column

Assume that there is a uniform one-dimensional flow with an average pore velocity V in the sand column. Before $t=0$ there was no radioactive tracer

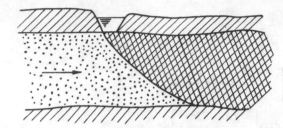

FIGURE 3.5. A test of releasing tracer in a one-dimensional flow field.

in the column. Starting at $t = 0$, the water containing the tracer with concentration C_0 is continuously injected into the sand column at point $x = 0$. This situation may be approximately simulated in the tracer-releasing test in the field as shown in Figure 3.5.

The fundamental solution of the plane source problem, Eq. (3.1.24), has been derived using the transformation in Eq. (3.1.20), therefore

$$C^0(x, t) = \frac{\mu/\theta}{\sqrt{4\pi D_L t}} \exp\left[-\frac{(x - Vt)^2}{4D_L t} \right] \qquad (3.2.16)$$

should be the fundamental solution of the following equation

$$\frac{\partial C^0}{\partial t} = D_L \frac{\partial^2 C^0}{\partial x^2} - V \frac{\partial C^0}{\partial x}, \qquad (3.2.17)$$

as long as there is a point source at $x = 0$ and $t = 0$. On the other hand, one can directly verify that Eq. (3.2.1) has the fundamental solution

$$C = C^0 e^{-\lambda t}. \qquad (3.2.18)$$

Substituting Eq. (3.2.16) into (3.2.18) yields

$$C = \frac{\mu/\theta}{\sqrt{4\pi D_L t}} \exp\left[-\frac{(x - Vt)^2}{4D_L t} - \lambda t \right]. \qquad (3.2.19)$$

According to the method mentioned in Section 3.1.3, continuous injection can be looked upon as the superposition of a series of transient injections. Let us consider a point source into which the mass of tracer $\dot\mu\, dt'$ is injected instantaneously at time t'. In terms of Eq. (3.2.19), the concentration difference at point x caused by the point source should be

$$dC = \frac{(\dot\mu/\theta)\, dt'}{\sqrt{4\pi D_L (t - t')}} \exp\left[-\frac{[x - V(t - t')]^2}{4D_L(t - t')} - \lambda(t - t') \right].$$

Integrating t' from 0 to t, we have

$$C(x, t) = \frac{\dot\mu/\theta}{\sqrt{4\pi D_L}} \exp\left(\frac{Vx}{2D_L} \right) \int_0^t \frac{1}{\sqrt{\tau}} \exp\left[-\frac{(x - V\tau)^2}{4D\tau} - \lambda\tau \right] d\tau. \qquad (3.2.20)$$

If the injection rate is equal to Darcy's velocity, $\dot\mu/\theta = C_0 V$, then Eq. (3.2.20) may be integrated to obtain

$$C(x,t) = \frac{C_0}{2} \exp\left(\frac{Vx}{2D_L}\right) \left\{ \exp\left[-\frac{x}{2D_L}\sqrt{V^2 + 4\lambda D_L} \right] \text{erfc}\left[\frac{x - \sqrt{V^2 + 4\lambda D_L}\,t}{2\sqrt{D_L t}} \right] \right.$$

$$\left. - \exp\left[\frac{x}{2D_L}\sqrt{V^2 + 4\lambda D_L} \right] \text{erfc}\left[\frac{x + \sqrt{V^2 + 4\lambda D_L}\,t}{2\sqrt{D_L t}} \right] \right\}. \qquad (3.2.21)$$

This is the solution of Problem 2. When $\lambda = 0$, it may be simplified to the form

$$C(x,t) = \frac{C_0}{2} \left\{ \text{erfc}\left[\frac{x - Vt}{2\sqrt{D_L t}} \right] - \exp\left(\frac{Vx}{D_L}\right) \text{erfc}\left[\frac{x + Vt}{2\sqrt{D_L t}} \right] \right\}. \qquad (3.2.22)$$

Comparing the solution of Problem 1 (Eq. (3.2.15)) with that of Problem 2 (Eq. (3.2.22)), we can see that only the second terms on the right-hand sides of the equations have different signs. Generally, the second term may be neglected in comparison with the first term. In this case, the same approximate solution as given below can be adopted for both Problem 1 and Problem 2:

$$C(x,t) = \frac{C_0}{2} \text{erfc}\left[\frac{x - Vt}{2\sqrt{D_L t}} \right]. \qquad (3.2.23)$$

We would like to mention some other results on analytic solutions for one-dimensional dispersion problems. Selim and Mansell (1976) found the analytic solutions for one-dimensional dispersion in a finite sand column including consideration of adsorption and chemical reactions. Basak and Murty (1981) studied the analytic solution of the nonlinear one-dimensional diffusion equation, with the diffusion coefficient being proportional to the concentration. Van Genuchten (1981) made a more comprehensive study of the analytic solutions for one-dimensional dispersion problems in a semi-infinite sand column. He considered the generation and decay of the tracer and the treatments of the second type of boundary conditions. Meanwhile, chemical transport with combined biological transformation and mass exchange between different phases in a soil column was studied by Mironenko and Pachepsky (1984). Lindstrom and Boersma (1989) gave a very general analytic solution of the one-dimensional advection-dispersion equation, which includes an arbitrary initial concentration distribution, time-dependent boundary conditions, and source or sink terms.

3.2.2 Two- and Three-Dimensional Dispersion Problems

Problem 3. Two-dimensional Dispersion of the Tracer in a Unidirectional Flow Field (Transient Injection)

Suppose that there is a unidirectional flow in the x direction with Darcy's velocity $q = nV$ in an aquifer which is infinite, homogeneous, and of constant thickness. At $t = 0$, a certain amount of tracer, m, is injected instantaneously into the aquifer at the origin. Let us find the time-space distribution of the concentration of the tracer. The mathematical model of this problem is

$$
\begin{cases}
\dfrac{\partial C}{\partial t} = D_L \dfrac{\partial^2 C}{\partial x^2} + D_T \dfrac{\partial^2 C}{\partial y^2} - V\dfrac{\partial C}{\partial x}, \\[2mm]
C(x, y, 0) = 0, \quad (x, y) \neq (0, 0), \\[2mm]
\displaystyle\int_{-\infty}^{\infty}\int_{-\infty}^{\infty} nC\,dx\,dy = m, \\[2mm]
C(\pm\infty, y, t) = 0, \quad t \geq 0, \\[2mm]
C(x, \pm\infty, t) = 0, \quad t \geq 0.
\end{cases}
\tag{3.2.24}
$$

where D_L and D_T are the longitudinal and transverse dispersion coefficients, respectively; n is the porosity. If the transformation $Y = y\sqrt{D_L/D_T}$ is adopted, and the observation is made in the moving coordinate system $\xi = x - Vt$, this problem becomes the transient line source problem mentioned earlier. The solution, therefore, can be directly obtained from Eq. (3.1.23),

$$
C(x, y, t) = \frac{m/n}{4\pi t\sqrt{D_L D_T}} \exp\left[-\frac{(x - Vt)^2}{4D_L t} - \frac{y^2}{4D_T t} \right].
\tag{3.2.25}
$$

If the effect of the molecular diffusion is neglected, then $D_L = \alpha_L V$ and $D_T = \alpha_T V$ can be substituted into this equation to obtain

$$
C(x, y, t) = \frac{m/n}{4\pi V t\sqrt{\alpha_L \alpha_T}} \exp\left[-\frac{(x - Vt)^2}{4\alpha_L V t} - \frac{y^2}{4\alpha_T V t} \right],
\tag{3.2.26}
$$

where α_L and α_T are the longitudinal and transverse dispersivities, respectively. Figure 3.6 shows that the concentration contours plotted according to Eq. (3.2.26) vary with time. The plume of the tracer moves forward along the x axis and its velocity is equal to the average velocity, V. As time proceeds, the areal extent of the plume increases. The rates at which the plume expands

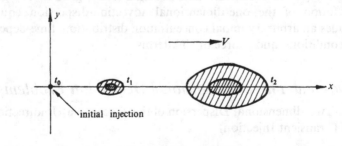

FIGURE 3.6. Variations of concentration contours with time in the case of transient injection.

in the longitudinal and transverse directions depend on the dispersion coefficients D_L and D_T.

Problem 4. Two-dimensional Dispersion of the Tracer in a Unidirectional
Flow Field with Continuous Injection

The basic assumptions in Problem 4 are the same as that in Problem 3,
except that starting at $t = 0$ the fluid containing the tracer with concentration C_0 is continuously injected into the aquifer at the origin with rate q.
Since continuous injection may be considered as a series of transient injections, we can integrate the solution of Problem 3, Eq. (3.2.25), over time t to
obtain the solution of Problem 4 as follows:

$$C(x, y, t) = \frac{C_0 q}{4\pi\sqrt{D_L D_T}} \int_0^t \exp\left[-\frac{(x - V\eta)^2}{4D_L\eta} - \frac{y^2}{4D_T\eta}\right]\frac{d\eta}{\eta}. \qquad (3.2.27)$$

Letting

$$a = \frac{x^2}{D_L} + \frac{y^2}{D_T}; \quad b = \frac{V^2}{4D_L}, \qquad (3.2.28)$$

and using the variable transformation $u = a/4\eta$, Eq. (3.2.27) is transformed
into

$$C(x, y, t) = \frac{C_0 q}{4\pi\sqrt{D_L D_T}} \exp\left(\frac{Vx}{2D_L}\right)\int_{a/4t}^{\infty} \exp\left(-u - \frac{ab}{u}\right)\frac{du}{u}. \qquad (3.2.29)$$

The above equation can be rewritten as

$$C(x, y, t) = \frac{C_0 q}{4\pi\sqrt{D_L D_T}} \exp\left(\frac{Vx}{2D_L}\right)[W(0, \sqrt{ab}) - W(bt, \sqrt{ab})], \qquad (3.2.30)$$

where

$$W(u, r) = \int_u^{\infty} \exp\left(-\xi - \frac{r^2}{4\xi}\right)\frac{d\xi}{\xi} \qquad (3.2.31)$$

is the well-known Hantush leaky well function. Letting $t \to \infty$, the asymptotic
concentration distribution of this case is

$$C(x, y) = \frac{C_0 q}{2\pi\sqrt{D_L D_T}} \exp\left(\frac{Vx}{2D_L}\right)K_0(ab)$$

$$= \frac{C_0 q}{2\pi\sqrt{D_L D_T}} \exp\left(\frac{Vx}{2D_L}\right)K_0\left(\frac{x^2 V^2}{4D_L^2} + \frac{y^2 V^2}{4D_L D_T}\right), \qquad (3.2.32)$$

where K_0 is the zero-order modified Bessel function of the second kind.

Problem 5. Two- and Three-dimensional Dispersions of a Tracer in a Semi-infinite Region

Leij and Dane (1990) considered a *two-dimensional dispersion problem* of a tracer in a semi-plane, which is as follows:

$$
\begin{cases}
\dfrac{\partial C}{\partial t} = D_L \dfrac{\partial^2 C}{\partial x^2} + D_T \dfrac{\partial^2 C}{\partial y^2} - V \dfrac{\partial C}{\partial x}, \\[2mm]
C(x, y, 0) = 0, \qquad 0 < x < \infty, \quad -\infty < y < \infty, \\[2mm]
\left. \dfrac{\partial C}{\partial x} \right|_{x=\infty} = 0, \qquad -\infty < y < \infty, \quad t > 0, \\[2mm]
C(0, y, t) = g(y), \quad -\infty < y < \infty, \quad t > 0, \\[2mm]
\left. \dfrac{\partial C}{\partial y} \right|_{y=\pm\infty} = 0, \qquad 0 < x < \infty, \quad t > 0,
\end{cases}
\qquad (3.2.33)
$$

where

$$
g(y) = \begin{cases}
C_L, & y < 0, \\[2mm]
\dfrac{1}{2}(C_L + C_R), & y = 0, \\[2mm]
C_R, & y > 0,
\end{cases}
\qquad (3.2.34)
$$

and C_L and C_R are given constants. Using the Laplace and Fourier transforms successively, Eq. (3.2.33) may be translated into a boundary value problem of an ordinary differential equation with parameters α and p:

$$
\begin{cases}
D_L \dfrac{d^2 \hat{C}}{dx^2} - V \dfrac{d\hat{C}}{dx} - (\alpha^2 D_T + p)\hat{C} = 0, \\[2mm]
\left. \dfrac{d\hat{C}}{dx} \right|_{x=\infty} = 0, \\[2mm]
\hat{C}(0, \alpha, p) = \hat{g}/p,
\end{cases}
\qquad (3.2.35)
$$

where $\hat{C}$ is the *Fourier transform* of C^*, i.e.,

$$
\hat{C}(x, \alpha, p) = F[C^*] = \frac{1}{\sqrt{2\pi}} \int_{-\infty}^{\infty} \exp(i\alpha y) C^*(x, y, p)\, dy; \qquad (3.2.36)
$$

and C^* is the *Laplace transform* of solution C of the original problem, i.e.,

$$
C^* = L[C] = \int_0^{\infty} e^{-pt} C(x, y, t)\, dt, \qquad (3.2.37)
$$

and $\hat{g}$ is the Fourier transform of g. After the solution of Eq. (3.2.35) is obtained, the Laplace and Fourier inverse transforms are used successively, and the solution of the original problem is finally obtained as:

$$C(x, y, t) = \frac{x}{\sqrt{4\pi D_L}} \int_0^t \tau^{-3/2} \left\{ \frac{C_L}{2} \operatorname{erfc}\left(\frac{y}{\sqrt{4D_T\tau}} \right) \right.$$

$$\left. + \frac{C_R}{2} \operatorname{erfc}\left(\frac{-y}{\sqrt{4D_T\tau}} \right) \right\} \exp\left[-\left(\frac{x - Vt}{\sqrt{4D_T\tau}} \right)^2 \right] d\tau. \quad (3.2.38)$$

The detailed derivation can be found in the paper by Leij and Dane (1990).

Similarly, the following three-dimensional dispersion problem in a semi-infinite region may be expressed as follows:

$$\begin{cases} \dfrac{\partial C}{\partial t} = D_L \dfrac{\partial^2 C}{\partial x^2} + D_T \left(\dfrac{\partial^2 C}{\partial y^2} + \dfrac{\partial^2 C}{\partial z^2} \right) - V \dfrac{\partial C}{\partial x}, \\[2mm] C(x, y, z, 0) = 0, \qquad 0 \le x < \infty, \quad -\infty < y < \infty, \quad -\infty < z < \infty, \\[2mm] C(0, y, z, t) = g(y, z), \quad -\infty < y < \infty, \quad -\infty < z < \infty, \quad t > 0, \\[2mm] \left. \dfrac{\partial C}{\partial x} \right|_{x=\infty} = 0, \qquad -\infty < y < \infty, \quad -\infty < z < \infty, \quad t > 0, \\[2mm] \left. \dfrac{\partial C}{\partial y} \right|_{y=\pm\infty} = 0, \qquad 0 \le x < \infty, \quad -\infty < z < \infty, \quad t > 0, \\[2mm] \left. \dfrac{\partial C}{\partial z} \right|_{z=\pm\infty} = 0, \qquad 0 \le x < \infty, \quad -\infty < y < \infty, \quad t > 0. \end{cases} \quad (3.2.39)$$

When $g(y, z)$ is given by

$$g(y, z) = \begin{cases} C_0, & |y| < a \quad \text{and} \quad |z| < b, \\ 0, & \text{otherwise}, \end{cases} \quad (3.2.40)$$

the solution of Eq. (3.2.39) is

$$C(x, y, z, t) = \frac{xC_0}{8\sqrt{\pi D_L}} \int_0^t \tau^{-3/2} \left\{ \operatorname{erfc}\left(\frac{a+y}{2\sqrt{D_T\tau}} \right) + \operatorname{erfc}\left(\frac{a-y}{2\sqrt{D_T\tau}} \right) \right\}$$

$$\cdot \left\{ \operatorname{erfc}\left(\frac{b+z}{2\sqrt{D_T\tau}} \right) + \operatorname{erfc}\left(\frac{b-z}{2\sqrt{D_T\tau}} \right) \right\} \exp\left[-\left(\frac{x - Vt}{2\sqrt{D_T\tau}} \right)^2 \right] d\tau. \quad (3.2.41)$$

The solutions of the two- and three-dimensional problems contain complicated integrals as shown in Eq. (3.2.38) and Eq. (3.2.41). The concentrations corresponding to any given point (x, y, z) and time t can be calculated by means of numerical integration.

The three-dimensional dispersion problem in unidirectional flow fields was further considered by Leij et al. (1991). In their model, adsorption decay, and flux boundary conditions are included. Tang and Aral (1992) presented an advection-dispersion model for the contaminant transport in layered aquifers and obtained associated analytical solutions. In their model, both horizontal and vertical flows are taken into account. First-order reaction and retardation effects are also considered.

3.2.3 Radial Dispersion Problems

Problem 6. Radial Dispersion of a Tracer in an Aquifer with No Natural Groundwater Velocity

It is assumed that there is a fully penetrating well with radius r_0 in a confined aquifer, which is horizontal, of constant thickness, infinite extent, homogeneous, and isotropic. Water with tracer concentration C_0 is continuously injected at constant rate Q into the well. A nearly steady two-dimensional radial flow will be formed very soon around the well when no natural flow exists. Meanwhile, the flux passing through any circle centered around the well and with an arbitrary radius r is

$$2\pi KBr\frac{\partial h}{\partial r} = -Q,$$

where K is the hydraulic conductivity, and B is the thickness of the aquifer. Based on this equation and Darcy's law, the average velocity is

$$V(r) = \frac{q}{n} = -\frac{K}{n}\frac{\partial h}{\partial r} = \frac{Q}{2\pi Bnr} = \frac{A}{r}, \qquad (3.2.42)$$

where $A = \dfrac{Q}{2\pi Bn}$. Substituting Eq. (3.2.42) into Eq. (2.6.12), the mathematical model of this problem is obtained as follows:

$$\begin{cases} \dfrac{\partial C}{\partial t} = \dfrac{\alpha_L A}{r}\dfrac{\partial^2 C}{\partial r^2} - \dfrac{A}{r}\dfrac{\partial C}{\partial r}, \\[2mm] C(r,0) = 0, \qquad r \geq r_0, \\[2mm] C(r_0,t) = C_0, \qquad t > 0, \\[2mm] C(\infty,t) = 0, \qquad t > 0. \end{cases} \qquad (3.2.43)$$

Tang and Babu (1979) found the exact analytic solution of this problem by Laplace transform method. The result is

$$\frac{C(r,t)}{C_0} = 1 - \left(\frac{r}{r_0}\right)^{1/2}\exp\left(\frac{r - r_0}{2\alpha_L}\right)(I_1 + I_2 + I_3), \qquad (3.2.44)$$

where

$$I_1 = \int_0^{\sqrt{a/r}} e^{-\rho^2 t}\left[\frac{\alpha/r - \rho^2}{\alpha/r_0 - \rho^2}\right]^{1/2}$$

$$\cdot \left\{\frac{K_{1/3}(arg\,\beta)I_{1/3}(arg\,\gamma) - K_{1/3}(arg\,\gamma)I_{1/3}(arg\,\beta)}{\frac{1}{4}[K_{1/3}(arg\,\gamma)]^2 + \left[\frac{\sqrt{3}}{2}K_{1/3}(arg\,\gamma) + \pi I_{1/3}(arg\,\gamma)\right]^2}\right\}\frac{d\rho}{\rho}, \qquad (3.2.45)$$

in which

$$\alpha = \frac{A}{4\alpha_L}, \quad \beta = \frac{2}{3}\left(\frac{r^3}{\alpha_L A}\right)^{1/2}, \quad \gamma = \frac{2}{3}\left(\frac{r_0^3}{\alpha_L A}\right)^{1/2},$$

$$arg\,\beta = \beta\left[\frac{\alpha}{r} - \rho^2\right]^{3/2}/\rho^2, \quad arg\,\gamma = \gamma\left[\frac{\alpha}{r_0} - \rho^2\right]^{3/2}/\rho^2,$$

$$I_2 = \int_{\sqrt{\alpha/r_0}}^{\sqrt{\alpha/r}} e^{-\rho^2 t}\left[\frac{\rho^2 - \alpha/r}{\alpha/r_0 - \rho^2}\right]^{1/2}\left\{\frac{-K_{1/3}(arg\,\gamma)J_{1/3}(arg\,\beta) + \pi I_{1/3}(arg\,\gamma)}{\frac{1}{4}[K_{1/3}(arg\,\gamma)]^2 + \left[\frac{\sqrt{3}}{2}K_{1/3}(arg\,\gamma)\right.}\right.$$

$$\left.\frac{\left[-\frac{\sqrt{3}}{2}J_{1/3}(arg\,\beta) + \frac{1}{2}Y_{1/3}(arg\,\beta)\right]}{\left. + \pi I_{1/3}(arg\,\gamma)\right]^2}\right\}\frac{d\rho}{\rho}, \tag{3.2.46}$$

where the arguments for I_2 are

$$arg\,\beta = \beta\left[\rho^2 - \frac{\alpha}{r}\right]^{3/2}/\rho^2, \quad arg\,\gamma = \gamma\left[\frac{\alpha}{r_0} - \rho^2\right]^{3/2}/\rho^2;$$

and

$$I_3 = \frac{2}{\pi}\int_{\sqrt{\alpha/r_0}}^{\infty} e^{-\rho^2 t}\left[\frac{\rho^2 - \alpha/r}{\rho^2 - \alpha/r_0}\right]^{1/2}$$

$$\cdot\left\{\frac{J_{1/3}(arg\,\gamma)Y_{1/3}(arg\,\beta) - J_{1/3}(arg\,\beta)Y_{1/3}(arg\,\gamma)}{[J_{1/3}(arg\,\gamma)]^2 + [Y_{1/3}(arg\,\gamma)]^2}\right\}\frac{d\rho}{\rho}, \tag{3.2.47}$$

where the arguments for I_3 are

$$arg\,\beta = \beta\left[\rho^2 - \frac{\alpha}{r}\right]^{3/2}/\rho^2, \quad arg\,\gamma = \gamma\left[\rho^2 - \frac{\alpha}{r_0}\right]^{3/2}/\rho^2.$$

In Eq. (3.2.45) to Eq. (3.2.47), $J_{1/3}$, $Y_{1/3}$, $I_{1/3}$ and $K_{1/3}$ are the Bessel functions of the first, the second kind, and the modified Bessel functions, respectively.

As shown by the above expression, the calculation of the solution of Eq. (3.2.44) is troublesome. The calculation will be especially difficult when time t is relatively small or large. Therefore, the following asymptotic solutions for short time and long time may be used separately. When t is small, we use

$$\frac{C(r,t)}{C_0} \approx \left(\frac{r}{r_0}\right)^{-1/4}\exp\left(\frac{r - r_0}{2\alpha_L}\right)\text{erfc}\left(\frac{\beta - \gamma}{2\sqrt{t}}\right). \tag{3.2.48}$$

When t is quite large, we adopt

$$\frac{C(r,t)}{C_0} \approx 1 - \exp\left(\frac{r - r_0}{2\alpha_L}\right)\frac{(I_4 + I_5)}{\pi}, \tag{3.2.49}$$

where

$$I_4 = \int_{\alpha/r}^{\alpha/r_0} \left[1 + \frac{(r - r_0)x}{4\alpha}\right] \exp\left[-tx - \frac{\gamma}{x}\left(\frac{\alpha}{r_0} - x\right)^{3/2}\right] \sin\left[\frac{\beta}{x}\left(x - \frac{\alpha}{r}\right)^{3/2}\right] \frac{dx}{x},$$

$$I_5 = \int_{\alpha/r_0}^{\infty} \left[1 + \frac{(r - r_0)x}{4\alpha}\right] \exp(-tx) \sin\left[\frac{\beta(x - \alpha/r)^{3/2} - \gamma(x - \alpha/r_0)^{3/2}}{x}\right] \frac{dx}{x}.$$

Moench and Ogata (1981) calculated the inverse Laplace transform by a numerical method to avoid calculation of these complex expressions. With this algorithm, the computational effort is reduced and the results obtained are accurate enough. A simpler analytic solution including the Airy function was given by Hsieh (1986). Chen (1987) changed the boundary condition of the constant concentration into the Cauchy boundary condition at the injecting well, and obtained the corresponding analytic solution.

In researching radial dispersion, Rasmuson (1981) considered the two-dimensional dispersion problem on the plane (r, z). An analytic solution for the case that there is a disc-shaped pollution source ($C(r, 0, t) = C_0$, $0 \leq r \leq a$) on the plane $z = 0$ was obtained. This solution includes the effect of adsorption. A problem of three-dimensional radial dispersion, as shown in Figure 3.7, was considered by Yates (1988). In the axisymmetric case, with adsorption and radioactive decay taken into consideration, Eq. (2.6.8) was rewritten as

$$R_d\frac{\partial C}{\partial t} = \frac{1}{r}\frac{\partial}{\partial r}\left(D_L r \frac{\partial C}{\partial r}\right) + \frac{\partial}{\partial z}\left(D_T \frac{\partial C}{\partial z}\right) - V\frac{\partial C}{\partial r} - \lambda R_d C, \qquad (3.2.50)$$

where velocity V is determined from Eq. (3.2.42), R_d is the retardation factor, and λ is the coefficient of radioactive decay.

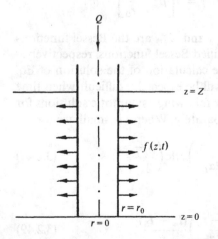

FIGURE 3.7. Three-dimensional axisymmetric radial dispersion.

The additional initial and boundary conditions are:

$$\begin{cases} C(r,z,0) = 0, & r > r_0, \quad 0 \le z \le Z, \\ C(r_0,z,t) = f(z,t), & 0 \le z \le Z, \quad t > 0, \\ C(r,z,t)|_{r=\infty} = 0, & 0 \le z \le Z, \quad t > 0, \\ \left.\dfrac{\partial C}{\partial z}\right|_{z=0} = 0, & r > r_0, \quad t > 0, \\ \left.\dfrac{\partial C}{\partial z}\right|_{z=Z} = 0, & r > t_0, \quad t > 0, \end{cases} \tag{3.2.51}$$

where $f(z,t)$ is an arbitrary function that can be used to simulate any testing condition in situ. Yates (1988) obtained the analytic solution of this problem using a technique similar to that used by Tang and Babu (1979). Chen (1989) considered the axisymmetric dispersion problem in a leaky aquifer. Waste water is injected through a fully penetrating well into the aquifer, but because of the leakage effect, the waste water may enter into the aquitard where the hydraulic conductivity is smaller than that in the aquifer. Under the assumptions that both the longitudinal dispersion in the aquifer and the transverse dispersion in the aquitard can be ignored, Chen (1989) derived a mathematical model for the problem, and obtained its analytic solution by the Laplace transformation method.

3.2.4 Dispersion Problems in Fractured Rock

The existence of *fractures* has a major influence on mass transport in groundwater. Since the structures of fracture systems are very complex, it is not easy to describe the mechanism of mass transport in fractures or to build the corresponding mathematical models. In this section, several analytic solutions, which were obtained under some ideal conditions, will be given.

Problem 7. Hydrodynamic Dispersion in a Single Fracture

Tang et al. (1981) studied advection-dispersion in a single fracture and the diffusion in the porous matrix around the fracture.

It is assumed that there is a narrow semi-infinite long fracture in a saturated porous medium, see Figure 3.8. The width of the fracture and the velocity of groundwater in the z direction are constant and equal to $2b$ and V, respectively. The solute concentration in the whole system is initially equal to zero. The solute concentration at the end point $z = 0$ remains $C = C_0$ starting from $t = 0$. It is assumed that the hydrodynamic dispersion in the fracture is one-dimensional and the solute in the fracture will enter the neighboring porous matrix through the walls of the fracture. The velocity of the porous matrix is assumed to be very slow so that the solute transport in

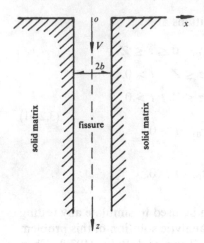

FIGURE 3.8. Mass transport in a single fracture.

the porous matrix depends mainly on the molecular diffusion. Based on these assumptions, the governing equation of the one-dimensional advection-dispersion in the fracture is

$$\frac{\partial C}{\partial t} = D \frac{\partial^2 C}{\partial z^2} - V \frac{\partial C}{\partial z} - \frac{q}{b}, \quad 0 \leq z < \infty, \quad (3.2.52)$$

where C is the concentration distribution in the fracture, D is the longitudinal dispersion coefficient, and q is the diffusive flux through the fracture walls $(M/L^2 T)$.

The diffusion equation in the porous matrix is

$$\frac{\partial C'}{\partial t} = D' \frac{\partial^2 C'}{\partial x^2}, \quad b \leq x < \infty, \quad (3.2.53)$$

where C' is the concentration distribution, and D' is the diffusion coefficient in the porous matrix. Eqs. (3.2.52) and (3.2.53) will be coupled by the diffusive flux q. According to the law of the linear diffusion, we have

$$q = -nD' \frac{\partial C'}{\partial x}\bigg|_{x=b}. \quad (3.2.54)$$

Substituting this equation into Eq. (3.2.54), we obtain the governing equation for a fracture as

$$\frac{\partial C}{\partial t} = D \frac{\partial^2 C}{\partial z^2} - V \frac{\partial C}{\partial z} + \frac{nD'}{b} \frac{\partial C'}{\partial x}\bigg|_{x=b}. \quad (3.2.55)$$

This equation can be solved simultaneously with Eq. (3.2.53). The subsidiary conditions for Eq. (3.2.55) are

$$
\begin{cases}
C(0, t) = C_0, \\
C(\infty, t) = 0, \\
C(z, 0) = 0,
\end{cases}
\tag{3.2.56}
$$

and the subsidiary conditions for Eq. (3.2.53) are

$$
\begin{cases}
C'(b, z, t) = C(z, t), \\
C'(\infty, z, t) = 0, \\
C'(x, z, 0) = 0,
\end{cases}
\tag{3.2.57}
$$

where the first condition is the boundary condition along the walls of the fracture. This boundary condition can also be used to connect the two equations. Tang et al. (1981) obtained the analytic solution of this problem by the Laplace transform. Under the condition $D \neq 0$, the concentration distribution in the fracture is

$$
\frac{C(z, t)}{C_0} = \frac{2 \exp(Vz/2D)}{\sqrt{\pi}} \int_{z/2\sqrt{Dt}}^{\infty} \exp\left[-\xi^2 - \frac{V^2 z^2}{16 D^2 \xi^2}\right] \mathrm{erfc}[f(\xi)]\, d\xi, \tag{3.2.58}
$$

where

$$
f(\xi) = \frac{nz^2}{4b} \sqrt{\frac{D'}{D}} \frac{1}{\xi \sqrt{4 D \xi^2 t - z^2}},
$$

and the concentration distribution in the pores is

$$
\frac{C'(x, z, t)}{C_0} = \frac{2 \exp(Vz/2D)}{\sqrt{\pi}} \int_{z/2\sqrt{Dt}}^{\infty} \exp\left[-\xi^2 - \frac{V^2 z^2}{16 D^2 \xi^2}\right] \mathrm{erfc}[g(\xi)]\, d\xi,
$$
$$
\tag{3.2.59}
$$

where

$$
g(\xi) = \left[\frac{nz^2}{4b} \sqrt{\frac{D'}{D}} \frac{1}{\xi} + \sqrt{\frac{D}{D'}}(x - b)\xi\right] \Big/ \sqrt{4 D \xi^2 t - z^2}. \tag{3.2.60}
$$

Adsorption and radioactive decay were also considered by Tang et al. (1981). The results, however, are more complex than those of the above two equations.

In addition, other studies following this orientation were made by Sudicky and Frind (1982). They obtained the analytic solution corresponding to a system of parallel fractures, and later they extended their results from a single species to a decay chain created during the solute transport (Subdicky and Frind, 1984). Chen (1986) studied the problem of radioactive mass entering into a single fracture from an injected well. Lowell (1989) considered the case that the concentration at the entrance varied periodically with time.

The limitation of analytic solutions is obvious. The conditions of existing an anlytic solution are very strict: the flow region must be regular, the structure of governing equation must be simple, all model parameters must be constant. Although these requirements can never be satisfied in practice, the analytic solution is still important in the study of advection-dispersion problems. It has at least the following two major uses:

1. Since the analytic solutions are exact solutions, we can use it to verify the accuracy of a numerical method and test the correctness of a program.
2. Analytic solutions can be used to simulate some simple tests conducted in the field and in the laboratory for better understanding the mechanics of dispersion and identifying dispersion parameters.

Exercises

3.1. Assume that there are two instantaneous point sources located at (x_1, y_1, z_1) and (x_2, y_2, z_2), respectively. Find the concentration distribution by superposition.

3.2. Assume that there is an instantaneous point source located at the center of two parallel walls in the x-y plane. The lengths of the walls are infinite. Find the concentration distribution by the image method.

3.3. Derive Eq. (3.1.21) from Eq. (3.1.19).

3.4. Write a subroutine for calculating the concentration distribution of Eq. (3.1.31) for the continuous injection problem with input parameters: $D_{11}, D_{22}, D_{33}, U, \dot{M}$ and θ.

3.5. Write a subroutine to calculate $C(x, t)$ in Eq. (3.2.15) with input parameters C_0, D_L, and V. Let $V = 1$, change D_L from 0.01 to 10, draw the breakthrough curves for $t = 30$.

3.6. Draw a figure to display the model given in Eqs. (3.2.33) and (3.2.34).

3.7. Derive the radial model (3.2.43) from the general advection-dispersion Eq. (2.6.10). What assumptions are used in deriving the radial model (3.2.43)?

3.8. Extend the mass transport model presented in Section 3.2.4, from the case of a single fracture to the case of a system of parallel fractures.

4
Finite Difference Methods and the Method of Characteristics for Solving Hydrodynamic Dispersion Equations

4.1 Finite Difference Methods

4.1.1 Finite Difference Approximations of Derivatives

We have known the basic idea of *finite difference methods* (FDM) in the study of groundwater flow problems. FDM includes three major steps. First, the flow region is divided by a grid and the time interval into time steps. Second, the partial derivatives involved in the PDE are replaced by their finite difference approximations. As a result, the PDE is transformed into a system of algebraic equations. Third, the algebraic system is solved and the nodal values of the unknown function are obtained. These discrete values approximately describe the time-space distribution of the unknown variable. We will see that exactly the same steps can be used to solve advection-dispersion problems.

Consider that there are three adjacent nodes along the x direction, $(x - \Delta x, y, z)$, (x, y, z) and $(x + \Delta x, y, z)$ in the center of each cube, as shown in Figure 4.1. If the nodes are numbered in accordance with the number of grid lines, they can be denoted by $(i - 1, j, k)$, (i, j, k) and $(i + 1, j, k)$. The spatial distances between the mesh planes in the three directions are assumed to be Δx, Δy, and Δz. The Taylor expansions centered around node (x, y, z, t) are

$$C(x + \Delta x, y, z, t) = C(x, y, z, t) + \frac{\partial C}{\partial x}\Delta x + \frac{\partial^2 C}{\partial x^2}\frac{(\Delta x)^2}{2} + O[(\Delta x)^3] \quad (4.1.1)$$

and

$$C(x - \Delta x, y, z, t) = C(x, y, z, t) - \frac{\partial C}{\partial x}\Delta x + \frac{\partial^2 C}{\partial x^2}\frac{(\Delta x)^2}{2} + O[(\Delta x)^3]. \quad (4.1.2)$$

In these equations, the values of the partial derivatives on the right-hand sides are all evaluated at point (x, y, z, t). From Eq. (4.1.1), we have

$$\frac{\partial C}{\partial x} = \frac{C(x + \Delta x, y, z, t) - C(x, y, z, t)}{\Delta x} + O(\Delta x), \quad (4.1.3)$$

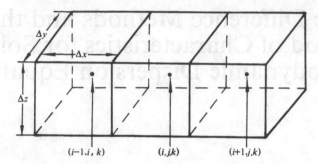

$(i-1,i,k)$ (i,j,k) $(i+1,j,k)$

FIGURE 4.1. Three adjacent nodes along the x direction.

and from Eq. (4.1.2), we have

$$\frac{\partial C}{\partial x} = \frac{C(x,y,z,t) - C(x - \Delta x, y, z, t)}{\Delta x} + O(\Delta x). \tag{4.1.4}$$

Subtraction of Eq. (4.1.2) from Eq. (4.1.1) results in

$$\frac{\partial C}{\partial x} = \frac{C(x + \Delta x, y, z, t) - C(x - \Delta x, y, z, t)}{2\Delta x} + O[(\Delta x)^2]. \tag{4.1.5}$$

By neglecting the second term on the right-hand side of Eqs. (4.1.3), (4.1.4), and (4.1.5), three approximate equations are obtained. These equations are called the *forward*, *backward*, and *central difference formulas*, respectively. The second term on the right-hand side is called the truncation error, which denotes the possible order of error when the derivative on the left-hand side is expressed by the first term on the right-hand side.

The summation of Eq. (4.1.1) and Eq. (4.1.2) generates the finite difference approximation of the second order derivative:

$$\frac{\partial^2 C}{\partial x^2} = \frac{C(x + \Delta x, y, z, t) - 2C(x, y, z, t) + C(x - \Delta x, y, z, t)}{(\Delta x)^2}. \tag{4.1.6}$$

The truncation error of this approximation is $O[(\Delta x)^2]$. Using the node numbers, this equation can be rewritten as

$$\left.\frac{\partial^2 C}{\partial x^2}\right|_{(i,j,k)} = \frac{C_{i+1,j,k} - 2C_{i,j,k} + C_{i-1,j,k}}{(\Delta x)^2}. \tag{4.1.7}$$

For the mixed partial derivative, we have

$$\left.\frac{\partial^2 C}{\partial x \partial y}\right|_{(i,j,k)} = \frac{C_{i+1,k} - C_{i-1,j+1,k} - C_{i+1,j-1,k} + C_{i-1,j-1,k}}{4\Delta x \Delta y}, \tag{4.1.8}$$

and its truncation error is $O[(\Delta x)^2 + (\Delta y)^2]$.

In the same way, we can obtain the finite difference approximations of other partial derivatives, such as $\partial C/\partial t$, $\partial C/\partial y$, $\partial^2 C/\partial y^2, \ldots, \partial^2 C/\partial z^2$, and so on. Thus. all first order and second order partial derivatives of the concentration at any node can be represented by linear combinations of the concentration values at that node and its neighboring nodes.

4.1.2 Finite Difference Solutions for One-Dimensional Dispersion Problems

Consider the following one-dimensional advection-dispersion equation:

$$\frac{\partial C}{\partial t} = D\frac{\partial^2 C}{\partial x^2} - V\frac{\partial C}{\partial x}, \quad 0 \le x \le L, \ 0 \le t \le T. \tag{4.1.9}$$

The time range $[0, T]$ and space region $[0, L]$ are both divided into a uniform grid with time and space intervals Δt and Δx, respectively. The concentration value of the i-th node located at x_i and at time t_n may be written as $C_{i,n}$. Now, let us derive a finite difference equation at any internal node to replace Eq. (4.1.9). The left-hand side of the equation is often approximated by the forward difference:

$$\frac{\partial C}{\partial t} = \frac{C_{i,n+1} - C_{i,n}}{\Delta t}. \tag{4.1.10}$$

On the right-hand side of Eq. (4.1.9), $\partial^2 C/\partial x^2$ is replaced by Eq. (4.1.6), and $\partial C/\partial x$ may be replaced by forward, backward, or central formulas, i.e., Eqs. (4.1.3), (4.1.4) or (4.1.5). If t in Eq. (4.1.5) and Eq. (4.1.6) is replaced by t_n, t_{n+1} or the middle time between t_n and t_{n+1}, three finite difference schemes can be obtained, that is, the *explicit*, *implicit* and *Crank-Nicolson schemes*.

(1) The explicit difference equation of Eq. (4.1.9) is

$$\frac{C_{i,n+1} - C_{i,n}}{\Delta t} = D\frac{C_{i+1,n} - 2C_{i,n} + C_{i-1,n}}{(\Delta x)^2} - V\frac{C_{i+1,n} - C_{i-1,n}}{2\Delta x}. \tag{4.1.11}$$

From this equation, we can directly obtain

$$C_{i,n+1} = \left[\frac{D\Delta t}{(\Delta x)^2} + \frac{V\Delta t}{2\Delta x}\right]C_{i-1,n} + \left[1 - \frac{2D\Delta t}{(\Delta x)^2}\right]C_{i,n} + \left[\frac{D\Delta t}{(\Delta x)^2} - \frac{V\Delta t}{2\Delta x}\right]C_{i+1,n}. \tag{4.1.12}$$

Therefore, the concentration values of all nodes at time t_{n+1} can be calculated directly without solving any equations, provided that their values at t_n are known.

(2) The implicit difference scheme of Eq. (4.1.9) is

$$\frac{C_{i,n+1} - C_{i,n}}{\Delta t} = D\frac{C_{i+1,n+1} - 2C_{i,n+1} + C_{i-1,n+1}}{(\Delta x)^2} - V\frac{C_{i+1,n+1} - C_{i-1,n+1}}{2\Delta x}, \tag{4.1.13}$$

where the unknown concentrations $C_{i-1,n+1}$, $C_{i,n+1}$ and $C_{i+1,n+1}$ at t_{n+1} are included. After rearrangement, we have

$$-\left[\frac{D\Delta t}{(\Delta x)^2} + \frac{V\Delta t}{2\Delta x}\right]C_{i-1,n+1} + \left[1 + \frac{2D\Delta t}{(\Delta x)^2}\right]C_{i,n+1} - \left[\frac{D\Delta t}{(\Delta x)^2} - \frac{V\Delta t}{2\Delta x}\right]C_{i+1,n+1}$$

$$= C_{i,n}. \qquad (4.1.14)$$

We can write similar equations for all nodes. When these equations are combined with the boundary conditions, a set of tridiagonal equations are formed. Thus, the concentration values of all nodes can be obtained by using the *Thomas algorithm*.

(3) The explicit and implicit schemes can be averaged to yield the Crank-Nicolson difference scheme, i.e.,

$$\frac{C_{i,n+1} - C_{i,n}}{\Delta t} = \frac{1}{2}\left(D\frac{C_{i+1,n} - 2C_{i,n} + C_{i-1,n}}{(\Delta x)^2} - V\frac{C_{i+1,n} - C_{i-1,n}}{2\Delta x}\right.$$

$$\left. + D\frac{C_{i+1,n+1} - 2C_{i,n+1} + C_{i-1,n+1}}{(\Delta x)^2} - V\frac{C_{i+1,n+1} - C_{i-1,n+1}}{2\Delta x}\right). \qquad (4.1.15)$$

After rearranging, we have

$$-\left[\frac{D\Delta t}{(\Delta x)^2} + \frac{V\Delta t}{2\Delta x}\right]C_{i-1,n+1} + 2\left[1 + \frac{D\Delta t}{(\Delta x)^2}\right]C_{i,n+1} - \left[\frac{D\Delta t}{(\Delta x)^2} - \frac{V\Delta t}{2\Delta x}\right]C_{i+1,n+1}$$

$$= \left[\frac{D\Delta t}{(\Delta x)^2} + \frac{V\Delta t}{2\Delta x}\right]C_{i-1,n} + 2\left[1 - \frac{D\Delta t}{(\Delta x)^2}\right]C_{i,n} + \left[\frac{D\Delta t}{(\Delta x)^2} - \frac{V\Delta t}{2\Delta x}\right]C_{i+1,n}. \qquad (4.1.16)$$

The right-hand side of this equation includes only the known concentrations at t_n, therefore the forms of both Eq. (4.1.16) and Eq. (4.1.14) are exactly the same. We can still use the Thomas algorithm to solve the tridiagonal equations and obtain the concentration values at t_{n+1}.

The truncation errors of explicit Eq. (4.1.11) and implicit Eq. (4.1.13) are both $O[(\Delta t) + (\Delta x)^2]$, but in the Crank-Nicolson scheme, the truncation error is $O[(\Delta t)^2 + (\Delta x)^2]$. Since these three schemes are all similar to those for groundwater flow, it is not necessary to explain the solution procedures in detail.

The *convergence* and *stability* of the three schemes will be stated below. If the solution of the difference equation tends to the solution of the original partial differential equation as $\Delta t \to 0$ and $\Delta x \to 0$, the difference scheme is called a convergent one. If the calculation error produced at a certain time level decreases, or at least does not increase when it propagates to the next time level, the difference scheme is called a stable one. One can prove that the implicit scheme of Eq. (4.1.13) and the Crank-Nicolson scheme of Eq. (4.1.16) are unconditionally convergent and stable, while the explicit scheme of Eq. (4.1.11) is not. Convergence and stability of an explicit scheme depend on the step sizes of both time and space. For Eq. (4.1.11), the restrictive conditions are

$$\Delta x < \frac{2D}{V}, \quad \Delta t < \frac{(\Delta x)^2}{2D}. \tag{4.1.17}$$

If the PDE is discretized by an explicit scheme, and the first order partial derivative $\partial C/\partial x$ is approximated by the backward difference, that is,

$$\frac{C_{i,n+1} - C_{i,n}}{\Delta t} = D\frac{C_{i+1,n} - 2C_{i,n} + C_{i-1,n}}{(\Delta x)^2} - V\frac{C_{i,n} - C_{i-1,n}}{\Delta x}, \tag{4.1.18}$$

the restrictive condition would be

$$\Delta t < \frac{(\Delta x)^2}{2D + V\Delta x}. \tag{4.1.19}$$

When the diffusion coefficient, D, is relatively small, or velocity, V, is relatively large, the explicit scheme is unacceptable because of the restrictive condition of Eq. (4.1.17) or Eq. (4.1.19). Let us derive condition (4.1.19) as an example. The accurate solution $\overline{C}$ of the partial differential equation (4.1.9) satisfies the following equation:

$$\frac{\overline{C}_{i,n+1} - \overline{C}_{i,n}}{\Delta t} + O(\Delta t) = D\frac{\overline{C}_{i+1,n} - 2\overline{C}_{i,n} + \overline{C}_{i-1,n}}{(\Delta x)^2} + O[(\Delta x)^2]$$

$$- V\frac{\overline{C}_{i,n} - \overline{C}_{i-1,n}}{\Delta x} + O(\Delta x). \tag{4.1.20}$$

The difference between the accurate solution $\overline{C}$ and the approximate solution C of the difference equation (4.1.18) is indicated by ε. Subtracting Eq. (4.1.20) from Eq. (4.1.18), we have

$$\frac{\varepsilon_{i,n+1} - \varepsilon_{i,n}}{\Delta t} + O(\Delta t) = D\frac{\varepsilon_{i+1,n} - 2\varepsilon_{i,n} + \varepsilon_{i-1,n}}{(\Delta x)^2} + O[(\Delta x)^2]$$

$$- V\frac{\varepsilon_{i,n} - \varepsilon_{i-1,n}}{\Delta x} + O[\Delta x]. \tag{4.1.21}$$

For simplicity, we introduce the following dimensionless parameters:

$$\tilde{D} = \frac{D\Delta t}{(\Delta x)^2}, \quad \tilde{V} = \frac{V\Delta t}{\Delta x}. \tag{4.1.22}$$

Eq. (4.1.21) can then be rewritten as

$$\varepsilon_{i,n+1} = \tilde{D}\varepsilon_{i+1,n} + (1 - 2\tilde{D} - \tilde{V})\varepsilon_{i,n} + (\tilde{D} + \tilde{V})\varepsilon_{i-1,n}$$

$$+ O[(\Delta t)^2 + \Delta t \cdot \Delta x]. \tag{4.1.23}$$

From Eq. (4.1.22), we have $\tilde{D} > 0$, $\tilde{V} > 0$. If the following condition

$$2\tilde{D} + \tilde{V} < 1 \tag{4.1.24}$$

is satisfied, then in terms of Eq. (4.1.23), we can obtain the estimation:

$$E_{n+1} \leq \tilde{D}E_n + (1 - 2\tilde{D} - \tilde{V})E_n + (\tilde{D} + \tilde{V})E_n + O[(\Delta t)^2 + \Delta t \cdot \Delta x]$$

$$= E_n + O[(\Delta t)^2 + \Delta t \cdot \Delta x] \tag{4.1.25}$$

where E_{n+1} and E_n are used to indicate the maximum absolute values of ε at time levels $n + 1$ and n, respectively. Since the initial condition used in the difference equation is the same as in the original equation, we have $E_0 = 0$. Thus, from Eq. (4.1.25), we have

$$E_1 \leq O[(\Delta t)^2 + \Delta t \cdot \Delta x]$$

$$E_2 \leq 2 \cdot O[(\Delta t)^2 + \Delta t \cdot \Delta x]$$

$$\cdots\cdots\cdots\cdots\cdots$$

$$E_n \leq n \cdot O[(\Delta t)^2 + \Delta t \cdot \Delta x]$$

$$\cdots\cdots\cdots\cdots\cdots$$

These equations show that $E_n \to 0$ as long as $\Delta t \to 0$, $\Delta x \to 0$ at any time level n. They also show that the solutions of the difference equations tend to the exact solution of the differential equation at the nodes, if condition (4.1.24) is satisfied. Substituting Eq. (4.1.22) into Eq. (4.1.24), we obtain:

$$\frac{2D\Delta t}{(\Delta x)^2} + \frac{V\Delta t}{\Delta x} = \Delta t \frac{2D + V\Delta x}{(\Delta x)^2} < 1,$$

which is just the condition given in Eq. (4.1.19). The error-enlargement factor of the difference equation from one time level to the next can be calculated by von Neumann's stability analysis, we have,

$$|\xi| = \left\{ \left(1 - 4\frac{D\Delta t}{(\Delta x)^2} \sin^2 \frac{\beta \Delta t}{\Delta x} \right)^2 + \left(\frac{V\Delta t}{\Delta x} \right)^2 \sin^2 \beta \Delta t \right\}^{1/2}. \tag{4.1.26}$$

A detailed analysis of the convergence and stability of various difference schemes can be found in the book written by Lapidus and Pinder (1982).

For the one-dimensional advection-dispersion equation with a variable dispersion coefficient:

$$\frac{\partial C}{\partial t} = \frac{\partial}{\partial x}\left(D\frac{\partial C}{\partial x} \right) - V\frac{\partial C}{\partial x}, \tag{4.1.27}$$

we can use the following difference formula

$$\frac{\partial}{\partial x}\left(D\frac{\partial C}{\partial x} \right)\bigg|_i = \frac{\left(D\frac{\partial C}{\partial x} \right)_{i+1/2} - \left(D\frac{\partial C}{\partial x} \right)_{i-1/2}}{\Delta x}$$

$$= \frac{D_{i+1/2}C_{i+1} - (D_{i+1/2} + D_{i-1/2})C_i + D_{i-1/2}C_{i-1}}{(\Delta x)^2}, \tag{4.1.28}$$

where $D_{i+1/2}$ and $D_{i-1/2}$ are the dispersion coefficients between node i and its adjacent nodes $i + 1$ and $i - 1$, respectively. The corresponding difference equations can be derived without difficulty. For example, we have the following implicit difference scheme:

$$\frac{C_{i,n+1} - C_{i,n}}{\Delta t} = \frac{D_{i+1/2}C_{i+1,n+1} - (D_{i+1/2} + D_{i-1/2})C_{i,n+1} + D_{i-1/2}C_{i-1,n+1}}{(\Delta x)^2}$$

$$- V_i \frac{C_{i+1,n+1} - C_{i-1,n+1}}{2\Delta x}. \qquad (4.1.29)$$

The node values of D may be used to express $D_{i+1/2}$ and $D_{i-1/2}$, e.g., we may use the harmonic mean, i.e., let

$$D_{i+1/2} = \frac{2D_i D_{i+1}}{D_i + D_{i+1}},$$

$$D_{i-1/2} = \frac{2D_i D_{i-1}}{D_i + D_{i-1}}.$$

The method mentioned above can also be used to solve the radial dispersion problem:

$$\frac{\partial C}{\partial t} = \frac{\alpha_L A}{r} \frac{\partial^2 C}{\partial r^2} - \frac{A}{r} \frac{\partial C}{\partial r}.$$

If the implicit scheme is adopted, we have the following finite difference approximation:

$$\frac{C_{i,n+1} - C_{i,n}}{\Delta t} = \frac{A}{r_i} \left[\alpha_L \frac{C_{i+1,n+1} - 2C_{i,n+1} + C_{i-1,n+1}}{(\Delta r)^2} - \frac{C_{i+1,n+1} - C_{i-1,n+1}}{2\Delta r} \right].$$

After rearranging, we have

$$- \left[\alpha_L + \frac{\Delta r}{2} \right] C_{i-1,n+1} + \left[2\alpha_L + \frac{r_i(\Delta r)^2}{A\Delta t} \right] C_{i,n+1} - \left[\alpha_L - \frac{\Delta r}{2} \right] C_{i+1,n+1}$$

$$= \frac{r_i(\Delta r)^2}{A\Delta t} C_{i,n}, \qquad (4.1.30)$$

where r_i is the radial distance of node i. Eq. (4.1.30) can be written for all internal nodes. After combining with boundary conditions, a set of tri-diagonal equations are formed. The concentrations of all nodes at the next time step can be obtained using the Thomas algorithm. If the sizes of Δt and Δr are appropriately controlled, the result obtained by this algorithm is very close to the exact solution. The program based on this algorithm is simple and efficient, and can be used for simulating the field tests of tracer injection in order to identify the longitudinal dispersion coefficient.

Now let us study the difficulties and problems which may be encountered when FDM is used for solving the advection-dispersion equations.

4.1.3 Numerical Dispersion and Overshoot

Through the above discussions, the reader may find that the traditional difference methods used for solving the problems of groundwater flow can basically be copied to solve the problems of advection-dispersion. The computational programs can also be written accordingly. In most cases, the calculated results are satisfactory, when compared with the accurate solutions. However, for a problem with a small dispersion coefficient and relatively large velocity, common finite difference methods will not be satisfactory because the transition zone formed by the hydrodynamic dispersion will not be calculated accurately. In other words, when the Peclet number of the problem is large, computational difficulties will be encountered. In order to illustrate this difficulty, let us consider the canonical one-dimensional advection-dispersion equation as follows:

$$\frac{\partial C}{\partial t} = D\frac{\partial^2 C}{\partial x^2} - V\frac{\partial C}{\partial x}, \tag{4.1.31}$$

where D and V are constants, and the initial and boundary conditions are

$$\begin{cases} C(x,0) = 0, & x > 0, \\ C(0,t) = C_0, & t \geq 0, \\ C(\infty,t) = 0, & t \geq 0. \end{cases} \tag{4.1.32}$$

The analytic solution of this problem has been obtained in paragraph 3.2.1. If finite difference methods are used, the grid Peclet number is defined as

$$Pe = \frac{V\Delta x}{D}.$$

When Δx is constant, the greater the ratio between V and D, the larger the Peclet number will be. Figures 4.2 and 4.3 represent the numerical solutions of the finite difference in comparison with the analytic solutions at $Pe = 0.5$ and $Pe = 100$, respectively.

As shown in these figures, for a small Peclet number the numerical solution is very close to the analytic solution, but for a large Peclet number the transition zone of the numerical solution becomes wider, and *oscillations* appear around the concentration front.

For the convenience of analysis, we shall first study the extreme case of $D = 0$. In this case there is only advection without concrete physical dispersion. Thus, Eq. (4.1.31) can be reduced to

$$\frac{\partial C}{\partial t} = -V\frac{\partial C}{\partial x}. \tag{4.1.33}$$

Its analytic and numerical solutions subject to the subsidiary conditions of

FIGURE 4.2. Comparison between analytical and numerical solutions of the implicit difference when $Pe = 0.5$, $t = 40d$. —— = analytical solution; • = numerical solution.

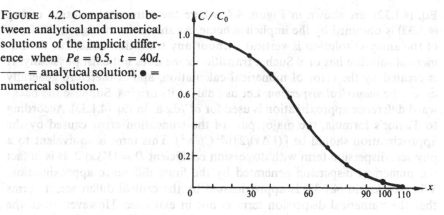

FIGURE 4.3. Comparison between analytical and numerical solutions of the implicit difference when $Pe = 100$, $t = 50d$. —— = analytical solution; o = numerical solution.

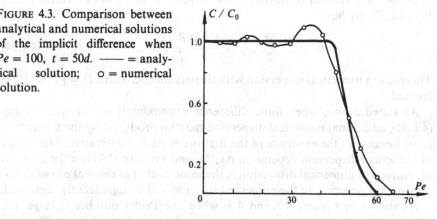

FIGURE 4.4. Comparison between analytical and numerical solutions in the case of pure convection, when $Pe = \infty$. —— = analytical solution; o = numerical solution.

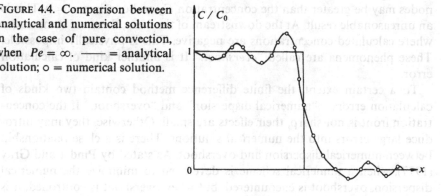

Eq. (4.1.32), are shown in Figure 4.4, where the numerical solution of Eq. (4.1.33) is obtained by the implicit scheme. As shown in the figure, the front of the analytic solution is vertical without any transition zone, but the numerical solution has one. Such a transition zone does not exist physically. It is created by the error of numerical calculation, and is therefore generally called the *numerical dispersion*. Let us analyze its origins. Suppose the backward difference approximation is used for $\partial C/\partial x$ as in Eq. (4.1.33). According to Taylor's formula, the major part of the truncation error caused by the approximation should be $[(V\Delta x)/2](\partial^2 C/\partial x^2)$. This term is equivalent to a physical dispersion term with dispersion coefficient $D = (V\Delta x)/2$. It is in fact the numerical dispersion generated by the finite difference approximation. When $\partial C/\partial x$ in (4.1.33) is approximated by the central difference, it seems that the numerical dispersion term is not in existence. However, since the difference approximation is also used in replacing $\partial C/\partial t$ on the left-hand side of Eq. (4.1.33), another truncation error $(\Delta t/2)(\partial^2 C/\partial t^2)$ will be created. Using Eq. (4.1.33), we have

$$\frac{\Delta t}{2}\frac{\partial^2 C}{\partial t^2} = \frac{V^2 \Delta t}{2}\frac{\partial^2 C}{\partial x^2}. \tag{4.1.34}$$

Therefore, a numerical dispersion with dispersion coefficient $D = [V^2 \Delta t]/2$ is created.

As stated above, when finite difference approximations are used to Eq. (4.1.31), additional numerical dispersion may be produced by the truncation error because of the existence of the first order partial derivative. The values of numerical dispersion depend on Δx, Δt, and velocity V. When the order of magnitude of numerical dispersion is the same as that of the real physical dispersion, the accuracy of the numerical solution will be significantly decreased.

As shown in Figures 4.2 and 4.3, when the Peclet number is large, the oscillation phenomenon occurs in the numerical solution of finite difference. At the upstream of the front, the calculated concentrations at some internal nodes may be greater than the concentration C_0 on boundary $x = 0$. This is an unreasonable result. At the downstream of the front, there are some nodes where calculated concentrations are negative. This is physically impossible. These phenomena are called "*overshoot*." It is another kind of calculation error.

To a certain extent, the finite difference method contain two kinds of calculation errors—"numerical dispersion" and "overshoot." If the concentration front is not sharp, their effects are small; Otherwise, they may introduce large errors into the numerical solutions. There is a close relationship between numerical dispersion and overshoot. As stated by Pinder and Gray (1977), when a numerical scheme is developed to minimize the numerical dispersion, overshoot is encountered; but when overshoot is controlled, it is generally at the expense of increased numerical dispersion. This problem will be discussed in detail in Chapter 6.

4.1.4 Finite Difference Solutions for Two- and Three-Dimensional Dispersion Problems

To extend the FDM from its one-dimensional form to two- and three-dimensional forms is straightforward. Finite difference approximations can be used to discretize each term in two- and three-dimensional advection-dispersion equations. However, we should note that the hydrodynamic dispersion coefficient is always a tensor even for isotropic porous media. Thus the discretization of dispersion terms may involve more neighboring nodes than that in the solution of groundwater flow problems. For a two-dimensional problem, we have four dispersion terms:

$$\frac{\partial}{\partial x_\alpha}\left(D_{\alpha\beta}\frac{\partial C}{\partial x_\beta}\right), \quad (\alpha, \beta = 1, 2). \tag{4.1.35}$$

When Eqs. (4.1.7) and (4.1.8) are used to discretize these terms, nine nodal concentrations: $C_{i-1,j-1}$, $C_{i-1,j}$, $C_{i-1,j+1}$, $C_{i,j-1}$, $C_{i,j}$, $C_{i,j+1}$, $C_{i+1,j-1}$, $C_{i+1,j}$ and $C_{i+1,j+1}$ will be involved in the finite difference equation of node (i,j). For a three-dimensional problem, the number of dispersion terms is nine, and 27 nodal values of concentration will be involved in the finite difference equation of a node.

The numerical dispersion problem associated with two- and three-dimensional finite element methods can also be analyzed. Let us consider the following two-dimensional advection-dispersion equation:

$$\frac{\partial C}{\partial t} + V_x\frac{\partial C}{\partial x} + V_y\frac{\partial C}{\partial y} = D_{xx}\frac{\partial^2 C}{\partial x^2} + 2D_{xy}\frac{\partial^2 C}{\partial x\partial y} + D_{yy}\frac{\partial^2 C}{\partial y^2}, \tag{4.1.36}$$

where V_x and V_y are components of the flow velocity in the x and y directions, respectively; D_{xx}, D_{xy} and D_{yy} are components of the dispersion coefficient tensor. If the backward difference is used for the first order partial derivatives in the equation, the resulting numerical dispersion will be

$$\frac{1}{2}\left[(V_x\Delta x - V_x^2\Delta t)\frac{\partial^2 C}{\partial x^2} - 2V_xV_y\Delta t\frac{\partial^2 C}{\partial x\partial y} + (V_y\Delta y - V_y^2\Delta t)\frac{\partial^2 C}{\partial y^2}\right]. \tag{4.1.37}$$

The derivation process of Eq. (4.1.37) is similar to the one-dimensional problem discussed above, and can be found in the paper written by Cheng et al. (1984). As expressed in Eq. (4.1.37), numerical dispersion may affect all components of the dispersion coefficient tensor.

Besides the numerical dispersion problem, it is difficult to use the FDM to describe irregular geometry and heterogeneous structure of an aquifer, as we have known in the solution of groundwater flow problems. For solving practical groundwater quality problems, therefore, we don't recommend use of the FDM.

4.2 The Method of Characteristics

4.2.1 Basic Idea of the Method of Characteristics

As stated in the previous section, the numerical dispersion increases with velocity. When the advection term dominates the advection-dispersion equation, the effect of numerical dispersion will become notable. In this case, the nature of the advection-dispersion equation is close to the pure-advection one. Without the dispersion term, the latter is a hyperbolic equation of the first order. This fact leads us to use the method of characteristics for hyperbolic equations to solve the advection-dispersion equation. The concept of this method is simple, and the method is capable of handling various complex conditions. Therefore, many researchers adopted it in water quality simulation in the early period (e.g., Garder et al. (1964), Pinder and Cooper (1970), and Konikow and Bredehoeft (1974)). Konikow and Bredehoeft (1978) gave a systematic exposition, which includes its principle, details of usage and a general program written in FORTRAN.

Let us consider the following two-dimensional advection-dispersion equation:

$$\frac{\partial C}{\partial t} + V_x \frac{\partial C}{\partial x} + V_y \frac{\partial C}{\partial y} = \frac{\partial}{\partial x}\left(D_{xx}\frac{\partial C}{\partial x} + D_{xy}\frac{\partial C}{\partial y}\right) + \frac{\partial}{\partial y}\left(D_{yx}\frac{\partial C}{\partial x} + D_{yy}\frac{\partial C}{\partial y}\right) + I.$$

$$(4.2.1)$$

For an arbitrary moving particle in the flow field, its coordinates are functions of time, i.e.,

$$x = x(t), \quad y = y(t). \tag{4.2.2}$$

According to the definition of velocity, we have

$$\frac{dx}{dt} = V_x \tag{4.2.3}$$

and

$$\frac{dy}{dt} = V_y. \tag{4.2.4}$$

In general, a curve satisfying Eq. (4.2.3) and Eq. (4.2.4) is called a characteristic. Along the characteristic, the total derivative of concentration with respect to time, or the material derivative, may be represented by

$$\frac{dC}{dt} = \frac{\partial C}{\partial t} + \frac{\partial C}{\partial x}\frac{dx}{dt} + \frac{\partial C}{\partial y}\frac{dy}{dt} = \frac{\partial C}{\partial t} + V_x \frac{\partial C}{\partial x} + V_y \frac{\partial C}{\partial y}. \tag{4.2.5}$$

This is just the left-hand side of Eq. (4.2.1). Therefore, along the characteristic, Eq. (4.2.1) can be turned into another equation, which contains only the dispersion terms, i.e.,

$$\frac{dC}{dt} = \frac{\partial}{\partial x}\left(D_{xx}\frac{\partial C}{\partial x} + D_{xy}\frac{\partial C}{\partial y}\right) + \frac{\partial}{\partial y}\left(D_{xy}\frac{\partial C}{\partial y} + D_{yy}\frac{\partial C}{\partial y}\right) + I. \quad (4.2.6)$$

The meaning of Eq. (4.2.6) is easily understood from the Lagrangian viewpoint. When we observe the process of advection-dispersion, not in a stationary coordinate system, but following a moving particle with an average velocity, what we see is not the advection, but only the dispersion in association with the particle. The solution of Eq. (4.2.6), $C = C(t)$, represents the time-dependent concentrations of the particles moving along the characteristic. Therefore, if we set a series of characteristics in the flow region, and find the changes of concentration along these characteristics caused by dispersion effect only, then the advection-dispersion problem is solved under the Lagrangian viewpoint. Since the solution under the Lagrangian viewpoint is not intuitive, it is required to establish a fixed coordinate system to describe the space distribution of the concentration. A family of characteristics will be depicted through a group of moving particles in the fixed coordinate system, meanwhile the changes of concentrations of moving particles are transformed into the fixed nodes of finite difference. Thus, we can obtain the conventional concentration distribution under the Eulerian viewpoint. In this case, the *method of characteristics* (MOC), which will be stated in this section, is actually a Lagrangian-Eulerian solution. The details of this technique will be presented in the following three sections.

4.2.2 Computation of the Advection Part

First, the domain being considered is divided into a common finite difference grid. The initial concentrations at all nodes are determined according to the given initial conditions. Let the initial concentration at node (i,j) be $C_{i,j,0}$. Next, place some particles into each grid square. It is suggested to put two, four, or nine particles in each grid square, see Figure 4.5. The initial concentration of each particle is prescribed to be equal to the initial concentration

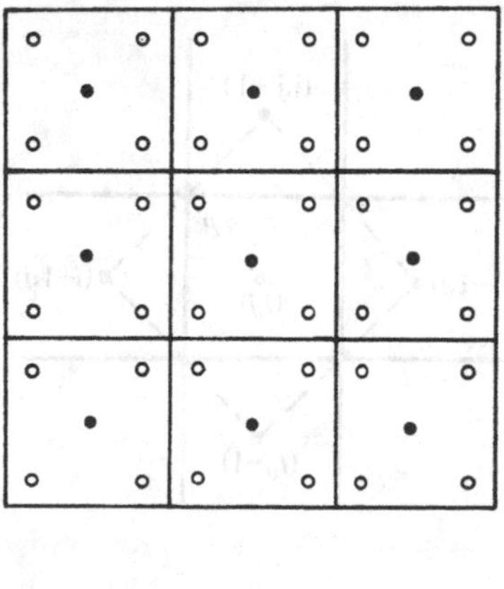

FIGURE 4.5. Particles put in the difference grid at the initial time. ● = node; o = moving point.

of the grid square that it occupies, e.g., if particle P is located in grid square (i,j), its initial concentration is prescribed to be $C_{p,0} = C_{i,j,0}$.

Our problem is how to determine the concentration of each grid square corresponding to a series of discrete instants in the future. The concentrations at all nodes are determined by both advection and dispersion effects. According to the basic idea of the method of characteristics mentioned in the previous paragraph, the two effects can be computed separately, where the advection part may be computed by tracing the particle positions. It is assumed that the concentration of any particle P at time t_k is known as $C_{p,k}$, its position is $(x_{p,k}, y_{p,k})$, and its velocity components are $V_{x,p,k}$ and $V_{y,p,k}$.

The position of particle P at $t_{k+1} = t_k + \Delta t$ can be calculated approximately as

$$x_{p,k+1} = x_{p,k} + V_{x,p,k} \cdot \Delta t, \tag{4.2.7}$$

$$y_{p,k+1} = y_{p,k} + V_{y,p,k} \cdot \Delta t. \tag{4.2.8}$$

In fact, it is equivalent to using the simple method of broken lines to approximately solve the characteristics of Eqs. (4.2.3) and (4.2.4). The velocity components in the equations are easily determined based on the given flow field. In terms of the flow field at a given time t_k, the node values of velocity components $V_{x,i,j,k}$ and $V_{y,i,j,k}$ can be obtained from Darcy's law. If particle P is located in grid square (i,j), it must be contained in a quadrilateral constituted by four vertices $(i-1,j)$, $(i+1,j)$, $(i,j-1)$ and $(i,j+1)$, see Figure 4.6. If the coordinates of particle P are known, the velocity components $V_{x,p,k}$ and $V_{y,p,k}$ of particle P can be determined by means of bilinear interpolation using known velocity components at the four vertices. Goode (1990) pointed out that the bilinear interpolation cannot describe the velocity change very well in the case of heterogeneous media, where the hydraulic conductivities are different among the grid squares. For this reason, he proposed a modified method for the velocity interpolation, where the coefficients of interpolation formula depend on the hydraulic conductivities of the grid squares. Once the velocity components of particle P are known, we can use Eq. (4.2.7) and Eq.

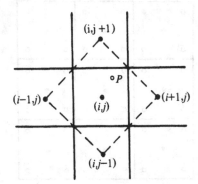

FIGURE 4.6. Velocity components of particles determined by bilinear interpolation.

(4.2.8) to find the new location of particle P at time t_{k+1}. After the same computation is completed for all particles, a new distribution of the particles will be obtained.

The next step is to calculate how many particles have entered each grid square. The concentrations of particles that have fallen into the same grid square are averaged, and the obtained mean concentration is assigned to the node of this grid square. Thus, each node (i,j) at t_{k+1}, has a mean concentration denoted by $C^*_{i,j,k+1}$. The superscript asterisk indicates that it is not the final concentration of node (i,j) at time t_{k+1}, but only the result of advection.

As for the estimation of $C^*_{i,j,k+1}$, Konikow and Bredehoeft (1978) simply let it be equal to the arithmetic mean of concentrations of the particles in the grid associated with node (i,j). Huyakorn and Pinder (1983) pointed out that the mean calculated in this way is not very smooth, in other words, the means of two adjacent grid squares may be significantly changed when there is one particle passing through their boundary. Huyakorn and Pinder (1983) suggested that each particle should have its exclusive acting area, and that the area should move with it, see Figure 4.7. For instance, particle 1 moves from square I to square IV, but all grids (I, II, III, and IV) have a part of its acting area. When we calculate the average concentrations for these grid squares, the influence of particle 1 should be considered according to the proportion of the acting area in each grid square.

We now propose another suggestion, which is to find out all particles P around the considered node (i,j) within a horizontal distance of $1.5\Delta x$ and a vertical distance of $1.5\Delta y$, see Figure 4.8. Then, we can calculate $C^*_{i,j,k+1}$ according to the following formula of weighted averaging:

$$C^*_{i,j,k+1} = \left(\sum_p \frac{1}{d_{i,j,p}} C_{p,k} \right) \Big/ \left(\sum_p \frac{1}{d_{i,j,p}} \right), \tag{4.2.9}$$

where $d_{i,j,p}$ indicates the distance between particle P and node (i,j), and $\sum_p$ is

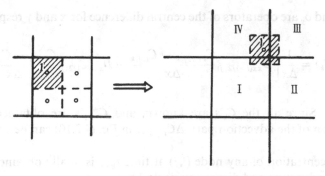

FIGURE 4.7. The acting area of the particle under consideration in calculating the mean concentration.

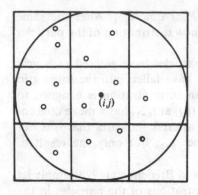

FIGURE 4.8. $C^*_{i,j,k+1}$ calculated by the weighted distance average.

the sum of all the particles P satisfying the conditions stated above. This treatment is simple to program and can also improve the accuracy of the solution.

4.2.3 Computation of the Dispersion Part

To calculate the effect of dispersion, we shall solve Eq. (4.2.6) along the characteristics. As in the computation of the advection part, the explicit finite difference formulas of Eqs. (4.2.7) and (4.2.8) have been used, which means that the time step size from t_k to t_{k+1} must be small enough. Therefore, we would also use an explicit scheme to solve Eq. (4.2.6). If the concentration on the right-hand side of Eq. (4.2.6) is replaced by the mean of $C_{i,j,k}$ and $C^*_{i,j,k+1}$, we then have

$$\Delta C_{i,j,k+1} = \frac{\Delta t}{2}[\delta_x(D_{xx}\delta_x C_k + D_{xy}\delta_y C_k) + \delta_y(D_{yx}\delta_x C_k + D_{yy}\delta_y C_k)]$$

$$+ \frac{\Delta t}{2}[\delta_x(D_{xx}\delta_x C^*_{k+1} + D_{xy}\delta_y C^*_{k+1}) + \delta_y(D_{yx}\delta_x C^*_{k+1} + D_{yy}\delta_y C^*_{k+1})]$$

$$+ \Delta t \cdot I, \tag{4.2.10}$$

where δ_x and δ_y are operators of the central difference for x and y respectively, e.g.,

$$\delta_x(D_{xx}\delta_x C_k) = \frac{1}{\Delta x}\left[D_{xx(i+1/2,j)}\frac{C_{i+1,j,k} - C_{i,j,k}}{\Delta x} - D_{xx(i-1/2,j)}\frac{C_{i,j,k} - C_{i-1,j,k}}{\Delta x}\right], \tag{4.2.11}$$

and so on. Since all the C_k's are known, and C^*_{k+1}'s are obtained in the computation of the advection part, $\Delta C_{i,j,k+1}$ in Eq. (4.2.10) can be calculated explicitly.

The concentration of any node (i,j) at time t_{k+1} is finally obtained as the sum of the advection and dispersion parts, i.e.,

$$C_{i,j,k+1} = C^*_{i,j,k+1} + \Delta C_{i,j,k+1}. \tag{4.2.12}$$

In order to continue the above computation procedure for the next time step, it is necessary to update concentrations of all the particles caused by dispersion. Let

$$C_{p,k+1} = C_{p,k} + \Delta C_{p,k+1},$$

where $\Delta C_{p,k+1}$ results from dispersion. In general, this concentration variation is assumed to be equal to the concentration variation of the grid square in which the particle P is located, i.e.,

$$\Delta C_{p,k+1} = \Delta C_{i,j,k+1}. \tag{4.2.13}$$

During the practical calculation, to avoid the occurrence of negative values of concentration, the following modification is often adopted. When $\Delta C_{i,j,k+1} < 0$, Eq. (4.2.13) will no longer be used to calculate $\Delta C_{p,k+1}$, but the following proportional relationship will be used instead:

$$\Delta C_{p,k+1} = \Delta C_{i,j,k+1} \frac{C_{p,k}}{C^*_{i,j,k+1}}. \tag{4.2.14}$$

As long as $|\Delta C_{i,j,k+1}| < C^*_{i,j,k+1}$, we will have $|\Delta C_{p,k+1}| < C_{p,k}$, so that $C_{p,k+1}$ cannot be a negative value.

After the concentration values of all particles at time t_{k+1} have been obtained, we can repeat the above procedure to get the concentration distribution at time t_{k+2}. This is the major procedure for solving advection-dispersion problems by the method of characteristics.

As stated above, Eq. (4.2.3), Eq. (4.2.4), and Eq. (4.2.6) are all solved by the method of explicit finite difference. In order to maintain the stability of the calculation procedure, the time step size must be strictly limited.

4.2.4 Treatment of Boundary Conditions and Source/Sink Terms

When the method of characteristics is used to solve advection-dispersion problems, various boundary conditions and source/sink terms require special treatments.

When the particles placed on the inflow boundary move out of the grid squares connected with the boundary, new particles should be added with concentrations equal to the concentrations of the inflow fluid at that time. On the outflow boundary, no special treatments are required as the particles move out of the boundary. On a no-flow boundary, new locations of the computed particles may cross the boundary. In this case we can use the method of mirror images to move the particles to return to the region, see Figure 4.9. It shows that the particles move along the no-flow boundary.

In grid squares having solute sources, new particles must be continuously added according to the strength of the sources, while in the grid squares that have solute sinks we have to remove a number of particles according to the

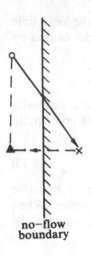

FIGURE 4.9. Mirror image on a no-flow boundary. o = previous location of the particle; × = new location of the particle; ▲ = point of the mirror image.

no-flow
boundary

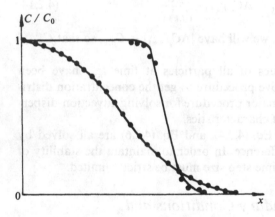

FIGURE 4.10. Comparison between the analytic and the numerical solutions of the characteristics method in a one-dimensional dispersion problem. —— = analytic solution; ● = numerical solution.

strength of the sinks. In a divergent flow field, the farther out the locations are, the smaller the density of the particles will be. Therefore, in some grid squares there may be no particles falling in. If this situation happens, new particles must be added.

From the above explanation, we find that the method of characteristics roughly treats both boundary fluxes and sink/source terms. As a result, large mass balance errors may occur.

As stated above, the principle of the method of characteristics is simple and easy to understand, but there are many special cases that must be treated in order to write a general code. Konikow and Bredehoeft (1978) gave a general program for the MOC written in FORTRAN. Several one-dimensional dispersion and radial dispersion problems were used for verification of their program. Figure 4.10 shows a comparison between the analytical and the numerical solutions obtained by the method of characteristics for a one-dimensional dispersion problem (stated in section 3.2.1, Problem 1). Konikow (1977) used the method of characteristics to simulate the migration of chloride ions in an alluvial aquifer located in the Rocky Mountains of the United States of America.

To extend the MOC for solving three-dimensional advection-dispersion problems is straight. Some modified techniques of the MOC will be given in Chapter 6.

4.2.5 The Random-Walk Model

Prickett et al. (1981) proposed a Random-Walk model for solving the problems of one-dimensional advection-dispersion. Its basic idea is that the mass transport in porous media may be looked upon as an average result of the movements of a large number of tracer particles. Each particle is engaged in two kinds of movements: one is advection, represented by the movement of the tracer particle with an average velocity in the flow field; the other is dispersion, which may be seen as a random fluctuation around the average movement. The development of modern computational techniques enables us to trace the two movements of a huge number of tracer particles by using a computer. As long as the number of tracer particles is large enough, mass transport in porous media can be described.

The Random-Walk method is the same as the method of characteristics in dealing with the advection of tracer particles, but is different with regard to the treatment of dispersion. In this method, the movement of each particle is constituted by the advection and an additional random movement, so that the new location of a particle can be simply determined without solving any equation in association with the dispersion. It is unnecessary to assign concentrations to a difference grid system within each time step and to modify the concentrations of the particles themselves. Thus, the computation is greatly simplified.

Similar to the method of characteristics, the Random-Walk considers the movements of a group of particles. If the location of a particle P at time t_k is $(x_{p,k}, y_{p,k})$, its location based on the advection movement after time Δt will be

$$x^*_{p,k+1} = x_{p,k} + V_{x,p,k} \cdot \Delta t, \qquad (4.2.15)$$

$$y^*_{p,k+1} = y_{p,k} + V_{y,p,k} \cdot \Delta t. \qquad (4.2.16)$$

These two equations have the same meaning as Eq. (4.2.7) and Eq. (4.2.8).

The superscript * is used to denote that $(x^*_{p,k+1}, y^*_{p,k+1})$ is just the result of advection, and not the final location of the particle at time $t_{k+1} = t_k + \Delta t$.

Its final location $(x_{p,k+1}, y_{p,k+1})$ may be looked upon as the result of the advection of particle P plus a dispersion, i.e.,

$$x_{p,k+1} = x^*_{p,k+1} + \delta x_{p,k+1}, \qquad (4.2.17)$$

$$y_{p,k+1} = y^*_{p,k+1} + \delta y_{p,k+1}, \qquad (4.2.18)$$

where $\delta x_{p,k+1}$ and $\delta y_{p,k+1}$ are additional components caused by dispersion. According to the statistical theory of dispersion phenomena, the dispersion results in a normal distribution of tracer particle locations around their average locations. This can be verified by the solution of one-dimensional advection-dispersion problems.

For the equation,

$$\frac{\partial C}{\partial t} + V\frac{\partial C}{\partial x} = D_L\frac{\partial^2 C}{\partial x^2},$$

the solution with a transient source was given in paragraph 3.1.3, i.e.,

$$C(x,t) = \frac{\mu/\theta}{2\sqrt{\pi D_L t}}\exp\left\{-\frac{(x-Vt)^2}{4D_L t}\right\}. \tag{4.2.19}$$

On the other hand, the density function of normal distribution is

$$n(x) = \frac{1}{\sqrt{2\pi}\sigma}\exp\left\{-\frac{(x-m)^2}{2\sigma^2}\right\}, \tag{4.2.20}$$

where m is the mathematical expectation, and σ is the variance. After comparing the above two equations we may find that for each given t, the concentration $C(x,t)$ is commensurate to a density function of normal distribution within a constant factor. Its corresponding mathematical expectation and variance are respectively:

$$m = Vt; \quad \sigma = \sqrt{2D_L t}. \tag{4.2.21}$$

For each time step, we assume that particle P first moves a distance $V\Delta t$ according to the advection, i.e., it arrives at the location of the mathematical expectation. The particle then makes a random walk along the $+x$ or $-x$ directions. The distance of the walk is

$$\sqrt{2D_L t}\cdot\text{ANORM}(0), \tag{4.2.22}$$

where ANORM(0) is a value of the normal distribution with mathematical expectation 0 and variance 1. ANORM(0) may be limited to the range of $(-6, +6)$ during calculation, because the probability of being more than 6 standard deviations from the mean is very small. Thus, the total displacement of particle P is

$$V\cdot\Delta t + \sqrt{2D_L t}\cdot\text{ANORM}(0). \tag{4.2.23}$$

This procedure is shown in Figure 4.11.

The procedure is repeated for a large number of tracer particles. Their displacements are different because the second term of Eq. (4.2.23) is stochastic. The more particles there are, the smaller the difference between their distribution and normal distribution will be. It is assumed that a total number of particles N_0 are instantaneously injected into the origin $(x = 0)$ at the initial time. After moving many time steps (the time is from 0 to t), the

FIGURE 4.11. New location of the particle in one-dimensional longitudinal dispersion.

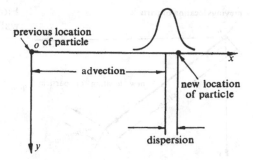

FIGURE 4.12. Transverse random walk of the tracer particle.

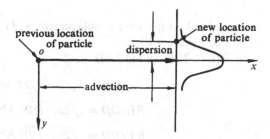

number of particles falling in a small area with point x as its center and with a distance of Δx is N_x. Based on the law of normal distribution, we have

$$\frac{N_x}{\Delta x} = \frac{N_0}{\sqrt{2\pi}\sqrt{2D_L t}} \exp\left\{-\frac{(x - Vt)^2}{4D_L t}\right\}. \tag{4.2.24}$$

Let N_0 be proportional to μ/θ, and the right-hand side of this equation is identical to Eq. (4.2.19). The concentration distribution $C(x, t)$ can therefore be obtained from the statistical distribution of N_x after using Eq. (4.2.23) for a large number of tracer particles and time steps. The same idea can be used to compute the transverse dispersion as shown in Figure 4.12. The displacement of a particle caused by transverse dispersion is

$$RT \cdot DX = \sqrt{2\alpha_T \cdot DX} \cdot \text{ANORM}(0), \tag{4.2.25}$$

where α_T is the transverse dispersivity and $DX = V_x \cdot \Delta t$ is the displacement caused by advection.

For the general case of Eq. (4.2.15) and Eq. (4.2.16), the following equations should be used

$$x_{p,k+1} = x_{p,k} + DX + RL \cdot DX + RT \cdot DY, \tag{4.2.26}$$

$$y_{p,k+1} = y_{p,k} + DY + RL \cdot DY + RT \cdot DX, \tag{4.2.27}$$

that is, New Location = Previous Location + Advection Displacement +

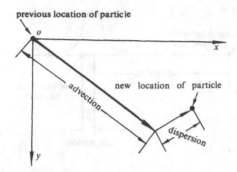

previous location of particle

FIGURE 4.13. Two-dimensional random walk of the tracer particle.

Longitudinal Dispersion Displacement + Transverse Dispersion Displacement (see Figure 4.13).

In Eqs. (4.2.26) and (4.2.27)

$$DX = V_x \cdot \Delta t, \quad DY = V_y \cdot \Delta t,$$

$$RL \cdot DD = \sqrt{2\alpha_L \cdot DD} \cdot \text{ANORM}(0),$$

$$RT \cdot DD = \sqrt{2\alpha_T \cdot DD} \cdot \text{ANORM}(0),$$

where DD is either DX or DY.

The method of Random-Walk is very similar to the method of characteristics in the treatment of boundary conditions and source/sink terms, but the treatment is more intuitive for the Random-Walk because the transport procedure of a group of tracer particles is simulated directly. On the no-flow boundary, the method of mirror reflection is still used to force the particles to stay in the region. On the inflow boundary and at injection wells, the number of particles to be added is proportional to the inflow solute quantity, and, on the outflow boundary or at sink points, particles are allowed to move out of the region. At a given concentration boundary, a definite number of particles should be maintained. All of these are easy to achieve in the program.

In the Random-Walk method, all particles move continuously throughout the whole region, and it is not necessary for them to be associated with a finite difference grid. During the computation procedure, the concentration is computed only for the times and subdomains of interest, while the method of characteristics requires computation of the concentration for all nodes and all time steps.

The accuracy of numerical solutions obtained by the Random-Walk method mainly depends on the number of tracer particles. If the number of particles is not large enough, the solution will become unsmooth. Consequently, the required accuracy may not be reached Prikett et al. (1981) provided a general program in the FORTRAN language with several examples. Figure 4.14 shows the numerical solution of a one-dimensional advection-

FIGURE 4.14. Comparison between the analytic solution and the
approximate solutions obtained
by the Random-Walk method.
● = approximate solution; —— =
analytic solution.

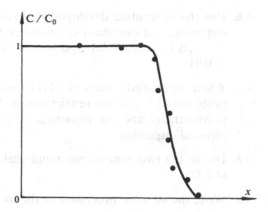

dispersion problem in a semi-infinite sand column obtained by the Random-Walk method and the corresponding analytic solution. From this figure, it is evident that the numerical solution is relatively rough. If we require that the numerical solution be identical to the analytic solution to three significant figures, the number of tracer particles simulated must be enormous.

Due to the rapid development of computation techniques, it is possible to trace enormous particles to maintain the accuracy of the approximate solution, and thus, the Random Walk method becomes more and more practical. Recently, Moltyaner et al. (1993) reported that the problem of numerical dispersion was completely eliminated by using the Random-Talk method to reproduce the observation results of a three-dimensional field experiment.

Exercises

4.1. Derive the expressions of truncation errors for the finite difference approximations in Eqs. (4.1.6) and (4.1.8).

4.2. When the central-finite difference approximation (4.1.5) is used to replace the first order derivative $\partial C/\partial x$, what is the condition of convergence for the explicit scheme Eq. (4.1.11)?

4.3. Prove that the implicit scheme in Eq. (4.1.13) is convergent without any restrictive conditions.

4.4. Write a subroutine to solve Problem 1 in Section 3.2.1 by using the explicit and implicit schemes. Let $D_L = 10$, $V = 1$, $\Delta X = 5$. Change the size of Δt to observe the variation of the computed concentration distribution.

4.5. Write the finite difference approximation for the 2-D problem given by Eq. (3.2.33).

4.6. Use the subroutine developed in Exercise 4.4 to observe numerical dispersion and overshoot errors (when the implicit scheme is used, let $V = 1, \Delta X = 5, \Delta t = 0.1$. Systematically change the value of D_L from 10 to 0.01).

4.7. When the explicit scheme (4.1.11) is used to solve the one-dimensional problem (4.1.9), find the restrictions on Δx and Δt such that the scheme is convergent and the numerical dispersion is less than 10% of the physical dispersion.

4.8. Derive the two-dimensional numerical dispersion terms given in Eq. (4.1.37).

4.9. Write the solution procedure of the method of characteristics (MOC) step by step.

4.10. What are the differences between the MOC and the Random-Walk method?

4.11. Present the key steps of using the Random-Walk method to solve three-dimensional advection-dispersion problems.

4.12. When the Random-Walk method is used to solve an advection-dispersion problem, how can the radioactive decay and adsorption be taken into account?

5
Finite Element Methods for Solving Hydrodynamic Dispersion Equations

5.1 Finite Element Methods for Two-Dimensional Problems

5.1.1 The Weighted Residual Method

Consider the following two-dimensional advection-dispersion equation

$$L(C) \equiv \frac{\partial C}{\partial t} + V_x \frac{\partial C}{\partial x} + V_y \frac{\partial C}{\partial y} - \frac{\partial}{\partial x}\left(D_{xx}\frac{\partial C}{\partial x} + D_{xy}\frac{\partial C}{\partial y}\right)$$

$$- \frac{\partial}{\partial y}\left(D_{xy}\frac{\partial C}{\partial x} + D_{yy}\frac{\partial C}{\partial y}\right) + QC - I = 0, \qquad (5.1.1)$$

which is subject to the initial condition

$$C(x, y, 0) = f, \quad (x, y) \in (R), \qquad (5.1.2)$$

boundary conditions

$$C(x, y, t) = g_1, \quad (x, y) \in (\Gamma_1), \qquad (5.1.3)$$

and

$$\left(D_{xx}\frac{\partial C}{\partial x} + D_{xy}\frac{\partial C}{\partial y}\right)n_x + \left(D_{yx}\frac{\partial C}{\partial x} + D_{yy}\frac{\partial C}{\partial y}\right)n_y = -g_2$$

$$(x, y) \in (\Gamma_2), \qquad (5.1.4)$$

where (R) is the flow domain, (Γ_1) and (Γ_2) are boundary sections of (R), f is a given function in (R), g_1 and g_2 are given functions along (Γ_1) and (Γ_2), respectively, and n_x and n_y are components of the unit outer normal vector to the boundary (Γ_2). Equation (5.1.3) expresses the boundary condition of given concentration, i.e., the first-type boundary condition, while Eq. (5.1.4) expresses the boundary condition of given dispersion flux, i.e., the second-type boundary condition.

In the *weighted residual method*, the following *trial function* is adopted:

$$\hat{C} = \sum_{j=1}^{N} C_j(t)\phi_j(x, y). \tag{5.1.5}$$

As an approximate solution of Eq. (5.1.1), $\hat{C}$ must satisfy the boundary condition (5.1.3) on (Γ_1), and when $N \to \infty$, it should tend towards the exact solution.

In Eq. (5.1.5),

$$\phi_1(x, y), \phi_2(x, y), \ldots, \phi_N(x, y) \tag{5.1.6}$$

is a set of linearly independent *basis functions*. They are the first N functions of a complete function system in the domain (R). $C_1(t), C_2(t), \ldots, C_N(t)$ are N undetermined coefficients related to time t. In what follows, we will see that this set of coefficients can be determined according to the requirement that $\hat{C}$ is an approximate solution of Eq. (5.1.1).

Because $\hat{C}$ is not an exact solution of Eq. (5.1.1), $L(\hat{C}) \neq 0$. Function

$$\hat{R}(x, y) = L(\hat{C}) \tag{5.1.7}$$

is called the *residual* produced by substituting the approximate solution $\hat{C}$ into Eq. (5.1.1). Naturally, the residual is required to be as small as possible in the whole domain. We can use the weighted mean

$$r = \iint_{(R)} W\hat{R}\, dx\, dy \Big/ \iint_{(R)} W\, dx\, dy \tag{5.1.8}$$

to measure the residual on domain (R), where $W(x, y)$ is a *weighting function*. If N weighting functions,

$$W_1(x, y), W_2(x, y), \ldots, W_N(x, y), \tag{5.1.9}$$

are selected by a certain method and all residuals r corresponding to these weighting functions are required to be equal to zero, we then have N equations:

$$\iint_{(R)} W_i L(\hat{C})\, dx\, dy = 0, \quad (i = 1, 2, \ldots, N). \tag{5.1.10}$$

Substituting Eq. (5.1.1) into Eq. (5.1.10), and using Green's formula to eliminate the terms with second-order derivatives, we arrive at the equation:

$$\iint_{(R)} \left\{ \frac{\partial \hat{C}}{\partial t} W_i + V_x \frac{\partial \hat{C}}{\partial x} W_i + V_y \frac{\partial \hat{C}}{\partial y} W_i + D_{xx} \frac{\partial \hat{C}}{\partial x} \frac{\partial W_i}{\partial x} + D_{xy} \frac{\partial \hat{C}}{\partial y} \frac{\partial W_i}{\partial x} \right.$$

$$\left. + D_{yx} \frac{\partial \hat{C}}{\partial x} \frac{\partial W_i}{\partial y} + D_{yy} \frac{\partial \hat{C}}{\partial y} \frac{\partial W_i}{\partial y} + Q\hat{C}W_i - IW_i \right\} dx\, dy$$

$$+ \int_{(\Gamma_2)} g_2 W_i\, d\Gamma = 0. \quad (i = 1, 2, \ldots, N). \tag{5.1.11}$$

The last term on the left-hand side of the above equation is obtained from the second-type boundary condition in Eq. (5.1.4).

Substituting Eq. (5.1.5) for $\hat{C}$ in Eq. (5.1.11), a set of *ordinary differential equations* (*ODE*) written in matrix form is obtained as follows:

$$[\mathbf{A}]\mathbf{C} + [\mathbf{B}]\frac{d\mathbf{C}}{dt} + \mathbf{F} = 0, \qquad (5.1.12)$$

where

$$\mathbf{C} = (C_1, C_2, \ldots, C_N)^T,$$

$$\frac{d\mathbf{C}}{dt} = \left(\frac{dC_1}{dt}, \frac{dC_2}{dt}, \ldots, \frac{dC_N}{dt}\right)^T.$$

The elements of coefficient matrices [**A**], [**B**] and vector **F** are, respectively:

$$A_{ij} = \iint_{(R)} \left(D_{xx}\frac{\partial W_i}{\partial x}\frac{\partial \phi_j}{\partial x} + D_{xy}\frac{\partial W_i}{\partial x}\frac{\partial \phi_j}{\partial y} + D_{yx}\frac{\partial W_i}{\partial y}\frac{\partial \phi_j}{\partial x} \right.$$

$$\left. + D_{yy}\frac{\partial W_i}{\partial y}\frac{\partial \phi_j}{\partial y} + V_x W_i\frac{\partial \phi_j}{\partial x} + V_y W_i\frac{\partial \phi_j}{\partial y} + Q W_i\phi_j \right) dx\,dy, \quad (5.1.13)$$

$$B_{ij} = \iint_{(R)} W_i\phi_j\,dx\,dy, \qquad (5.1.14)$$

$$F_i = -\iint_{(R)} I W_i\,dx\,dy + \int_{(\Gamma_2)} g_2 W_i\,d\Gamma. \qquad (5.1.15)$$

It is quite evident that once the system of basis functions (5.1.6) and weighting functions (5.1.9) are selected, all these elements can be calculated.

When the *finite difference approximation* is used to replace the time derivative in Eq. (5.1.12), the system of *ODE*s will become a system of *algebraic equations* in each time step. For example, assume that $\mathbf{C}_t$ is the value of **C** at time t. In order to find $\mathbf{C}_{t+\Delta t}$ at time $t + \Delta t$, we use the following finite difference approximation:

$$\frac{d\mathbf{C}}{dt} = \frac{\mathbf{C}_{t+\Delta t} - \mathbf{C}_t}{\Delta t} \qquad (5.1.16)$$

and substitute it into Eq. (5.1.12). The result is

$$([\mathbf{A}] + [\mathbf{B}]/\Delta t)\mathbf{C}_{t+\Delta t} = ([\mathbf{B}]/\Delta t)\mathbf{C}_t - \mathbf{F},$$

or

$$[\mathbf{E}]\mathbf{C}_{t+\Delta t} = \mathbf{G}, \qquad (5.1.17)$$

where

$$[\mathbf{E}] = [\mathbf{A}] + [\mathbf{B}]/\Delta t; \quad \mathbf{G} = \left(\frac{[\mathbf{B}]}{\Delta t}\right)\mathbf{C}_t - \mathbf{F}. \qquad (5.1.18)$$

Equation (5.1.17) is a system of linear algebraic equations. It can be solved by either the iteration method or the direct method. At the end of this chapter we will discuss these methods. Note that because of the existence of advection terms in the advection-dispersion equation, coefficient matrix [A] is not symmetric. This fact can be clearly seen from the structure of Eq. (5.1.13). Therefore, [E] in Eq. (5.1.17) is not a symmetric matrix either. This is the major difference between the finite element equations for mass transport problems and for groundwater flow problems. When [E] is an asymmetric matrix, the direct solution method requires more storage space and computational effort.

After solving $C_{t+\Delta t}$ from Eq. (5.1.17), the approximate solution $\hat{C}$ at $t + \Delta t$ can be obtained by substituting $C_{t+\Delta t}$ into the right-hand side of Eq. (5.1.5).

These are the main steps of solving advection-dispersion equations with the method of weighted residuals. The remaining problem is how to choose the basis and weighting functions for simplifying the calculation of the coefficient matrices.

Galerkin Method

The Galerkin method uses the basis functions in Eq. (5.1.6) as the weighting functions. We shall explain this method in detail in the following sections.

Subdomain Method

The method of subdomain is also called *the method of element collocation.* Subdivide the region (R) into several elements, and denote as (R_i), $i = 1, 2, \dots, N$. Let each element be associated with a weighting function, and define the weighting functions as:

$$W_i(x, y) = \begin{cases} 1 & \text{when } (x, y) \in (R_i), \\ 0 & \text{when } (x, y) \notin (R_i). \end{cases} \tag{5.1.19}$$

Under such a selection, Eq. (5.1.13) to Eq. (5.1.15), which were used to calculate coefficient matrices for Eq. (5.1.12), can be simplified to

$$A_{ij} = \iint_{(R_i)} \left(V_x \frac{\partial \phi_j}{\partial x} + V_y \frac{\partial \phi_j}{\partial y} + Q\phi_j \right) dx\, dy - \int_{(l_i)} \left[\left(D_{xx} \frac{\partial \phi_j}{\partial x} + D_{xy} \frac{\partial \phi_j}{\partial y} \right) n_x \right.$$
$$\left. + \left(D_{xy} \frac{\partial \phi_j}{\partial x} + D_{yy} \frac{\partial \phi_j}{\partial y} \right) n_y \right] dl, \tag{5.1.20}$$

$$B_{ij} = \iint_{(R_i)} \phi_j\, dx\, dy, \tag{5.1.21}$$

$$F_i = -\iint_{(R_i)} I\, dx\, dy \tag{5.1.22}$$

where (l_i) is the boundary of subdomain (R_i), and n_x and n_y are components of its unit outer normal vector.

Collocation Method

In this method, N points (x_i, y_i), $i = 1, 2, \ldots, N$, are selected from the domain (R) and called *collocation points*. The weighting functions are then defined as

$$W_i(x, y) = \delta(x - x_i) \cdot \delta(y - y_i)$$

$$(i = 1, 2, \ldots, N) \tag{5.1.23}$$

where $\delta(x - x_i)$ and $\delta(y - y_i)$ are Dirac-δ functions. It implies that the value of $W_i(x, y)$ is non-zero only at the collocation point (x_i, y_i), and also, for any function $a(x, y)$, we have

$$\iint_{(R)} a(x, y) W_i(x, y) \, dx \, dy = a(x_i, y_i). \tag{5.1.24}$$

Under these circumstances, Eq. (5.1.13) to Eq. (5.1.15) can be simplified to

$$A_{ij} = \left[V_x \frac{\partial \phi_j}{\partial x} + V_y \frac{\partial \phi_j}{\partial y} + Q\phi_j - \frac{\partial}{\partial x} \left(D_{xx} \frac{\partial \phi_j}{\partial x} + D_{xy} \frac{\partial \phi_j}{\partial y} \right) \right.$$

$$\left. - \frac{\partial}{\partial y} \left(D_{yx} \frac{\partial \phi_j}{\partial x} + D_{yy} \frac{\partial \phi_j}{\partial y} \right) \right]_{(x_i, y_i)}, \tag{5.1.25}$$

$$B_{ij} = \phi_j(x_i, y_i), \tag{5.1.26}$$

$$F_i = -I(x_i, y_i). \tag{5.1.27}$$

Note that no integration is needed to obtain these coefficients. Therefore, the method of collocation is a numerical method which requires relatively less computational effort. Its accuracy depends on both the selection of the basis functions and the locations of the collocation points. Pinder and Shapiro (1979) put forward a method of orthogonal collocation for advection-dispersion equations. The basis points of Gauss' quadratic formula are selected as the collocation points.

5.1.2 Finite Element Discretization and Basis Functions

Now let us return to the Galerkin FEM. Divide the region (R) into several elements (R_m), $m = 1, 2, \ldots, M$, and take the vertices, points on the sides, and sometimes the internal points of elements as nodes. Suppose that there are N nodes in the whole region, we can then rewrite Eq. (5.1.13) to Eq. (5.1.15) as follows:

$$A_{ij} = \sum_{m=1}^{M} \iint_{(R_m)} \left(D_{xx} \frac{\partial \phi_i}{\partial x} \frac{\partial \phi_j}{\partial x} + D_{xy} \frac{\partial \phi_i}{\partial x} \frac{\partial \phi_j}{\partial y} + D_{yx} \frac{\partial \phi_i}{\partial y} \frac{\partial \phi_j}{\partial x} \right.$$

$$\left. + D_{yy} \frac{\partial \phi_i}{\partial y} \frac{\partial \phi_j}{\partial y} + V_x \phi_i \frac{\partial \phi_j}{\partial x} + V_y \phi_i \frac{\partial \phi_j}{\partial y} + Q\phi_i \phi_j \right) dx \, dy, \tag{5.1.28}$$

$$B_{ij} = \sum_{m=1}^{M} \iint_{(R_m)} \phi_i \phi_j \, dx \, dy, \tag{5.1.29}$$

$$F_i = - \sum_{m=1}^{M} \iint_{(R_m)} I\phi_i \, dx \, dy + \sum_{m=1}^{M} \int_{(\Gamma_{2,m})} g_2 \phi_i \, d\Gamma, \tag{5.1.30}$$

where $(\Gamma_{2,m})$ is the common boundary of (R_m) and (Γ_2).

Let the ith basis function $\phi_i(x, y)$ be associated with the ith node and let

$$\phi_i(x_j, y_j) = \delta_{ij} = \begin{cases} 1 & \text{when } j = i, \\ 0 & \text{when } j \neq i, \end{cases} \tag{5.1.31}$$

where (x_j, y_j) are the coordinates of node j. Equation (5.1.31) means that the value of $\phi_i(x, y)$ is defined to be 1 at node i, and zero at other nodes. Furthermore, we define $\phi_i(x, y)$ as non-zero only in those elements which link up with node i, and to be always zero in other elements.

These two stipulations have several advantages. First, because A_{ij} and B_{ij} given in Eqs. (5.1.28) and (5.1.29) are non-zero only when nodes i and j are in the same element, coefficient matrices $[A]$ and $[B]$ now become highly sparse. Second, we do not need to integrate on the whole domain (R) for calculating A_{ij} and B_{ij}. To calculate them we only need to integrate on the elements that have common nodes i and j. Third, from the definition of Eq. (5.1.31), the solutions $C_{t+\Delta t}$ of Eq. (5.1.12) are just the unknown nodal values of approximate solution $\hat{C}$ at the time $t + \Delta t$, so we do not need go back to Eq. (5.1.5). Therefore, when we combine the finite element discretization with the Galerkin method, the calculation process demonstrated in Section 5.1.1 will be much simplified. It is quite evident that even though the above stipulations are satisfied, the shape of elements, the arrangement of nodes and the expressions of basis functions are still undetermined. In other words, we still have a great deal of flexibility in selecting a Galerkin finite element method.

Two commonly-used options for solving advection-dispersion equations are given below.

Triangular Elements and Linear Basis Functions

Divide domain (R) into triangular elements and take all vertices as nodes. Figure 5.1 shows an arbitrary triangular element (Δ).

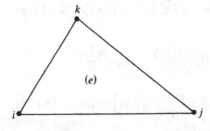

FIGURE 5.1. A linear triangular element and its nodes.

Suppose that the nodes at the three vertices are numbered as i, j, k, with coordinates (x_i, y_i), (x_j, y_j) and (x_k, y_k), respectively. In the element, the basis functions related to the three nodes are defined as linear functions:

$$\phi_l(x, y) = \frac{1}{2\Delta}(a_l + b_l x + c_l y),$$

$$l = i, j, k; \quad (x, y) \in (\Delta), \tag{5.1.32}$$

where

$$a_i = x_j y_k - x_k y_j, \quad b_i = y_j - y_k, \quad c_i = x_k - x_j;$$

$$a_j = x_k y_i - x_i y_k, \quad b_j = y_k - y_i, \quad c_j = x_i - x_k; \tag{5.1.33}$$

$$a_k = x_i y_j - x_j y_i, \quad b_k = y_i - y_j, \quad c_k = x_j - x_i;$$

and Δ is the area of triangular element (Δ).

For the basis functions given in Eq. (5.1.32), we have

$$\iint_{(\Delta)} \phi_i \, dx \, dy = \frac{\Delta}{3}, \quad \iint_{(\Delta)} \phi_i^2 \, dx \, dy = \frac{\Delta}{6};$$

$$\iint_{(\Delta)} \phi_i \phi_j \, dx \, dy = \frac{\Delta}{12}, \quad (i \neq j). \tag{5.1.34}$$

Substituting these expressions into Eq. (5.1.28) through Eq. (5.1.30), we can directly compute the parts of A_{ij}, B_{ij} and F_i in element (Δ), respectively, i.e.,

$$A_{ij}^{(\Delta)} = \iint_{(\Delta)} \left(D_{xx} \frac{\partial \phi_i}{\partial x} \frac{\partial \phi_j}{\partial x} + D_{xy} \frac{\partial \phi_i}{\partial x} \frac{\partial \phi_j}{\partial y} + D_{yx} \frac{\partial \phi_i}{\partial y} \frac{\partial \phi_j}{\partial x} \right.$$

$$\left. + D_{yy} \frac{\partial \phi_i}{\partial y} \frac{\partial \phi_j}{\partial y} + V_x \phi_i \frac{\partial \phi_j}{\partial x} + V_y \phi_i \frac{\partial \phi_j}{\partial y} + Q \phi_i \phi_j \right) dx \, dy$$

$$= \frac{1}{4\Delta} [D_{xx} b_i b_j + D_{xy}(b_i c_j + c_i b_j) + D_{yy} c_i c_j]$$

$$+ \frac{1}{6}(V_x b_i + V_y c_j) + B_{ij}^{(\Delta)} Q, \tag{5.1.35}$$

$$B_{ij}^{(\Delta)} = \iint_{(\Delta)} \phi_i \phi_j \, dx \, dy = \begin{cases} \dfrac{\Delta}{6} & i = j, \\[2mm] \dfrac{\Delta}{12} & i \neq j, \end{cases} \tag{5.1.36}$$

$$F_i^{(\Delta)} = -\iint_{(\Delta)} I \phi_i \, dx \, dy + \int_{(\Gamma_{2,\Delta})} g_2 \phi_i \, d\Gamma = -\frac{\Delta}{3} I + \frac{g_2}{2} \Gamma_{2,\Delta}, \tag{5.1.37}$$

where $\Gamma_{2,\Delta}$ is the side length of element (Δ) in common with Γ_2. When they have no overlapping part, $\Gamma_{2,\Delta} = 0$.

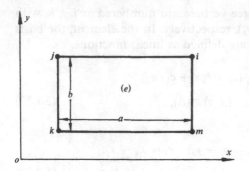

FIGURE 5.2. A rectangular element and its nodes.

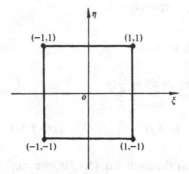

FIGURE 5.3. A square element and its nodes in a local coordinate system.

The next step is the same as that for solving the problems of groundwater flow, that is, to form the global matrices $[A]$, $[B]$ and vector $\mathbf{F}$ through assembling their components of each element. Thus, the system of Eq. (5.1.12) is built up. From Eq. (5.1.35), it is clear that $[A]$ is an asymmetric matrix because of the existence of advection terms.

Rectangular Elements and Bilinear Basis Functions

In this case, domain (R) is divided into a net of rectangles, and all grid points are taken as nodes as they were in the finite difference method. Figure 5.2 shows an arbitrary rectangular element (e) with length a and width b. Its four nodes are numbered as i, j, k, and m, respectively.

Using the transformation

$$\begin{cases} \xi = 2\dfrac{x - x_0}{a}, \\[2mm] \eta = 2\dfrac{y - y_0}{b}, \end{cases} \tag{5.1.38}$$

where (x_0, y_0) are the coordinates of the central point of the rectangular element, the element can be transformed into a standard square element in the local coordinate system, as shown in Figure 5.3.

In the standard element, the coordinates of the four nodes are $\xi = \pm 1$, $\eta = \pm 1$. The basis functions in the standard element are defined as:

$$\phi_i(\xi, \eta) = \frac{1}{4}(1 + \xi)(1 + \eta),$$

$$\phi_j(\xi, \eta) = \frac{1}{4}(1 - \xi)(1 + \eta),$$

$$\phi_k(\xi, \eta) = \frac{1}{4}(1 - \xi)(1 - \eta),$$

$$\phi_m(\xi, \eta) = \frac{1}{4}(1 + \xi)(1 - \eta). \tag{5.1.39}$$

These basis functions are bilinear functions of ξ and η. Using the variable transformation formula for double integrals, the components of A_{ij} and B_{ij} in element (e) can be calculated, respectively, as follows:

$$
\begin{aligned}
A_{ij}^{(e)} &= \int_{-1}^{1}\int_{-1}^{1} \left(\frac{D_{xx}}{a^2}\frac{\partial\phi_i}{\partial\xi}\frac{\partial\phi_j}{\partial\xi} + \frac{D_{xy}}{ab}\frac{\partial\phi_i}{\partial\xi}\frac{\partial\phi_j}{\partial\eta} + \frac{D_{yx}}{ab}\frac{\partial\phi_i}{\partial\eta}\frac{\partial\phi_j}{\partial\xi} \right. \\
&\qquad \left. + \frac{D_{yy}}{b^2}\frac{\partial\phi_i}{\partial\eta}\frac{\partial\phi_j}{\partial\eta} + V_x\frac{\phi_i}{a}\frac{\partial\phi_j}{\partial\xi} + V_y\frac{\phi_i}{b}\frac{\partial\phi_j}{\partial\eta} + Q\phi_i\phi_j \right) ab\, d\xi\, d\eta \\
&= \frac{1}{12ab}(2D_{yy}a^2 + 4D_{xx}b^2 - V_y a^2 b - 2V_x ab^2) + B_{ij}^{(e)}Q, \tag{5.1.40}
\end{aligned}
$$

$$
B_{ij}^{(e)} = \int_{-1}^{1}\int_{-1}^{1} \phi_i\phi_j ab\, d\xi\, d\eta =
\begin{cases}
\dfrac{1}{9}ab & (j = i), \\[2mm]
\dfrac{1}{18}ab & (j \neq i).
\end{cases}
\tag{5.1.41}
$$

For $A_{ik}^{(e)}$, $B_{ik}^{(e)}$, $A_{im}^{(e)}$, $B_{im}^{(e)}$, ..., similar results can also be derived. After the computations are completed for all elements, the next is to assemble them to form global matrices $[A]$, $[B]$ and vector $\mathbf{F}$. Thus, the system of equations (5.1.12) is built up for rectangular elements.

5.1.3 High-Order Elements and Hermite Elements

The utilization of triangular elements and linear basis functions implies that the values of unknown concentration may be approximated by linear functions in each element, so the approximate solution $\hat{C}$ is a piecewise linear function. A natural way to improve the accuracy of the approximate solution is to use quadratic or high-order functions to approximate the unknown concentration. This introduces the problem of using *high-order elements*.

The complete quadratic function of x, y has six coefficients and thus six conditions are needed to define it. It is quite obvious that when the function

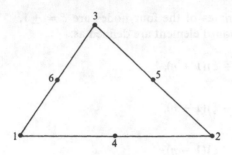

FIGURE 5.4. A quadratic triangle.

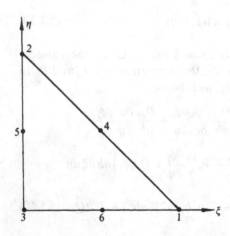

FIGURE 5.5. A standard quadratic triangle in a local coordinate system.

values on six points are given in each element, the conditions will be satisfied. Therefore, we can designate six nodes in each triangle element, which are respectively the three vertices and three middle points on the three sides, as shown in Figure 5.4.

Through the transformation

$$\begin{cases} x = x_1\xi + x_2\eta + x_3(1 - \xi - \eta), \\ y = y_1\xi + y_2\eta + y_3(1 - \xi - \eta), \end{cases} \tag{5.1.42}$$

the triangle (Δ) in the xy plane as shown in Figure 5.4 becomes a standard triangle (Δ') in the $\xi\eta$ plane in a local coordinate system, as shown in Figure 5.5.

In the standard element (Δ'), the complete quadratic function of ξ, η has the following form.

$$\phi(\xi, \eta) = a + b\xi + c\eta + d\xi^2 + e\xi\eta + f\eta^2. \tag{5.1.43}$$

If the value of quadratic basis function $\phi_i(\xi, \eta)$ associated with node i within element (Δ') is required to be 1 at node i and zero at the other five points, we then have six conditions. With the six conditions, the six coefficients of $\phi_i(\xi, \eta)$ can be determined, as listed in Table 5.1.

TABLE 5.1. Basis functions in the local coordinate system.

Linear element	Quadratic element	Cubic element
$\phi_1 = \xi$	$\phi_1 = 2\xi^2 - \xi$	$\phi_1 = \frac{1}{2}\xi(3\xi - 1)(3\xi - 2)$
$\phi_2 = \eta$	$\phi_2 = 2\eta^2 - \eta$	$\phi_2 = \frac{1}{2}\eta(3\eta - 1)(3\eta - 2)$
$\phi_3 = 1 - \xi - \eta$	$\phi_3 = (1 - \xi - \eta)(1 - 2\xi - 2\eta)$	$\phi_3 = \frac{1}{2}(1 - \xi - \eta)(2 - 3\xi - 3\eta)(1 - 3\xi - 3\eta)$
	$\phi_4 = 4\xi\eta$	$\phi_4 = \frac{9}{2}\xi\eta(3\xi - 1)$
	$\phi_5 = 4(1 - \xi - \eta)\eta$	$\phi_5 = \frac{9}{2}\xi\eta(3\eta - 1)$
	$\phi_6 = 4(1 - \xi - \eta)\xi$	$\phi_6 = \frac{9}{2}\eta(1 - \xi - \eta)(3\eta - 1)$
		$\phi_7 = \frac{9}{2}\eta(1 - \xi - \eta)(2 - 3\xi - 3\eta)$
		$\phi_8 = \frac{9}{2}\xi(1 - \xi - \eta)(2 - 3\xi - 3\eta)$
		$\phi_9 = \frac{9}{2}\xi(1 - \xi - \eta)(3\xi - 1)$
		$\phi_{10} = 27\xi\eta(1 - \xi - \eta)$

FIGURE 5.6. The third-order triangle.

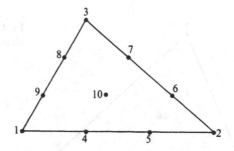

Using Eq. (5.1.42), the Jacobian of the transformation can be calculated and all the coefficients of Eqs. (5.1.28) to (5.1.30) can be integrated in the local coordinate system. The integrands are polynomial expressions with ξ and η lower than the third order.

The *cubic finite element* may be derived in a manner similar to the quadratic element. In each triangle we define 10 nodes, and use a complete cubic polynomial to approximately express the value of the unknown function in the element, see Figure 5.6. Using the transformation in Eq. (5.1.42), element (Δ) becomes a standard triangular element (Δ') in the $\xi\eta$ plane within the local coordinate system. The relevant nodes are shown in Figure 5.7. The third-order Galerkin basis functions in the local coordinate system are also listed in Table 5.1.

With the expressions of basis functions, the values of coefficients A_{ij}, B_{ij} and F_i in element (Δ) can be obtained by numerical integration. The integrands are all fourth-order polynomials of ξ and η in the local coordinate system. The Gauss quadratic formula is often employed for this purpose.

Besides using 10 nodal values, we have other methods to determine the 10 coefficients of a cubic polynomial. For instance, take three corner points and a center point of the triangle as nodes and assume that function values, as well as the partial derivatives with respect to x and y at the three vertices, are

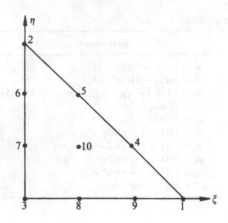

FIGURE 5.7. The third order standard triangle in a local coordinate system.

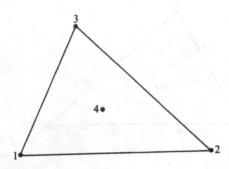

FIGURE 5.8. The Hermite triangular element.

given. If the value of the function at the center node is also given, then we have ten conditions to determine the ten coefficients. Such an element is called a *Hermite element*, which is shown in Figure 5.8.

Let the unknown function C in element (Δ) be replaced approximately by a cubic polynomial $\hat{C}$, which has the following form:

$$\hat{C} = \sum_{i=1}^{10} C_i(t)\phi_i(x, y), \quad (x, y) \in (\Delta); \qquad (5.1.44)$$

Each vertex of the triangle is related to three basis functions and three undetermined coefficients. The center point is related to $C_4(t)$ and $\phi_4(x, y)$ see Table 5.2. All basis functions are cubic polynomials of x and y.

According to the principles of selecting basis functions mentioned previously, we specify that the values of ϕ_1, $\partial\phi_5/\partial x$, $\partial\phi_8/\partial y$ at node 1 are equal to 1, and ϕ_l ($l = 2, 3, \ldots, 10$), $\partial\phi_l/\partial x$ ($l = 1, 2, \ldots, 10$, $l \neq 5$) and $\partial\phi_l/\partial y$ ($l = 1, 2, \ldots, 10$, $l \neq 8$) at node 1 are equal to zero. Basis functions for the other nodes are similarly defined. The basis functions defined in the local coordinate system as cubic polynomials of ξ and η are listed in Table 5.3. They are called *third-order Hermite basis functions*.

TABLE 5.2. Number of basis functions associated with each node.

Node	Undetermined coefficients	Basis functions
1	C_1, C_5, C_8	ϕ_1, ϕ_5, ϕ_8
2	C_2, C_6, C_9	ϕ_2, ϕ_6, ϕ_9
3	C_3, C_7, C_{10}	$\phi_3, \phi_7, \phi_{10}$
4	C_4	ϕ_4

As we now have the expressions of basis functions for each element, the coefficients of Eq. (5.1.12) can easily be calculated. Three solutions associated with each vertex are obtained.

These solutions are the value of the unknown function, and the values of its partial derivatives with respect to x and y at the node. The solution associated with the center node is just the value of the unknown function at that point.

The solution obtained by the Hermite element method has continuous first-order partial derivatives at the vertices. This is the main reason of choosing the Hermite element method.

When the flow equation is solved by the FEM with linear elements, the flow velocity obtained by Darcy's law is discontinuous along the sides of elements, because the gradient field of the numerical solution is piecewise constant. In order to avoid this problem, some authors proposed using the third-order Hermite elements to solve the flow equation, so that the continuity of flow velocity can be maintained, and the accuracy of the numerical solution of the advection-dispersion equation may be improved.

Recently, Cordes and Kinzelbach (1992) presented an approach for generating highly-accurate velocity fields for mass transport calculation when linear, quadratic and cubic FEM are used.

5.1.4 Isoparametric Finite Elements

Figure 5.9 shows a *quadrilateral element* with curved sides and seven nodes. The sides connected with nodes 1 and 2, as well as 6 and 7, are linear; the side with nodes 1, 4, and 7 is quadratic; the side with nodes 2, 3, 5, and 6 is of the third order. It is obvious that elements of such a shape are very suitable for representing the boundaries of an irregularly shaped domain or the inner borders in a non-homogeneous medium. We now wish to transform the element into a quadrilateral in the $\xi\eta$ plane of a local coordinate system through the transformation $x = x(\xi, \eta)$, $y = y(\xi, \eta)$. The positions of related nodes are shown in Figure 5.9.

This transformation can be constructed using basis functions. When node i is located on the side of the element, as nodes 3, 4, 5 in Figure 5.9, its basis

TABLE 5.3. Hermite basis functions in the local coordinate system.

Nodes	Hermite basis functions*		
1	$\phi_1 = \xi^2(\xi + 3\eta + 3\zeta) - 7\xi\eta\zeta$	$\phi_5 = \xi^2(c_3\eta - c_2\zeta) + (c_2 - c_3)\xi\eta\zeta$	$\phi_8 = \xi^2(b_2\zeta - b_3\eta) + (b_3 - b_2)\xi\eta\zeta$
2	$\phi_2 = \eta^2(\eta + 3\xi + 3\zeta) - 7\xi\eta\zeta$	$\phi_6 = \eta^2(c_1\zeta - c_3\xi) + (c_3 - c_1)\xi\eta\zeta$	$\phi_9 = \eta^2(b_3\xi - b_1\zeta) + (b_1 - b_3)\xi\eta\zeta$
3	$\phi_3 = \zeta^2(\zeta + 3\xi + 3\eta) + 7\xi\eta\zeta$	$\phi_7 = \zeta^2(c_2\xi - c_1\eta) + (c_1 - c_2)\xi\eta\zeta$	$\phi_{10} = \zeta^2(b_1\eta - b_2\xi) + (b_2 - b_1)\xi\eta\zeta$
4	$\phi_4 = 27\xi\eta\zeta$		

* In this table, $\zeta = 1 - \xi - \eta$; b_i, c_i ($i = 1, 2, 3$) are determined by (5.1.33).

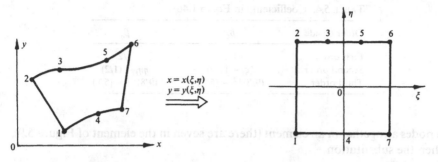

FIGURE 5.9. Transformation of the quadrilateral element with curved sides into the standard element in a local coordinate system.

functions can be defined as

$$\phi_i(\xi, \eta) = \begin{cases} \frac{1}{2}(1 - \xi^2)(1 + \eta\eta_i), & \text{when } \xi_i = 0, \eta_i = \pm 1; \\[2mm] \frac{1}{2}(1 + \xi\xi_i)(1 - \eta^2), & \text{when } \xi_i = \pm 1, \eta_i = 0; \\[2mm] \frac{9}{32}(1 - \xi^2)(1 + 9\xi\xi_i)(1 + \eta\eta_i), & \text{when } \xi_i = \pm\frac{1}{3}, \eta_i = \pm 1; \\[2mm] \frac{9}{32}(1 + \xi\xi_i)(1 - \eta^2)(1 + 9\eta\eta_i), & \text{when } \xi_i = \pm 1, \eta_i = \pm\frac{1}{3}. \end{cases}$$

$$(5.1.45)$$

The first expression can be used for node 4 in Figure 5.9, and the third expression for nodes 3 and 5.

The corner nodes may cause difficulties because they may be the intersections of curves with different orders. Let us consider the corner node i, and define its basis functions as follows:

$$\begin{cases} \phi_i(\xi, \eta) = \alpha_i\beta_i, \\[2mm] \alpha_i = \frac{1}{4}(1 + \xi\xi_i)(1 + \eta\eta_i), \\[2mm] \beta_i = \beta_\xi + \beta_\eta, \end{cases} \qquad (5.1.46)$$

where the values of β_ξ and β_η depend on the orders of sides $\xi = \pm 1, \eta = \pm 1$, see Table 5.4.

It is not difficult to verify that the value of basis function $\phi_i(\xi, \eta)$, thus defined, is equal to 1 at node i, and zero at the other nodes.

With the aid of these basis functions, the transformation between the global and local coordinate systems can also be found. Assume that there are

TABLE 5.4. Coefficients in Eq. (5.1.46).

Order of side	β_ξ	β_η
First order	1/2	1/2
Second order	$\xi\xi_i - (1/2)$	$\eta\eta_i - (1/2)$
Third order	$(9/8)\xi^2 - (5/8)$	$(9/8)\eta^2 - (5/8)$

n nodes altogether in an element (there are seven in the element of Figure 5.9), then the substitution

$$\begin{cases} x = \phi_1 x_1 + \phi_2 x_2 + \cdots + \phi_n x_n, \\ y = \phi_1 y_1 + \phi_2 y_2 + \cdots + \phi_n y_n, \end{cases} \tag{5.1.47}$$

transforms the element into a standard one in $\xi\eta$ coordinates, where (x_i, y_i), $(i = 1, 2, \ldots, n)$ are the coordinates of node i, and ϕ_i the corresponding basis function. When we take $\xi = \xi_i$, $\eta = \eta_i$ on the right-hand side of Eq. (5.1.47), we will have $x = x_i$, $y = y_i$, because the values of basis functions are all zero except $\phi_i = 1$. Thus, the transformation of Eq. (5.1.47) does provide the corresponding relation between points (ξ_i, η_i) and (x_i, y_i).

Since the parameters used in coordinate transformation (5.1.47) are identical to those used in determining the unknown concentration in Eq. (5.1.5), i.e., all of them are basis functions, so this method has the name of *isoparametric finite element method*.

After the basis functions and coordinate transformation are determined, the only problem left is calculating the elements of coefficient matrices in Eq. (5.1.12).

First of all, it is required to calculate the *Jacobian matrix* of the transformation in Eq. (5.1.47). We have:

$$[\mathbf{J}(\xi, \eta)] = \begin{bmatrix} \dfrac{\partial x}{\partial \xi} & \dfrac{\partial x}{\partial \eta} \\ \dfrac{\partial y}{\partial \xi} & \dfrac{\partial y}{\partial \eta} \end{bmatrix} = \begin{bmatrix} \dfrac{\partial \phi_1}{\partial \xi} & \dfrac{\partial \phi_2}{\partial \xi} & \cdots & \dfrac{\partial \phi_n}{\partial \xi} \\ \dfrac{\partial \phi_1}{\partial \eta} & \dfrac{\partial \phi_2}{\partial \eta} & \cdots & \dfrac{\partial \phi_n}{\partial \eta} \end{bmatrix} \begin{bmatrix} x_1 & y_1 \\ x_2 & y_2 \\ \cdots & \cdots \\ x_n & y_n \end{bmatrix}. \tag{5.1.48}$$

Since the nodal coordinates and the expressions of basis functions are known, the value of the Jacobian determinant, det $[\mathbf{J}]$, can be obtained from Eq. (5.1.48) directly. In order to transform the integrals (5.1.28) to (5.1.30) into the local coordinate system for calculation, it is also necessary to transfer $\partial \phi_i / \partial x$, $\partial \phi_j / \partial x$, $\partial \phi_i / \partial y$, $\partial \phi_j / \partial y$, which appear in the integrands, into the local coordinate system. By using the chain rule of differentiation, we arrive at

$$\begin{bmatrix} \dfrac{\partial \phi_i}{\partial x} \\ \dfrac{\partial \phi_i}{\partial y} \end{bmatrix} = [\mathbf{J}(\xi, \eta)]^{-1} \begin{bmatrix} \dfrac{\partial \phi_i}{\partial \xi} \\ \dfrac{\partial \phi_i}{\partial \eta} \end{bmatrix}, \tag{5.1.49}$$

where $[J(\xi, \eta)]^{-1}$ is the inverse matrix of $[J]$ in Eq. (5.1.48), and $\partial \phi_i / \partial \xi$ and $\partial \phi_i / \partial \eta$ can be calculated by Eqs. (5.1.45) and (5.1.46). As a result, Eq. (5.1.49) expresses $\partial \phi_i / \partial x$ and $\partial \phi_i / \partial y$ as functions of ξ and η. Using the same method, the expressions of $\partial \phi_j / \partial x$ and $\partial \phi_j / \partial y$ in terms of ξ and η can be obtained. Now, the calculation of A_{ij} associated with element (e) in Eq. (5.1.28) becomes the calculation of the following integral:

$$A_{ij}^{(e)} = \int_{-1}^{1} \int_{-1}^{1} f(\xi, \eta) \, d\xi \, d\eta. \tag{5.1.50}$$

The Gauss quadratic formula is often used to complete the numerical integration. The calculations of $B_{ij}^{(e)}$ and $F_i^{(e)}$ are similar. With these being completed, matrices $[A]$, $[B]$ and vector F can be formed.

5.1.5 Treatment of Boundary Conditions

In the formulas derived above, we have considered the first- and second-types of boundary conditions. However, to deal with the third-type of boundary condition, the derivation procedure in the Galerkin method should be modified slightly. The boundary condition with known solute flux can be given as follows:

$$(D \, \mathrm{grad} \, C - VC) \cdot n = -g_3, \quad (x, y) \in (\Gamma_3). \tag{5.1.51}$$

Consider the general advection-dispersion equation

$$\frac{\partial C}{\partial t} - \mathrm{div}(D \, \mathrm{grad} \, C - VC) = 0. \tag{5.1.52}$$

Its Galerkin equations are

$$\iint_{(R)} \left\{ \frac{\partial \hat{C}}{\partial t} - \mathrm{div}(D \, \mathrm{grad} \, \hat{C} - V\hat{C}) \right\} \phi_i \, dx \, dy = 0,$$

$$(i = 1, 2, \ldots, N) \tag{5.1.53}$$

where $\hat{C}$ is the trial function defined by Eq. (5.1.5).

Applying Green's formula to the second term (including both advection and dispersion) of the integrand in Eq. (5.1.53), we obtain

$$\iint_{(R)} \left\{ \frac{\partial \hat{C}}{\partial t} \phi_i + (D \, \mathrm{grad} \, \hat{C} - V\hat{C}) \cdot \mathrm{grad} \, \phi_i \right\} dx \, dy$$

$$- \int_{(\Gamma_3)} \phi_i (D \, \mathrm{grad} \, \hat{C} - V\hat{C}) \cdot n \, d\Gamma = 0,$$

$$(i = 1, 2, \ldots, N). \tag{5.1.54}$$

Substituting Eq. (5.1.5) and boundary condition (5.1.51) into the above line integral, we have:

$$
\iint_{(R)} \left\{ \sum_{j=1}^{N} \left(D_{xx} \frac{\partial \phi_j}{\partial x} \frac{\partial \phi_i}{\partial x} + D_{xy} \frac{\partial \phi_j}{\partial y} \frac{\partial \phi_i}{\partial x} + D_{yx} \frac{\partial \phi_j}{\partial x} \frac{\partial \phi_i}{\partial y} + D_{yy} \frac{\partial \phi_j}{\partial y} \frac{\partial \phi_i}{\partial y} \right. \right.
$$

$$
\left. \left. - V_x \phi_j \frac{\partial \phi_i}{\partial x} - V_y \phi_j \frac{\partial \phi_i}{\partial y} \right) C_j + \sum_{j=1}^{N} \phi_i \phi_j \frac{\partial C_j}{\partial t} \right\} dx\, dy + \int_{(\Gamma_3)} g_3 \phi_i \, d\Gamma = 0,
$$

$$
(i = 1, 2, \ldots, N). \tag{5.1.55}
$$

We can write this system of equations in the form of Eq. (5.1.12), in which the elements of coefficient matrices $[A]$, $[B]$ and vector $\mathbf{F}$ are, respectively:

$$
A_{ij} = \iint_{(R)} \left(D_{xx} \frac{\partial \phi_i}{\partial x} \frac{\partial \phi_j}{\partial x} + D_{xy} \frac{\partial \phi_i}{\partial x} \frac{\partial \phi_j}{\partial y} + D_{yx} \frac{\partial \phi_i}{\partial y} \frac{\partial \phi_j}{\partial x} \right.
$$

$$
\left. + D_{yy} \frac{\partial \phi_i}{\partial y} \frac{\partial \phi_j}{\partial y} - V_x \phi_j \frac{\partial \phi_i}{\partial x} - V_y \phi_j \frac{\partial \phi_i}{\partial y} \right) dx\, dy, \tag{5.1.56}
$$

$$
B_{ij} = \iint_{(R)} \phi_i \phi_j \, dx\, dy, \tag{5.1.57}
$$

$$
F_i = \int_{(\Gamma_3)} g_3 \phi_i \, d\Gamma. \tag{5.1.58}
$$

Therefore, the third-type of boundary condition can also be included in the Galerkin finite element formula.

It should be noted that during the derivation of Eq. (5.1.11), Green's formula is used only for the dispersion term. But, in the derivation of Eq. (5.1.55), Green's formula is used for both dispersion and advection terms. These two methods are equal in theory, but somewhat different in calculation. Galeati and Gambolati (1989) made a comparison between the two methods for a problem with pumping wells. Their examples show that it is better to use Green's formula for the dispersion term only.

Before ending this section, we would like to summarize the principal steps of the FEM:

Step 1. Divide the flow region into elements. Usually, we can select either triangular or quadrilateral elements. When the flow region has irregular geometry and complex structure (with faults, for example), for an accurate simulation, it is better to use the elements with two- or three-order curve sides.

Step 2. Locate nodes. Number and locations of nodes in an element are determined by the order of the element. A linear side requires two nodes located at its ends, a quadratic side requires another node located at its middle, while a cubic side requires two additional nodes located at 1/3 and 2/3 of the side as shown in Figure 5.9. Sometimes, extra nodes, which may be located at inner points of the element, are needed to make the number of nodes to be consistent with the order of the element.

Step 3. Define basis functions in the local coordinate system. According to the number and locations of each element, the basis function associated with each node in the local coordinate system can be found from Tables 5.1 and 5.3 and 5.4.

Step 4. Calculate the Jacobian of coordinate transformation. Once the basis functions are determined in the local coordinate system, we will have the coordinate transformation relation, Eq. (5.1.47), for each element. Then, we can calculate the Jacobian (5.1.48) and its inverse.

Step 5. Calculate local coefficients $A_{i,j}^{(e)}$, $B_{i,j}^{(e)}$, and $F_i^{(e)}$ for each element. The computation should be completed in the local coordinate system. When linear or bilinear elements are used, these coefficients are already given explicitly in Section 5.1.2.

Step 6. Calculate global coefficients $A_{i,j}$, $B_{i,j}$, and F_i These global values can be obtained by accumulating their local values in all relative elements and incorporating boundary conditions. Thus, the PDE is transformed into a system of ODE by spatial discretization.

Step 7. Discrete the time domain into time steps. In each time step, the time derivative in Eq. (5.1.12) is replaced by its finite difference approximation. Thus, the system of ODE in each time step is transformed into a system of algebraic equations as given in Eq. (5.1.17). The initial condition is used to determine the right-hand side of this equation in the first time step.

Step 8. Solve the algebraic system to obtain the concentration distribution for each time step. This procedure is repeated until the total simulation time is achieved. The right-hand of Eq. (5.1.17) depends on the solution of last time step. The solution methods of Eq. (5.1.17) will be discussed in Section 5.4.

5.2 The Multiple Cell Balance Method

5.2.1 Governing Equations

In a saturated aquifer with variable thickness, the two-dimensional advection-dispersion equation is

$$\frac{\partial(mC)}{\partial t} = \text{div}(m\mathbf{D}\,\text{grad}\,C) - \text{div}(\mathbf{V}mC) - \frac{C'W}{n}, \qquad (5.2.1)$$

where m is the saturated thickness of the aquifer, which varies with space and time; n the effective porosity of the aquifer; W the rate of extraction or injection per unit area of the aquifer, positive for extraction and negative for injection; C' the solute concentration contained in the source or sink, and the meaning of other notations are the same as before. Equation (5.2.1) is obtained by integrating the three-dimensional equation (2.6.13) along the vertical direction, and the concentration C in Eq. (5.2.1) is defined as the average concentration over the whole thickness of the aquifer.

Assume that (D) is a domain encircled by an arbitrary curve (L) in the flow region. Integrating Eq. (5.2.1) over (D) and using Green's formula, we then have:

$$-\int_{(L)} m\mathbf{D}\,\text{grad}\,C\cdot\mathbf{n}\,dl + \int_{(L)} \mathbf{V}mC\cdot\mathbf{n}\,dl$$

$$= \iint_{(D)} \left[\frac{\partial(mC)}{\partial t} + \frac{C'W}{n} \right] dx\,dy, \qquad (5.2.2)$$

where $\mathbf{n}$ is the unit internal normal vector of (L). The first term on the left-hand side of this equation represents the solute flux passing through (L) into (D) resulting from hydrodynamic dispersion, while the second term represents the solute flux resulting from advection. The right-hand side shows the increment of solute mass in domain (D). Therefore, Eq. (5.2.2) is just the mass conservation equation of solute in domain (D).

In Cartesian coordinates, the scalar form of Eq. (5.2.2) is

$$\int_{(L)} m\left[\left(D_{xx}\frac{\partial C}{\partial x} + D_{xy}\frac{\partial C}{\partial y} \right) dy - \left(D_{xy}\frac{\partial C}{\partial x} + D_{yy}\frac{\partial C}{\partial y} \right) dx \right]$$

$$+ \int_{(L)} mC(V_y\,dx - V_x\,dy)$$

$$= \iint_{(D)} \left[\frac{\partial(mC)}{\partial t} + \frac{C'W}{n} \right] dx\,dy. \qquad (5.2.3)$$

Sun and Yeh (1983) derived a system of discrete equations from the conservation equation (5.2.3) directly with the aid of basis functions used in the FEM. The mathematics associated with this method is very simple and the errors in local and global mass conservations are quite small. This method can be used for unconfined aquifers and the saturated-unsaturated zone, as well as for confined aquifers. This flexibility makes the method very useful in practice.

Bear (1979) named a numerical method, which is based on element mass conservation, as "*Multiple Cell Balance Method.*" We use the name because it vividly describes the essence of this kind of method.

5.2.2 An Algorithm Based on Multiple Cell Balance

First, partition the domain into a triangular net and consider an arbitrary element (e) (see Figure 5.1 in Section 5.1). Let the nodes be numbered as i, j, k, and their coordinates (x_i, y_i), (x_j, y_j), (x_k, y_k). In this element, the concentration of each point can be represented approximately by linear interpolation of its nodal values, i.e.,

$$C(x, y, t) = \phi_i(x, y)C_i(t) + \phi_j(x, y)C_j(t) + \phi_k(x, y)C_k(t),$$

$$(x, y) \in (e), \tag{5.2.4}$$

where interpolation functions

$$\phi_l(x, y) = \frac{1}{2\Delta_e}(a_l + b_l x + c_l y); \quad l = i, j, k \tag{5.2.5}$$

are just the basis functions of linear FEM (see Eq. (5.1.32)). $C_i(t)$, $C_j(t)$, $C_k(t)$ in Eq. (5.2.4) are the solute concentrations of nodes i, j, k at time t, respectively. Coefficients a_l, b_l, and c_l in the above equation are given by Eq. (5.1.33), Δ_e is the area of element (e).

From Eq. (5.2.4) we can directly obtain the following expressions for element (e):

$$\frac{\partial C}{\partial x} = \frac{1}{2\Delta_e}(b_i C_i + b_j C_j + b_k C_k), \tag{5.2.6}$$

$$\frac{\partial C}{\partial y} = \frac{1}{2\Delta_e}(c_i C_i + c_j C_j + c_k C_k), \tag{5.2.7}$$

$$\frac{\partial C}{\partial t} = \phi_i \frac{\partial C_i}{\partial t} + \phi_j \frac{\partial C_j}{\partial t} + \phi_k \frac{\partial C_k}{\partial t}. \tag{5.2.8}$$

Saturated thickness m, head h and velocity components V_x, V_y can also be represented by the same linear interpolation functions defined in Eq. (5.2.5).

Now let us consider all the elements with node i as one of their vertices. Link up the center and the midpoints of the sides of each triangle to form a subdomain surrounding node i (see Figure 5.10). This area is called the *exclusive subdomain* of node i and is taken as the domain (D) in Eq. (5.2.3). The part of element (e) in domain (D), i.e., the quadrilateral $iACB$ in Figure 5.10, is denoted by (e_i). Substituting Eqs. (5.2.6) and (5.2.7) into the first line

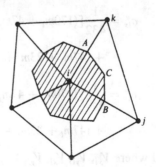

FIGURE 5.10. The exclusive subdomain for node i.

integral on the left-hand side of Eq. (5.2.3), we then obtain

$$\int_{(L)} m\left[\left(D_{xx}\frac{\partial C}{\partial x} + D_{xy}\frac{\partial C}{\partial y}\right)dy - \left(D_{xy}\frac{\partial C}{\partial x} + D_{yy}\frac{\partial C}{\partial y}\right)dy\right]$$

$$= \sum_{e_i} (\alpha_{ii}^e C_i + \alpha_{ij}^e C_j + \alpha_{ik}^e C_k), \tag{5.2.9}$$

where

$$\alpha_{il}^e = -\frac{m_{e_i}}{4\Delta_e}[D_{xx}^e b_i b_l + D_{xy}^e(c_i b_l + b_i c_l) + D_{yy}^e c_i c_l], \quad (l = i, j, k) \tag{5.2.10}$$

$$m_{e_i} = \frac{1}{24}(10m_i + 7m_j + 7m_k). \tag{5.2.11}$$

$\sum_{e_i}$ is the sum over all the elements with node i as one of their vertices, and $D_{xx}^e, D_{xy}^e, D_{yy}^e$ are values of dispersion coefficients of element (e).

For the second line integral on the left-hand side of Eq. (5.2.3), we have

$$\int_{(L)} mC(V_y\,dx - V_x\,dy) = -\iint_{(D)} \left[\frac{\partial}{\partial x}(V_x mC) + \frac{\partial}{\partial y}(V_y mC)\right]dx\,dy$$

$$= -\iint_{(D)} \left(mV_x\frac{\partial C}{\partial x} + mV_y\frac{\partial C}{\partial y}\right)dx\,dy$$

$$- \iint_{(D)} C\left[\frac{\partial(mV_x)}{\partial x} + \frac{\partial(mV_y)}{\partial y}\right]dx\,dy. \tag{5.2.12}$$

Calculating the first integral on the right-hand side, we get

$$\iint_{(D)} m\left[V_x\frac{\partial C}{\partial x} + V_y\frac{\partial C}{\partial y}\right]dx\,dy = \sum_{e_i} (\beta_{ii}^e C_i + \beta_{ij}^e C_j + \beta_{ik}^e C_k), \tag{5.2.13}$$

where

$$\beta_{il}^e = \frac{1}{6}(\sigma_{ix}^e b_l + \sigma_{iy}^e c_l); \quad l = i, j, k,$$

$$\sigma_{ix}^e = \frac{1}{432}[(170m_i + 47m_j + 47m_k)V_{xi}$$

$$+ (47m_i + 23m_j + 14m_k)V_{xj} + (47m_i + 14m_j + 23m_k)V_{xk}],$$

$$\sigma_{iy}^e = \frac{1}{432}[(170m_i + 47m_j + 47m_k)V_{yi}$$

$$+ (47m_i + 23m_j + 14m_k)V_{yj} + (47m_i + 14m_j + 23m_k)V_{jk}],$$

and where $V_{xi}, V_{yi}, V_{xj}, V_{yj}, V_{xk}, V_{yk}$ are the values of velocity components V_x, V_y at nodes i, j, k, respectively.

For the second integral on the right-hand side of Eq. (5.1.12), we may arrive at

$$-\iint_{(D)} C\left[\frac{\partial(mV_x)}{\partial x} + \frac{\partial(mV_y)}{\partial y}\right] dx\,dy = \iint_{(D)} \frac{C}{n}\left(S\frac{\partial h}{\partial t} + W\right) dx\,dy$$

$$= \sum_{e_i} (\gamma_{ii}^e C_i + \gamma_{ij}^e C_j + \gamma_{ik}^e C_k), \quad (5.2.14)$$

where

$$\gamma_{ii}^e = \frac{1}{n}\left[\frac{1}{432}\left(170\frac{\partial h_i}{\partial t} + 47\frac{\partial h_j}{\partial t} + 47\frac{\partial h_k}{\partial t}\right)S_e + \frac{22}{36}W_i\right],$$

$$\gamma_{ij}^e = \frac{1}{n}\left[\frac{1}{432}\left(47\frac{\partial h_i}{\partial t} + 23\frac{\partial h_j}{\partial t} + 14\frac{\partial h_k}{\partial t}\right)S_e + \frac{7}{36}W_i\right],$$

$$\gamma_{ik}^e = \frac{1}{n}\left[\frac{1}{432}\left(47\frac{\partial h_i}{\partial t} + 14\frac{\partial h_j}{\partial t} + 23\frac{\partial h_k}{\partial t}\right)S_e + \frac{7}{36}W_i\right],$$

and with S_e being the value of storage coefficient, S, in element (e).

In the calculation of Eqs. (5.2.13) and (5.2.14), the following formulas are used:

$$\iint_{(e_i)} \phi_i\,dx\,dy = \frac{\Delta_e}{3}\cdot\frac{22}{36}, \quad \iint_{(e_i)} \phi_j\,dx\,dy = \iint_{(e_i)} \phi_k\,dx\,dy = \frac{\Delta_e}{3}\cdot\frac{7}{36},$$

$$\iint_{(e_i)} \phi_i\phi_i\,dx\,dy = \frac{\Delta_e}{3}\cdot\frac{170}{432}, \quad \iint_{(e_i)} \phi_i\phi_j\,dx\,dy = \iint_{(e_i)} \phi_i\phi_k\,dx\,dy = \frac{\Delta_e}{3}\cdot\frac{47}{432},$$

$$\iint_{(e_i)} \phi_j\phi_j\,dx\,dy = \iint_{(e_i)} \phi_k\phi_k\,dx\,dy = \frac{\Delta_e}{3}\cdot\frac{23}{432}, \quad \iint_{(e_i)} \phi_j\phi_k\,dx\,dy = \frac{\Delta_e}{3}\cdot\frac{47}{432}.$$

$$(5.2.15)$$

With the aid of the variable substitution,

$$\begin{cases} x = x_i\xi + x_j\eta + x_k(1 - \xi - \eta), \\ y = y_i\xi + y_j\eta + y_k(1 - \xi - \eta), \end{cases} \quad (5.2.16)$$

all integrals in Eq. (5.2.15) can be computed in the $\xi\eta$ plane.

Dividing the right-hand side of Eq. (5.2.3) into three parts:

$$\iint_{(D)} \left[\frac{\partial(mC)}{\partial t} + \frac{C'W}{n}\right] dx\,dy = \iint_{(D)} m\frac{\partial C}{\partial t}\,dx\,dy + \iint_{(D)} C\frac{\partial m}{\partial t}\,dx\,dy$$

$$+ \iint_{(D)} \frac{C'W}{n}\,dx\,dy \quad (5.2.17)$$

and calculating each part on the right-hand side, we have

$$\iint_{(D)} m\frac{\partial C}{\partial t}\,dx\,dy = \sum_e \left(B_{ii}^e\frac{\partial C_i}{\partial t} + B_{ij}^e\frac{\partial C_j}{\partial t} + B_{ik}^e\frac{\partial C_k}{\partial t}\right), \quad (5.2.18)$$

where

$$B_{ii}^e = \frac{\Delta_e}{3} \cdot \frac{1}{432}(170m_i + 47m_j + 47m_k),$$

$$B_{ij}^e = \frac{\Delta_e}{3} \cdot \frac{1}{432}(47m_i + 23m_j + 14m_k),$$

$$B_{ik}^e = \frac{\Delta_e}{3} \cdot \frac{1}{432}(47m_i + 14m_j + 23m_k),$$

and

$$\iint_{(D)} C \frac{\partial m}{\partial t} dx\, dy = \sum_{e_i} (\delta_{ii}^e C_i + \delta_{ij}^e C_j + \delta_{ik}^e C_k), \qquad (5.2.19)$$

in which

$$\delta_{ii}^e = \frac{\Delta_e}{3} \cdot \frac{1}{432}\left(170\frac{\partial m_i}{\partial t} + 47\frac{\partial m_j}{\partial t} + 47\frac{\partial m_k}{\partial t}\right),$$

$$\delta_{ij}^e = \frac{\Delta_e}{3} \cdot \frac{1}{432}\left(47\frac{\partial m_i}{\partial t} + 23\frac{\partial m_j}{\partial t} + 14\frac{\partial m_k}{\partial t}\right),$$

$$\delta_{ik}^e = \frac{\Delta_e}{3} \cdot \frac{1}{432}\left(47\frac{\partial m_i}{\partial t} + 14\frac{\partial m_j}{\partial t} + 23\frac{\partial m_k}{\partial t}\right).$$

Furthermore, let

$$F_i = \iint_{(D)} \frac{C'W}{n} dx\, dy = \frac{C_i'}{n} W_i P_i, \qquad (5.2.20)$$

where C_i' is the solute concentration contained in the water extracted from, or injected into, the exclusive subdomain of node i. For extraction, C_i' is unknown. However, it is the same as the concentration C_i at node i. P_i is the area of the exclusive subdomain (D) of node i. $W_i P_i$ in Eq. (5.2.20) is the rate of extraction or injection for the subdomain.

Substituting Eqs. (5.2.9), (5.2.13), (5.2.14), and Eqs. (5.2.18) through (5.2.20) into Eq. (5.2.3), we then have:

$$\sum_{e_i} (A_{ii}^e C_i + A_{ij}^e C_j + A_{ik}^e C_k) + \sum_{e_i} \left(B_{ii}^e \frac{\partial C_i}{\partial t} + B_{ij}^e \frac{\partial C_j}{\partial t} + B_{ik}^e \frac{\partial C_k}{\partial t}\right) + F_i = 0,$$

$$(5.2.21)$$

where

$$A_{il}^e = -\alpha_{il}^e + \beta_{il}^e - \gamma_{il}^e + \delta_{il}^e, \quad (l = i, j, k). \qquad (5.2.22)$$

Eq. (5.2.21) can alternately be written as

$$A_{ii}' C_i + \sum_{j \neq i} A_{ij}' C_j + B_{ii}' \frac{\partial C_i}{\partial t} + \sum_{j \neq i} B_{ij}' \frac{\partial C_j}{\partial t} + F_i = 0, \qquad (5.2.23)$$

FIGURE 5.11. The balance domain of a boundary node.

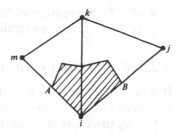

where $\sum_{j \neq i}$ is the summation over all adjacent nodes of node i, and

$$A'_{ii} = \sum_{e_i} A^e_{ii}, \quad B'_{ii} = \sum_{e_i} B^e_{ii},$$

$$A'_{ij} = \sum_{e_{ij}} A^e_{ij}, \quad B'_{ij} = \sum_{e_{ij}} B^e_{ij}, \tag{5.2.24}$$

where $\sum_{e_{ij}}$ denotes the summation over all elements with both nodes i and j as vertices.

We have discussed the situation of node i as an interior node. If node i is located on the boundary, as shown in Figure 5.11, we must take into consideration the given boundary conditions in setting up the relevant mass balance equations.

It is not necessary to build an equation for the boundary node i when the concentration at the node is given. In other equations where C_i appears, we can use the given value to substitute C_i. In the case of a given dispersion flux, we have

$$m\mathbf{D} \operatorname{grad} C \cdot \mathbf{n} = -g_2; \quad (x, y) \in A\hat{i}B,$$

where g_2 is a known function. Equation (5.2.1) will include the integral $\int_{(A\hat{i}B)} g_2 \, dl$, which can be regarded as a known value and merged into the right-hand side term F. Because the advection flux $\int_{(A\hat{i}B)} m\mathbf{V}C \cdot \mathbf{n} \, dl$ over $A\hat{i}B$ has been included in Eqs. (5.2.13) and (5.2.14), the boundary conditions of outflow or no-flow can be satisfied automatically.

Taking into account the boundary conditions, and listing the relevant equations (5.2.23) for every unknown concentration node, we obtain a system of equations. Using the symbols of Section 5.1.1, this set of equations can be written as

$$[\mathbf{A}']\mathbf{C} + [\mathbf{B}']\frac{d\mathbf{C}}{dt} + \mathbf{F}' = 0, \tag{5.2.25}$$

where the elements of matrices $[\mathbf{A}']$ and $[\mathbf{B}']$ are defined by Eq. (5.2.24), and the elements of vector $\mathbf{F}'$ are determined by Eq. (5.2.20) and relevant boundary conditions. Equation (5.2.25) is a system of ODEs whose form is the same as that of Eq. (5.1.12).

A FORTAN program of the MCB is given in Appendix B, in which all steps of the method are explained in detail.

5.2.3 Comparing with the Finite Element Method

It is interesting to make a comparison between Eqs. (5.2.25) and the discrete equations obtained from the linear Galerkin FEM.

Letting thickness $m =$ constant and div $V = 0$, Eq. (5.2.1) then reduces to Eq. (5.1.1), where $I = -(C'W)/(nm)$. The coefficients α_{il}^e of Eq. (5.2.10) are reduced to

$$\alpha_{il}^e = -\frac{1}{4\Delta_e}[D_{xx}^e b_i b_l + D_{xy}^e(c_i b_l + b_i c_l) + D_{yy}^e c_i c_l].$$

Furthermore, let velocity components V_x^e and V_y^e, be constant in each element. Coefficients β_{il}^e in Eq. (5.2.13) are then simplified to

$$\beta_{il}^e = \frac{1}{6}(V_x^e b_l + V_y^e c_l).$$

Since div $V = 0$, all γ_{il} in Eq. (5.2.14) are equal to zero. From $\partial m/\partial t = 0$, all δ_{il}^e in Eq. (5.2.19) are also equal to zero. Therefore, coefficients A_{il}^e, given by Eq. (5.2.22), are simplified to

$$A_{il}^e = \frac{1}{4\Delta_e}[D_{xx}^e b_i b_l + D_{xy}^e(b_i c_l + c_i b_l) + D_{yy}^e c_i c_l]$$

$$+ \frac{1}{6}(V_x^e b_l + V_y^e c_l). \tag{5.2.26}$$

Comparing this formula with Eq. (5.1.35), we find that matrix $[A']$ is exactly the same as matrix $[A]$ obtained by the Galerkin FEM using linear triangle elements.

Now let us study the relationship of coefficient matrices $[B']$ and $[B]$. When $m =$ constant, B_{ij}^e in Eq. (5.2.18) becomes

$$B_{ij}^e = \begin{cases} \dfrac{\Delta_e}{3} \cdot \dfrac{22}{36} & \text{when } i = j, \\[2mm] \dfrac{\Delta_e}{3} \cdot \dfrac{7}{36} & \text{when } i \neq j. \end{cases} \tag{5.2.27}$$

Comparing this equation with Eq. (5.1.36), it is apparent that they are identical in form, but have slightly different coefficients. When we build the equation for node i, the time derivatives of concentration should be weighted over the three nodes i, j, k in element (e). The weighting coefficients of FEM are 1/2, 1/4, 1/4, while those of the multiple cell balance method are 22/36, 7/36, 7/36. This set of coefficients is derived on the basis of maintaining local mass conservation.

Therefore, a program written for the multiple cell balance method can be turned into a program for Galerkin's FEM by changing only two constants. Coefficients B_{ij}^e can also take other values. For example, let

$$B_{ij}^e = \begin{cases} \dfrac{\Delta_e}{3} & \text{when} \quad i = j, \\[2ex] 0 & \text{when} \quad i \neq j. \end{cases} \tag{5.2.28}$$

This result comes from the approximation

$$\iint_{(D)} \frac{\partial C}{\partial t} \, dx \, dy \approx \frac{dC_i}{dt} \cdot P_i, \tag{5.2.29}$$

in calculating the integral in Eq. (5.2.18).

Under this condition, matrix $[\mathbf{B}']$ is simplified to a diagonal matrix, with the elements of each column concentrated onto the diagonal. This kind of method is called the *mass lumped FEM*. When the method is used to solve the flow equation, less computer memory is required and some advantages are gained in calculation. The mass lumped method has been used by some authors to solve water quality equations. In the next section, we are going to make a comparison among FEM, MCB and mass lumped FEM.

5.2.4 The Test of Numerical Solutions

The correctness and accuracy of numerical solutions can be tested through comparison with an analytic solution. In what follows, the problem of tracer transport in a semi-infinite sand-column will be used to test the numerical solutions.

The governing equation of this problem is

$$\frac{\partial C}{\partial t} = D \frac{\partial^2 C}{\partial x^2} - V \frac{\partial C}{\partial x}, \tag{5.2.30}$$

with initial and boundary conditions

$$C(x, 0) = 0, \quad C(\infty, t) = 0, \quad C(0, t) = C_0, \tag{5.2.31}$$

The analytic solution of this problem has been given in Section 3.2.1 as

$$C(x, t) = \frac{C_0}{2} \left\{ \text{erfc} \left[\frac{x - Vt}{2\sqrt{Dt}} \right] + \exp \left(\frac{Vx}{D} \right) \text{erfc} \left[\frac{x + Vt}{2\sqrt{Dt}} \right] \right\}. \tag{5.2.32}$$

The local Peclet number is defined as

$$Pe = \frac{V \Delta x}{D}, \tag{5.2.33}$$

where V is the mean flow velocity in pores, D the longitudinal hydrodynamic dispersion coefficient, and Δx the interval between grid lines in the x direction.

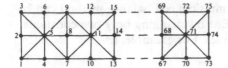

FIGURE 5.12. The grid used for Example 1.

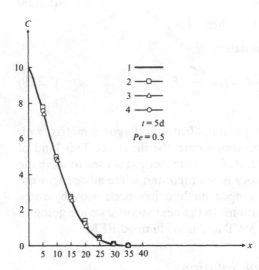

FIGURE 5.13. Comparison between numerical and analytic solutions when $Pe = 0.5$, $t = 5\,d$. (1) analytic solution; (2) FEM; (3) mass lumped FEM; (4) MCB.

First, let us consider the case of small Peclet number.

Let $C_0 = 10\,g/m^3$, $V = 1\,m/d$, $\Delta x = 5\,m$, $D = 10\,m^2/d$, so that $Pe = 0.5$.

We used 75 nodes and 96 triangles to solve this problem, as shown in Figure 5.12. The purpose of using three rows of elements is to keep the symmetry of the numerical solutions. For the time interval $0 \le t \le 50$ days, the length of the discretization model in the x direction is sufficient to approximately represent a semi-infinite region.

We have compared the results of linear FEM, MCB and mass lumped FEM. In the case of small Peclet number, all of these are satisfactory. For the solutions of short time period ($t = 5\,d$), the result of MCB is a little better than the other two (see Figure 5.13). For the solutions of long time period ($t = 50\,d$), they produce almost the same results, see Figure 5.14.

Now, turn to the case of large Peclet number. Again, let $C_0 = 10\,g/m^3$, $V = 1\,m/d$, and $\Delta x = 5\,m$, but set $D = 0.05\,m^2/d$, so $Pe = 100$. In this case, the front of the analytic solution becomes steep and numerical solutions oscillate significantly around the concentration front. It can be seen in Figure 5.15 that the front produced by the numerical method is smoother than that produced by the analytic solution. This example tells us that the solutions of FEM still contain both types of errors: numerical dispersion and overshoot (see Section 4.1.3).

In the next chapter, we will introduce various numerical methods which

FIGURE 5.14. Comparison between numerical and analytic solutions when $Pe = 0.5$, $t = 50\,d$. —— = analytic solution; ● = numerical solution.

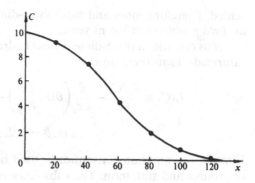

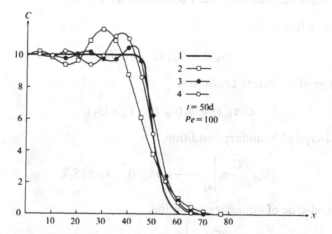

FIGURE 5.15. Comparison between numerical and analytic solutions when $Pe = 100$, $t = 50\,d$. (1) analytical solution; (2) FEM; (3) mass lumped FEM; (4) MCB.

may help to avoid the oscillations of numerical solutions and reduce the numerical dispersion.

5.3 Finite Element Methods for Three-Dimensional Problems

5.3.1 The Galerkin Finite Element Method

Most of the hydrodynamic dispersion problems encountered in reality are three-dimensional. For instance, the pollution source may be near the ground surface, the injection well of the waste water may be partially penetrated, there may exist hydraulic relations between layers, and so on. All these common cases are three-dimensional problems. If they are simplified into two-dimensional ones, the essential aspects of problems cannot be represented. Therefore, more and more three-dimensional models were developed for field problems in recent years.

Let us consider a three-dimensional hydrodynamic dispersion equation for saturated-unsaturated flow:

$$L(C) \equiv \frac{\partial(\theta C)}{\partial t} - \frac{\partial}{\partial x_\alpha}\left(\theta D_{\alpha\beta} \frac{\partial C}{\partial x_\beta}\right) + \frac{\partial}{\partial x_\alpha}(\theta V_\alpha C) + M = 0.$$

$$(\alpha, \beta = 1, 2, 3) \tag{5.3.1}$$

Einstein's summation convention is used here for brevity, where M denotes the source and sink term. The subsidiary conditions of Eq. (5.3.1) consist of the initial condition and boundary conditions as follows:

the initial condition,

$$C(\mathbf{x}, 0) = C_0(\mathbf{x}), \quad \mathbf{x} \in (R); \tag{5.3.2}$$

the first-type of boundary condition,

$$C(\mathbf{x}_B, t) = g_1(\mathbf{x}_B, t), \quad \mathbf{x}_B \in (S_1); \tag{5.3.3}$$

the second-type of boundary condition,

$$\theta D_{\alpha\beta} \frac{\partial C}{\partial x_B} n_\alpha \bigg|_{\mathbf{x}_B} = -g_2(\mathbf{x}_B, t), \quad \mathbf{x}_B \in (S_2); \tag{5.3.4}$$

and the third-type of boundary condition,

$$\theta \left(CV - D_{\alpha\beta} \frac{\partial C}{\partial x_B}\right) n_\alpha \bigg|_{\mathbf{x}_B} = -g_3(\mathbf{x}_B, t), \quad \mathbf{x}_B \in (S_3). \tag{5.3.5}$$

These equations have been given in Eqs. (2.6.30) to (2.6.33), where $\mathbf{x} = (x_1, x_2, x_3)$ represents point (x, y, z) in space; $\mathbf{x}_B$ is the point on the boundary of the region; C_0, g_1, g_2, g_3 are the known functions, and $(S_1), (S_2), (S_3)$ are boundary surfaces of the region, which together form the whole boundary surface (S) for region (R).

Let us take the trial solution

$$\hat{C}(x, t) = \sum_{i=1}^{N} C_i(t)\phi_i(\mathbf{x}), \tag{5.3.6}$$

where $\{\phi_i(\mathbf{x}), i = 1, 2, \ldots, N\}$ is a system of basis functions and $\{C_i(t), i = 1, 2, \ldots, N\}$ are the coefficients which depend upon time. According to the Galerkin method, these coefficients can be obtained by solving the following equations:

$$\iiint_{(R)} L(\hat{C})\phi_i(\mathbf{x}) \, dR = 0, \quad (i = 1, 2, \ldots, N). \tag{5.3.7}$$

Substituting Eq. (5.3.1) into Eq. (5.3.7), we have

$$\iiint_{(R)} \left\{ \frac{\partial(\theta\hat{C})}{\partial t} - \frac{\partial}{\partial x_\alpha}\left(\theta D_{\alpha\beta}\frac{\partial\hat{C}}{\partial x_\beta}\right) + \frac{\partial}{\partial x_\alpha}(\theta V_\alpha\hat{C}) + M \right\} \phi_i \, dR = 0,$$

$$(i = 1, 2, \ldots, N). \tag{5.3.8}$$

Using Green's formula to eliminate the second-order derivative terms from the integrands, we arrive at

$$\iiint_{(R)} \left\{ \frac{\partial(\theta\hat{C})}{\partial t} + \frac{\partial}{\partial x_\alpha}(\theta V_\alpha\hat{C}) + M \right\} \phi_i \, dR + \iiint_{(R)} \left(\theta D_{\alpha\beta}\frac{\partial\hat{C}}{\partial x_\beta} \right) \frac{\partial\phi_i}{\partial x_\alpha} \, dR$$

$$+ \iint_{(S_2)} g_2 \phi_i \, dS = 0. \tag{5.3.9}$$

The last term in this equation, which is a surface integral, is defined by the second-type of boundary condition given in Eq. (5.3.4). Here, we will temporarily assume that there exists only this kind of boundary condition. We will explain below how to deal with the third-type of boundary condition.

Substituting (5.3.6) into (5.3.9), we have

$$\iiint_{(R)} \sum_{j=1}^{N} \left\{ \frac{\partial\theta}{\partial t}\phi_i\phi_j + \frac{\partial}{\partial x_\alpha}(\theta V_\alpha)\phi_i\phi_j + \theta V_\alpha\phi_i\frac{\partial\phi_j}{\partial x_\alpha} + \theta D_{\alpha\beta}\frac{\partial\phi_i}{\partial x_\alpha}\frac{\partial\phi_j}{\partial x_\beta} \right\} C_j \, dR$$

$$+ \iiint_{(R)} \sum_{j=1}^{N} \theta\phi_i\phi_j\left(\frac{dC_j}{dt}\right) dR + \iiint_{(R)} M\phi_i \, dR + \iint_{(S_2)} g_2\phi_i \, dS = 0,$$

$$(i = 1, 2, \ldots, N). \tag{5.3.10}$$

This equation can be rewritten in vector-matrix form as follows:

$$[A]C + [B]\frac{dC}{dt} + F = 0, \tag{5.3.11}$$

where the elements of matrices [A], [B] and vector F are given by

$$A_{ij} = \iiint_{(R)} \left\{ \frac{\partial\theta}{\partial t}\phi_i\phi_j + \frac{\partial}{\partial x_\alpha}(\theta V_\alpha)\phi_i\phi_j \right.$$

$$\left. + \theta V_\alpha\phi_i\frac{\partial\phi_j}{\partial x_\alpha} + \theta D_{\alpha\beta}\frac{\partial\phi_i}{\partial x_\alpha}\frac{\partial\phi_j}{\partial x_\beta} \right\} dR, \tag{5.3.12}$$

$$B_{ij} = \iiint_{(R)} \theta\phi_i\phi_j \, dR, \tag{5.3.13}$$

$$F_i = \iiint_{(R)} M\phi_i \, dR + \iint_{(S_2)} g_2\phi_i \, dS. \tag{5.3.14}$$

Once the basis function ϕ_i is selected, these integrals can be easily calculated. Then, $C_i(t)$ can be obtained from Eq. (5.3.11). Substituting $C_i(t)$ into Eq. (5.3.6), an approximate solution $\hat{C}(x, t)$ can be obtained.

If there exists the third-type of boundary condition—Eq. (5.3.5), we must apply Green's formula to both the dispersion and advection terms of Eq. (5.3.8) simultaneously. The coefficients of Eq. (5.3.11) then become

$$A_{ij} = \iiint_{(R)} \left(\frac{\partial \theta}{\partial t} \phi_i \phi_j + \theta D_{\alpha\beta} \frac{\partial \phi_i}{\partial x_\alpha} \frac{\partial \phi_j}{\partial x_\beta} - \theta V_\alpha \phi_j \frac{\partial \phi_i}{\partial x_\alpha} \right) dR, \quad (5.3.15a)$$

$$B_{ij} = \iiint_{(R)} \theta \phi_i \phi_j \, dR, \quad (5.3.15b)$$

$$F_i = \iiint_{(R)} M \phi_i \, dR + \iint_{(S_3)} g_3 \phi_i \, dS. \quad (5.3.15c)$$

It should be noted that the three-dimensional Galerkin FEM is a direct extension of its two-dimensional form. The principles of selecting basis functions combined with finite element discretization are the same as those in the two-dimensional case. That is, let ϕ_i be equal to 1 at node i and zero at other nodes. Such a selection of basis functions will make the coefficient matrices [A] and [B] of Eq. (5.3.11) turn into highly-sparse matrices. Furthermore, coefficient $C_i(t)$ in Eq. (5.3.6) is just equal to the concentration of node i at time t.

The problem left is how to select the shape of elements and the order of basis functions. The simplest way is by adopting the tetrahedron element and linear basis functions. As it is inconvenient to input the geometrical information of tetrahedron elements, the combined elements such as triangular prism and hexahedron are often used. The former can be partitioned into three tetrahedrons, and the latter into five.

Gupta et al. (1975) used *isoparametric FEM* to solve the water quality equation. Using isoparametric FEM in three-dimensional problems can reduce the number of nodes, and achieve a higher accuracy.

For the standard cubic element in a local coordinate system, $-1 \le \xi \le 1$, $-1 \le \eta \le 1$, and $-1 \le \zeta \le 1$ (see Figure 5.16). We may select nodes in addition to the corners. For edges of second order, the center points are

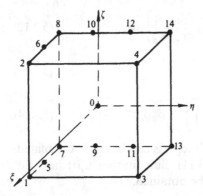

FIGURE 5.16. Standard element in the local coordinate system.

FIGURE 5.17. Isoparameter element in the global coordinate system.

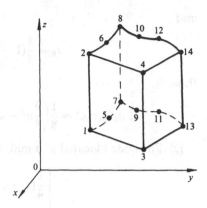

TABLE 5.5. Values of coefficient β in Eq. (5.3.17).

Order of sides	Values of β		
	β_ξ	β_η	β_ζ
First	1/3	1/3	1/3
Second	$\xi\xi_i - 2/3$	$\zeta\zeta_i - 2/3$	$\zeta\zeta_i - 2/3$
Third	$(9/8)\xi - 19/24$	$(9/8)\eta^2 - 19/24$	$(9/8)\zeta^2 - 19/24$

taken as nodes (nodes 5 and 6 in Figure 5.16). For edges of third order, points at one-third and two-third of the edges are taken as nodes (nodes 9, 10, 11, 12 in Figure 5.16). The corresponding nodes in the global coordinate system are shown in Figure 5.17. The associated basis functions can be defined as follows:

(1) For node i located at a corner point, we define

$$\phi_i(\xi, \eta, \zeta) = \alpha_i \beta_i, \tag{5.3.16}$$

where

$$\begin{cases} \alpha_i = \dfrac{1}{8}(1 + \xi\xi_i)(1 + \eta\eta_i)(1 + \zeta\zeta_i), \\ \beta_i = \beta_\xi + \beta_\eta + \beta_\zeta. \end{cases} \tag{5.3.17}$$

β_ξ, β_η, β_ζ depend on the orders of intersection sides of the node, as given in Table 5.5. For example, let us consider node 8 in Figure 5.16. Its local coordinates are $(-1, -1, 1)$. The orders of its three intersecting sides are 2, 3, and 1, respectively. From Table 5.5 we find

$$\beta_8 = \left(-\xi - \frac{2}{3}\right) + \left(\frac{9}{8}\eta^2 - \frac{19}{24}\right) + \frac{1}{3},$$

and

$$\alpha_8 = \frac{1}{8}(1 - \xi)(1 - \eta)(1 + \zeta),$$

thus,

$$\phi_8(\xi, \eta, \zeta) = \frac{1}{8}\left(\frac{9}{8}\eta^2 - \xi - \frac{9}{8}\right)(1 - \xi)(1 - \eta)(1 + \zeta).$$

(2) For node i located at a midpoint of a edge, we define

$$\phi_i(\xi, \eta, \zeta) = \begin{cases} \frac{1}{4}(1 - \xi^2)(1 + \eta\eta_i)(1 + \zeta\zeta_i), \\ \\ \quad \text{when} \quad \xi_i = 0, \eta_i = \pm 1, \zeta_i = \pm 1; \\ \\ \frac{1}{4}(1 + \xi\xi_i)(1 - \eta^2)(1 + \zeta\zeta_i), \\ \\ \quad \text{when} \quad \xi_i = \pm 1, \eta_i = 0, \zeta_i = \pm 1; \\ \\ \frac{1}{4}(1 + \xi\xi_i)(1 + \eta\eta_i)(1 - \zeta^2), \\ \\ \quad \text{when} \quad \xi_i = \pm 1, \eta_i = \pm 1, \zeta_i = 0. \end{cases} \qquad (5.3.18)$$

Node 5 has local coordinates $(0, -1, -1)$ and is an example of this case. Using the first formula of Eq. (5.1.18), we can find

$$\phi_5(\xi, \eta, \zeta) = \frac{1}{4}(1 - \xi^2)(1 - \eta)(1 - \zeta).$$

(3) For node i located at one-third of an edge, we define

$$\phi_i(\xi, \eta, \zeta) = \begin{cases} \frac{9}{64}(1 - \xi^2)(1 + 9\xi\xi_i)(1 + \eta\eta_i)(1 + \zeta\zeta_i), \\ \\ \quad \text{when} \quad \xi_i = \pm\frac{1}{3}, \eta_i = \pm 1, \zeta_i = \pm 1 \\ \\ \frac{9}{64}(1 + \xi\xi_i)(1 - \eta^2)(1 + 9\eta\eta_i)(1 + \zeta\zeta_i), \\ \\ \quad \text{when} \quad \xi_i = \pm 1, \eta_i = \pm\frac{1}{3}, \zeta_i = \pm 1 \\ \\ \frac{9}{64}(1 + \xi\xi_i)(1 + \eta\eta_i)(1 - \zeta^2)(1 + 9\zeta\zeta_i), \\ \\ \quad \text{when} \quad \xi_i = \pm 1, \eta_i = \pm 1, \zeta_i = \pm\frac{1}{3}. \end{cases} \qquad (5.3.19)$$

Node 10 has local coordinates $(-1, -\frac{1}{3}, 1)$ in Figure 5.1.6 and is an example of this case. Using the second formula of Eq. (5.3.19), we have

$$\phi_{10}(\xi, \eta, \zeta) = \frac{9}{64}(1 - \xi)(1 - \eta^2)(1 - 3\eta)(1 + \zeta).$$

Just as in the two-dimensional case of isoparameter FEM, the basis functions defined above can also serve as parameters of coordinate transformation which transforms the irregular element in the global coordinates into a standard element in the local coordinates. Suppose that there is a mixed element with sides of the first, second, and third orders and n nodes in the global coordinates (Figure 5.17). Using coordinate transformation

$$\begin{cases} x = \sum_{i=1}^{n} \phi_i(\xi, \eta, \zeta) x_i, \\ y = \sum_{i=1}^{n} \phi_i(\xi, \eta, \zeta) y_i, \\ z = \sum_{i=1}^{n} \phi_i(\xi, \eta, \zeta) z_i, \end{cases} \tag{5.3.20}$$

the irregular element can be transformed into a standard element in the local coordinates, where (x_i, y_i, z_i) are the coordinates of node i in the global coordinates. This conclusion can be verified directly from the definition of basis functions.

The Jacobian matrix of the transformation in Eq. (5.3.20) is

$$[J] = \begin{bmatrix} \dfrac{\partial x}{\partial \xi} & \dfrac{\partial x}{\partial \eta} & \dfrac{\partial x}{\partial \zeta} \\ \dfrac{\partial y}{\partial \xi} & \dfrac{\partial y}{\partial \eta} & \dfrac{\partial y}{\partial \zeta} \\ \dfrac{\partial z}{\partial \xi} & \dfrac{\partial z}{\partial \eta} & \dfrac{\partial z}{\partial \zeta} \end{bmatrix} = \begin{bmatrix} \dfrac{\partial \phi_1}{\partial \xi} & \dfrac{\partial \phi_2}{\partial \xi} & \cdots & \dfrac{\partial \phi_n}{\partial \xi} \\ \dfrac{\partial \phi_1}{\partial \eta} & \dfrac{\partial \phi_2}{\partial \eta} & \cdots & \dfrac{\partial \phi_n}{\partial \eta} \\ \dfrac{\partial \phi_1}{\partial \zeta} & \dfrac{\partial \phi_2}{\partial \zeta} & \cdots & \dfrac{\partial \phi_n}{\partial \zeta} \end{bmatrix} \begin{bmatrix} x_1 & y_1 & z_1 \\ x_2 & y_2 & z_2 \\ \cdots & \cdots & \cdots \\ \cdots & \cdots & \cdots \\ \cdots & \cdots & \cdots \\ x_n & y_n & z_n \end{bmatrix}, \tag{5.3.21}$$

and all partial derivatives $\partial \phi_i / \partial x$, $\partial \phi_i / \partial y$, $\partial \phi_i / \partial z$ can be represented as:

$$\begin{bmatrix} \dfrac{\partial \phi_i}{\partial x} \\ \dfrac{\partial \phi_i}{\partial y} \\ \dfrac{\partial \phi_i}{\partial z} \end{bmatrix} = [J]^{-1} \begin{bmatrix} \dfrac{\partial \phi_i}{\partial \xi} \\ \dfrac{\partial \phi_i}{\partial \eta} \\ \dfrac{\partial \phi_i}{\partial \zeta} \end{bmatrix}. \tag{5.3.22}$$

Thus, the coefficients of FEM equations, i.e., A_{ij}, B_{ij} and F_i in Eq. (5.3.12), can all be obtained through calculating triple integrals of the following form in the local coordinates:

$$\int_{-1}^{1}\int_{-1}^{1}\int_{-1}^{1} f(\xi, \eta, \zeta)\, d\xi\, d\eta\, d\zeta. \tag{5.3.23}$$

The Gaussian quadrature formula for calculating triple integrals is

$$\int_{-1}^{1}\int_{-1}^{1}\int_{-1}^{1} f(\xi, \eta, \zeta)\, d\xi\, d\eta\, d\zeta = \sum_{i=1}^{m}\sum_{j=1}^{m}\sum_{k=1}^{m} \omega_i \omega_j \omega_k f(\xi_i, \eta_j, \zeta_k), \tag{5.3.24}$$

where ω_i, ω_j, ω_k are weighting coefficients and ξ_i, η_j, ζ_k are basis points of the Gaussian quadrature formula, and m is the number of basis points used in each direction.

Gupta et al. (1975) developed a FORTRAN code which uses the 3-D isoparametric FEM to solve groundwater quality problems.

5.3.2 *Triangular Prism Elements and Associated Basis Functions*

Sun (1979) used the triangular prism element in the modeling of three-dimensional groundwater flow. Any irregular aquifer can be covered by a cylinder with the circumference of the maximum cross section of the aquifer as its generator, with the highest and lowest levels of the aquifer as its upper and lower bases, respectively (see Figure 5.18).

Subdivide the cylinder vertically into several horizontal layers, and partition the layer with the largest cross-sectional area into triangular elements. Then, make perpendiculars along the edges of the triangle to subdivide the cylinder into several layers of triangular prisms. The triangular prisms, thus subdivided, are aligned from top to bottom (see Figure 5.19.) If the flanks of the aquifer are not vertical, some nodes will certainly be located outside of the aquifer. They can be identified as such by inputting appropriate information for these nodes.

Therefore, one only need to input the following geometric information to generate all of the elements:

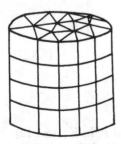

FIGURE 5.18. A special subdivision method.

1. Nodal numbers and coordinates of triangle elements at the largest cross section.
2. Number of layers and their intervals in the vertical direction.
3. Number of nodes that are located at the outside of the aquifer.

The amount of this input data is almost the same as that for two-dimensional models.

Let us consider an arbitrary triangular prism element shown in Figure 5.20. Its six vertices:

$$P_1(x_i, y_i, z_0), \quad P_2(x_j, y_j, z_0), \quad P_3(x_k, y_k, z_0);$$

$$P_4(x_i, y_i, z_1), \quad P_5(x_j, y_j, z_1), \quad P_6(x_k, y_k, z_1)$$

are defined as nodes. The associated basis functions of these nodes in the element are defined as:

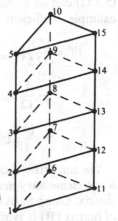

FIGURE 5.19. A string of triangular prism elements aligned from top to bottom.

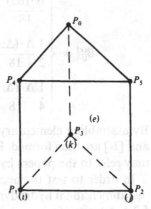

FIGURE 5.20. A triangular prism element (e) and its nodes.

$$\phi_1(x, y, z) = z^- \phi_i(x, y),$$
$$\phi_2(x, y, z) = z^- \phi_j(x, y),$$
$$\phi_3(x, y, z) = z^- \phi_k(x, y),$$
$$\phi_4(x, y, z) = z^+ \phi_i(x, y),$$
$$\phi_5(x, y, z) = z^+ \phi_j(x, y),$$
$$\phi_6(x, y, z) = z^+ \phi_k(x, y)$$

(5.3.25)

in which

$$z^- = \frac{z_1 - z}{\Delta z}, \quad z^+ = \frac{z - z_0}{\Delta z}, \quad \Delta z = z_1 - z_0,$$

and ϕ_i, ϕ_j, ϕ_k are the two-dimensional basis functions that have been given in Eq. (5.1.32).

With basis functions (5.3.25), all coefficients of finite element equations (5.3.11) can be directly calculated without using numerical integration. For example, coefficient A_{ii} in element (e) can be represented as:

$$A_{ii}^{(e)} = \frac{(\Delta z)}{12\Delta}[D_{xx}b_ib_i + D_{xy}b_ic_i] - \frac{1}{12}D_{xz}b_i + \frac{(\Delta z)}{12\Delta}[D_{yx}c_ib_i + D_{yy}c_ic_i]$$

$$- \frac{1}{12}D_{yz}c_i - \frac{1}{12}[D_{zx}b_i + D_{zy}c_i] + \frac{1}{6}D_{zz}\frac{\Delta}{(\Delta z)}$$

$$+ \frac{(\Delta z)}{18}[V_xb_i + V_yc_i] - \frac{1}{12}V_z \cdot \Delta. \tag{5.3.26}$$

We can write similar expressions for other A_{ij} $(i, j = 1, 2, \ldots, 6)$. Thus, a 6×6 elementary matrix $[\mathbf{A}]^{(e)}$ can be formed. Obviously, it is an asymmetric matrix. Using basis functions (5.3.25), we find that elementary matrix $[\mathbf{B}]^{(e)}$ of matrix $[\mathbf{B}]$ is symmetric with

$$B_{ij}^{(e)} = \begin{cases} \dfrac{\Delta \cdot (\Delta z)}{18}, & \text{when} \quad i = j, \\[3mm] \dfrac{1}{2}\dfrac{\Delta \cdot (\Delta z)}{18}, & \begin{array}{l}\text{when } i \text{ and } j \text{ have the same } (x, y) \text{ or } z \\ \text{coordinates,}\end{array} \\[3mm] \dfrac{1}{4}\dfrac{\Delta \cdot (\Delta z)}{18}, & \text{otherwise.} \end{cases} \tag{5.3.27}$$

By assembling elementary matrices of all elements, the global matrices $[\mathbf{A}]$ and $[\mathbf{B}]$ are then formed. For a detailed discussion of this method, the reader may refer to the papers by Sun et al. (1984).

In order to test the accuracy of the numerical solution, we again use the one-dimensional hydrodynamic dispersion problem introduced in paragraph 5.2.4. A series of triangular prismatic elements aligned in direction z has been

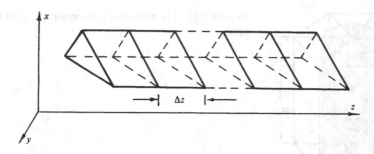

FIGURE 5.21. A series of triangular prismatic elements which represent a semi-infinite sand column.

TABLE 5.6. A comparison between numerical and analytical solutions.

Distance	0.0	5.0	10.0	15.0	30.0	45.0	50.0
Analytical solution	10.0000	9.8405	9.6037	9.2776	7.7202	5.4930	4.7038
Numerical solution	10.0000	9.8399	9.6023	9.2753	7.7162	5.4937	4.7075

used to represent a semi-infinite sand column as shown in Figure 5.21. A comparison between the numerical solution and analytical solution is shown in Table 5.6.

To obtain the solutions listed in Table 5.6, the following data are used:

$$C_0 = 10\,g/m^3, \quad V = 1\,m/d, \quad D = 10\,m^2/d, \quad \Delta z = 5\,m, \text{ and thus, } Pe = 0.5.$$

As shown in the table, when the Peclet number is relatively small, the accuracy of the numerical solution is very high. However, our numerical experiments also show that when the Peclet number is large, two kinds of numerical errors, numerical dispersion and overshoot, cannot be avoided.

5.3.3 A Mixed Finite Element–Finite Difference Method

A mixed finite element–finite difference (FE–FD) method presented by Sun (1979) uses two-dimensional mass lumped FEM in the horizontal direction and one-dimensional FDM in the vertical direction to obtain discretization equations for three-dimensional flow problems. Babu et al. (1984) developed a similar method and used it to solve three-dimensional groundwater quality problems. Instead of using the FEM, Sun (1981) used MCB in the horizontal direction. Next, we will describe the application of this method in the solution of three-dimensional advection-dispersion equations. This application is also discussed in Sun (1989).

The flow region is partitioned into triangular prism elements, as described in the previous section. All vertices of the elements are defined as nodes. The nodes are numbered in a two-dimensional array (i, m), where i represents the

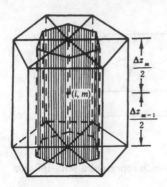

FIGURE 5.22. The exclusive subdomain $(R_{i,m})$ of node (i, m).

number of the node in the horizontal direction, and m the number in the vertical direction. The three-dimensional exclusive subdomain of a node (i, m) is defined as a column whose bottom is the two-dimensional exclusive subdomain of the node, as shown in Figure 5.10, and whose height is the distance Δz between two layers, as shown in Figure 5.22.

The integral form of the hydrodynamic dispersion equation, Eq. (2.4.31), can be written for the exclusive subdomain of a node as follows:

$$\iiint_{(R_{i,m})} \left(\frac{\partial C}{\partial t} + M \right) dR = \left\{ \iint_{(S_s)} + \iint_{(S_u)} + \iint_{(S_l)} \right\} (D\nabla C - \mathbf{V}C) \cdot \mathbf{n} \, ds,$$

(5.3.28)

where M is the source and sink term, and (S_s), (S_u), (S_l) represents the side, upper base, and lower base of the exclusive subdomain $(R_{i,m})$ of node (i, m), respectively.

The integral on the left-hand side of Eq. (5.3.28) can be calculated approximately by

$$\iiint_{(R_{i,m})} \left(\frac{\partial C}{\partial t} + M \right) dR \approx \left(\frac{dC_{i,m}}{dt} + M_{i,m} \right) P_i \Delta z,$$

(5.3.29)

where P_i is the base area of $(R_{i,m})$.

For the first surface integral on the right-hand side of Eq. (5.3.28), the following approximate expression can be used:

$$\iint_{(S_s)} [\cdot] \, dS \approx \left[\iint_{(L)} [\cdot] \, dt \right] \Delta z$$

$$= \left[\sum_{c_l} (A_{ii}^e C_{i,m} + A_{ij}^e C_{j,m} + A_{i,k}^e C_{k,m}) \right] \Delta z,$$

(5.3.30)

where (L) is the periphery of (S_s). Other symbols have the same meaning as those in Section 5.2.2, and A_{il}^e ($l = i, j, k$) are given by Eq. (5.2.22).

Assuming that direction z is a principal direction of the dispersion, i.e., $D_{xz} = D_{yz} = 0$, then

$$\iint_{(S_u)} [\cdot] dS \approx \left[D_{zz} \frac{\partial C}{\partial z} - V_z C \right]_{(i, m+(1/2))} P_i$$

$$= \left[D_{zz(i, m+(1/2))} \frac{C_{i, m+1} - C_{i, m}}{\Delta z} - V_{z(i, m+(1/2))} \frac{C_{i, m+1} + C_{i, m}}{2} \right] P_i$$

$$\tag{5.3.31}$$

in which

$$D_{zz(i, m+(1/2))} = \frac{1}{2}(D_{zz(i, m)} + D_{zz(i, m+1)}),$$

$$V_{z(i, m+(1/2))} = \frac{1}{2}(V_{z(i, m)} + D_{z(i, m+1)}).$$

Similarly, we have

$$\iint_{(S_l)} [\cdot] dS \approx -\left[D_{zz} \frac{\partial C}{\partial z} - V_z C \right]_{(i, m-(1/2))} P_i$$

$$= -\left[D_{zz(i, m-(1/2))} \frac{C_{i, m} - C_{i, m-1}}{\Delta z} - V_{z(i, m-(1/2))} \frac{C_{i, m} + C_{i, m-1}}{2} \right] P_i.$$

$$\tag{5.3.32}$$

Substitute Eqs. (5.3.29) to (5.3.32) into Eq. (5.3.28), and let

$$C_m = \{C_{1, m}, C_{2, m}, \dots, C_{N, m}\}^T,$$

$$\frac{dC_m}{dt} = \left\{ \frac{dC_{1, m}}{dt}, \frac{dC_{2, m}}{dt}, \dots, \frac{dC_{N, m}}{dt} \right\}^T,$$

where N is the number of nodes of layer m, the discretization equations for layer m is then obtained

$$[A_1]_m C_m + [A_2]_m C_{m+1} + [A_3]_m C_{m-1} + [B]_m \frac{dC_m}{dt} + F_m = 0, \quad (5.3.33)$$

where $[A_2]_m$, $[A_3]_m$ and $[B]_m$ are all diagonal matrices. We can use a layer-by-layer iteration method to obtain the concentration distribution of the whole region in each time step. Updated values of C_m can be obtained by solving Eq. (5.3.33), with C_{m-1} and C_{m+1} assigned their most recent values obtained in the iterative procedure.

Babu and Pinder (1984) adopted the following trial solution:

$$\hat{C}(x, y, z, t) = \sum_{i=1}^{N} C_i(z, t)\phi_i(x, y), \tag{5.3.34}$$

where $\{\phi_i(x, y)\}$ are two-dimensional basis functions. Using the Galerkin finite element for x-y directions and the finite difference for the z direction, they derived a system of equations similar to Eq. (5.3.33), but with non-diagonal matrices $[A_2]_m$ and $[A_3]_m$.

In Huyakorn et al. (1986), trial solution (5.3.34) is used for x-y directions, but for the z direction, they used the following:

$$\hat{C}_i(z, t) = \sum_{m=1}^{M} C_{i,m}(t)\phi_m(z), \tag{5.3.35}$$

where $\phi_m(z)$ is the one-dimensional basis function. This type of FEM is especially suitable for layered aquifers.

5.3.4 The Multiple Cell Balance Method

Sun et al. (1984) extended the MCB method from the two-dimensional case to the three-dimensional case. The 3-D MCB method also uses the triangular prism element and basis functions which are defined in Eq. (5.3.25) and introduced in Section 5.3.2. In an arbitrary element (e) shown in Figure 5.20, the unknown concentration $C(x, y, z, t)$ may be approximately expressed as

$$C(x, y, z, t) = \sum_{r=1}^{6} C_r^e(t)\phi_r(x, y, z), \quad (x, y, z) \in (e), \tag{5.3.36}$$

where $C_r^e(t)$ $(r = 1, 2, \ldots, 6)$ are concentrations at the six nodes of element (e). Approximate expressions for partial derivatives $\partial C/\partial x$, $\partial C/\partial y$, $\partial C/\partial z$ and $\partial C/\partial t$ in this element can be obtained through partial differentiation, as follows

$$\frac{\partial C}{\partial x} = \sum_{r=1}^{6} C_r^e(t)\frac{\partial \phi_r}{\partial x},$$

$$\frac{\partial C}{\partial y} = \sum_{r=1}^{6} C_r^e(t)\frac{\partial \phi_r}{\partial y},$$

$$\frac{\partial C}{\partial z} = \sum_{r=1}^{6} C_r^e(t)\frac{\partial \phi_r}{\partial z}, \tag{5.3.37}$$

$$\frac{\partial C}{\partial t} = \sum_{r=1}^{6} \frac{dC_r^e}{dt}\phi_r.$$

Using the basis functions defined in Eq. (5.3.25), we have

$$\frac{\partial \phi_1}{\partial x} = \frac{1}{2\Delta}z^- b_i, \quad \frac{\partial \phi_2}{\partial x} = \frac{1}{2\Delta}z^- b_j, \quad \frac{\partial \phi_3}{\partial x} = \frac{1}{2\Delta}z^- b_k,$$

$$\frac{\partial \phi_4}{\partial x} = \frac{1}{2\Delta}z^+ b_i, \quad \frac{\partial \phi_5}{\partial x} = \frac{1}{2\Delta}z^+ b_j, \quad \frac{\partial \phi_6}{\partial x} = \frac{1}{2\Delta}z^+ b_k.$$

Similar expressions can be obtained for $\partial \phi_r/\partial y$ $(r = 1, 2, \ldots, 6)$ by changing b_i, b_j, b_k to c_i, c_j, c_k.

For $\partial\phi_r/\partial z \; (r = 1, 2, \ldots, 6)$, we have

$$\frac{\partial\phi_1}{\partial z} = -\frac{1}{(\Delta z)}\phi_i, \quad \frac{\partial\phi_2}{\partial z} = -\frac{1}{(\Delta z)}\phi_j, \quad \frac{\partial\phi_3}{\partial z} = -\frac{1}{(\Delta z)}\phi_k,$$

$$\frac{\partial\phi_4}{\partial z} = \frac{1}{(\Delta z)}\phi_i, \quad \frac{\partial\phi_5}{\partial z} = \frac{1}{(\Delta z)}\phi_j, \quad \frac{\partial\phi_6}{\partial z} = \frac{1}{(\Delta z)}\phi_k,$$

where ϕ_i, ϕ_j, ϕ_k are two-dimensional linear basis functions. Substituting $\partial\phi_r/\partial x$, $\partial\phi_r/\partial y$, $\partial\phi_r/\partial z$ and ϕ_r into Eq. (5.3.37), we then have the approximate expressions for $\partial C/\partial x$, $\partial C/\partial y$, $\partial C/\partial z$, $\partial C/\partial t$ in element (e).

Now, let us go back to the integral equation (5.3.20), and rewrite it in the following form:

$$\iiint_{(R)} \left(\frac{\partial C}{\partial t} + M\right) dR = \iint_{(S)} D_{\alpha\beta}\frac{\partial C}{\partial x_\beta}n_\alpha dS - \iiint_{(R)} V_\alpha\frac{\partial C}{\partial x_\alpha} dR. \quad (5.3.38)$$

Einstein's summation convention has been used in the above equation, where (R) is the exclusive subdomain of the considered node and (S) is its boundary surface, see Figure 5.22. Note that all integrands in the equation only include first-order partial derivatives of the unknown concentration. When the approximate expressions of those partial derivatives given in Eq. (5.3.37) are substituted into Eq. (5.3.38) and integrated, we can obtain a discretized equation which connects the unknown concentrations of the node and its surrounding nodes. A system of discrete equations can then be formed by considering all nodes where the concentration is unknown.

The details of this method will be shown by considering the node P in Figure 5.23. The part of the exclusive subdomain of node P in element (e) is located at the upper-left corner of the element. The surface integral on the right-hand side of Eq. (5.3.38) in element (e) will be just the integrals over surfaces (S_1), (S_2) and (S_3) as shown in Figure 5.23. With Eq. (5.3.37), these

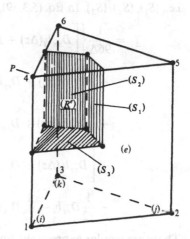

FIGURE 5.23. The part (R^e) of the exclusive subdomain of node P in element (e) under P (node P is just node 4 in the element).

integrals can be easily calculated. For surface (S_1), we have

$$\iint_{(S_1)} \frac{\partial C}{\partial x} \, dy \, dz = \frac{(\Delta z)}{96\Delta} [b_i \underline{C}_1^e + b_j \underline{C}_2^e + b_k \underline{C}_3^e) + 3(b_i \underline{C}_4^e + b_j \underline{C}_5^e + b_k \underline{C}_6^e)]$$
$$\cdot (b_j - b_i),$$

$$\iint_{(S_1)} \frac{\partial C}{\partial y} \, dy \, dz = \frac{(\Delta z)}{96\Delta} [c_i \underline{C}_1^e + c_j \underline{C}_2^e + c_k \underline{C}_3^e) + 3(c_i \underline{C}_4^e + c_j \underline{C}_5^e + c_k \underline{C}_6^e)]$$
$$\cdot (c_j - c_i),$$

$$\iint_{(S_1)} \frac{\partial C}{\partial z} \, dy \, dz = \frac{1}{2} [-5\underline{C}_1^e - 5\underline{C}_2^e - 2\underline{C}_3^e + 5\underline{C}_4^e + 5\underline{C}_5^e + 2\underline{C}_6^e]$$
$$\cdot (b_i c_j - c_i b_j)(c_j - c_i),$$

where, $\underline{C}_r^e$ $(r = 1, 2, \ldots, 6)$ represents the concentrations of six nodes in element $(\underline{e})$, and $b_i, b_j, b_k, c_i, c_j, c_k$ are determined by Eq. (5.1.33). Similar results can be obtained for (S_2). For the surface integral over (S_3), we have

$$\iint_{(S_3)} \frac{\partial C}{\partial x} \, dx \, dy = \frac{1}{12} (b_i \underline{C}_1^e + b_j \underline{C}_2^e + b_k \underline{C}_3^e + b_i \underline{C}_4^e + b_j \underline{C}_5^e + b_k \underline{C}_6^e),$$

$$\iint_{(S_3)} \frac{\partial C}{\partial y} \, dx \, dy = \frac{1}{12} (c_i \underline{C}_1^e + c_j \underline{C}_2^e + c_k \underline{C}_3^e + c_i \underline{C}_4^e + c_j \underline{C}_5^e + c_k \underline{C}_6^e),$$

$$\iint_{(S_3)} \frac{\partial C}{\partial z} \, dx \, dy = \frac{\Delta}{108(\Delta z)} (-22\underline{C}_1^e - 7\underline{C}_2^e - 7\underline{C}_3^e + 22\underline{C}_4^e + 7\underline{C}_5^e + 7\underline{C}_6^e).$$

Adding the above surface integrals together, we obtain

$$\iint_{(\underline{S}^e)} D_{\alpha\beta} \frac{\partial C}{\partial x_\beta} n_\alpha \, dS = \sum_{r=1}^{6} \underline{T}_r \underline{C}_r^e, \tag{5.3.39}$$

where $(\underline{S}^e)$ is the part of the boundary surface of exclusive subdomain in $(\underline{e})$, i.e., (S_1), (S_2), (S_3). In Eq. (5.3.39),

$$\underline{T}_1 = \frac{1}{96\Delta} \left\{ \left[D_{xx} b_i (\Delta z) + D_{xy} c_i (\Delta z) - \frac{5}{3} D_{xz} (b_i c_j - c_i b_j) \right] (b_j - b_i) \right.$$

$$+ \left[D_{xy} b_i (\Delta z) + D_{yy} c_i (\Delta z) - \frac{5}{3} D_{yz} (b_i c_j - c_i b_j) \right] (c_j - c_i)$$

$$+ \left[D_{xx} b_i (\Delta z) + D_{xy} c_i (\Delta z) - \frac{5}{3} D_{xz} (b_i c_k - c_i b_k) \right] (b_k - b_i)$$

$$\left. + \left[D_{xy} b_i (\Delta z) + D_{yy} c_i (\Delta z) - \frac{5}{3} D_{yz} (b_i c_k - c_i b_k) \right] (c_k - c_i) \right\}$$

$$- \frac{1}{12} D_{xz} b_i - \frac{1}{12} D_{yz} c_i + \frac{11}{54} \frac{\Delta}{(\Delta z)} D_{zz}.$$

There are similar expressions for other $\underline{T}_r$'s.

Using Eq. (5.3.37), the value of the second integral on the right-hand side of Eq. (5.3.38) can be calculated for element $(\underline{e})$. After some elementary integration, we have

$$\iiint_{(\underline{R}^e)} V_\alpha \frac{\partial C}{\partial x_\alpha} dR = \sum_{r=1}^{6} \underline{Q}_r \underline{C}_r^e, \qquad (5.3.40)$$

where $(\underline{R}^e)$ is the part of the exclusive subdomain of node P in $(\underline{e})$. In the above equation

$$\underline{Q}_1 = \frac{(\Delta z)}{48}(V_x b_i + V_y c_i) - \frac{11}{108}\Delta \cdot V_z.$$

For other $\underline{Q}_r$ $(r = 2, 3, \ldots, 6)$, we have similar expressions.

Using the approximate expression of $\partial C / \partial t$ in Eq. (5.3.37), the value of the integral on the left-hand side of Eq. (5.3.38) can be obtained for element $(\underline{e})$. It also has the form of a summation of weighted nodal concentrations, that is,

$$\iiint_{(\underline{R}^e)} \frac{\partial C}{\partial t} dR = \sum_{r=1}^{6} \underline{B}_r \frac{d}{dt}(\underline{C}_r^e), \qquad (5.3.41)$$

where $\underline{B}_1 = \frac{(\Delta z)}{12} \cdot \frac{11}{36}\Delta$, $\underline{B}_2 = \frac{(\Delta z)}{12} \cdot \frac{7}{36}\Delta$, and so on.

Now, let us consider the calculation of each term of Eq. (5.3.38) for the upper part of the exclusive subdomain of a node. Assume that $(\bar{e})$ is an element with node P as its vertex and located above P. The part of the exclusive subdomain of node P in this element is marked as $(\bar{R}^e)$ and the relevant boundary surface is $(\bar{S}^e)$ as in Figure 5.24. Node P in element $(\bar{e})$ corresponds to node 1. By the same method as the one mentioned above, it can be calculated as

$$\iint_{(\bar{S}^e)} D_{\alpha\beta} \frac{\partial C}{\partial x_\beta} n_\alpha dS = \sum_{r=1}^{6} \bar{T}_r \bar{C}_r, \qquad (5.3.42)$$

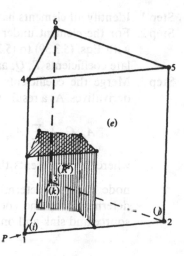

FIGURE 5.24. The part $(\bar{R}^e)$ of the exclusive subdomain of node P in element $(\bar{e})$ above P (node P is just node 1 in this element).

where $\bar{C}_r^e$ $(r = 1, 2, \ldots, 6)$ are the concentrations of six nodes in element $(\bar{e})$, and

$$
\begin{aligned}
\bar{T}_1 = \frac{1}{96\Delta}\Big\{ &\Big[3D_{xx}(\Delta z)b_i + 3D_{xy}(\Delta z)c_i - \frac{5}{3}D_{xz}(b_i c_j - c_i b_j) \Big](b_j - b_i) \\
&+ \Big[3D_{xy}(\Delta z)b_i + 3D_{yy}(\Delta z)c_i - \frac{5}{3}D_{yz}(b_i c_j - c_i b_j) \Big](c_j - c_i) \\
&+ \Big[3D_{xx}(\Delta z)b_i + 3D_{xy}(\Delta z)c_i - \frac{5}{3}D_{xz}(b_i c_k - c_i b_k) \Big](b_k - b_i) \\
&+ \Big[3D_{xy}(\Delta z)b_i + 3D_{yy}(\Delta z)c_i - \frac{5}{3}D_{yz}(b_i c_k - c_i b_k) \Big](c_k - c_i) \\
&+ \frac{1}{12}D_{xz}b_i + \frac{1}{12}D_{yz}c_i - \frac{11}{54}\frac{\Delta}{(\Delta z)}D_{zz}.
\end{aligned}
$$

The other expressions for other $\bar{T}_r$'s can also be obtained. Similarly, we can calculate the following integrals:

$$
\iiint_{(\bar{R}^e)} V_\alpha \frac{\partial C}{\partial x_\alpha} dR = \sum_{r=1}^{6} \bar{Q}_r \bar{C}_r^e, \tag{5.3.43}
$$

and

$$
\iiint_{(\bar{R}^e)} \frac{\partial C}{\partial t} dR = \sum_{r=1}^{6} \bar{B}_r \frac{d}{dt}(\bar{C}_r^e), \tag{5.3.44}
$$

in which $\bar{Q}_1 = \frac{(\Delta z)}{16}(V_x b_i + V_y C_i) - \frac{5}{36}\Delta \cdot V_z$, $\bar{B}_1 = \frac{(\Delta z)}{4} \cdot \frac{11}{36}\Delta$, $\bar{B}_2 = \frac{(\Delta z)}{4} \cdot \frac{7}{36}\Delta$, and so on.

With the above results, we may transfer Eq. (5.3.38) into a discrete equation connecting the unknown concentrations of node P and its neighboring nodes according to the following steps:

Step 1. Identify all elements having node P as one of their vertices;

Step 2. For the element under point P, calculate coefficients $\underline{T}_r$, $\underline{Q}_r$ and $\underline{B}_r$ with Eqs. (5.3.39) to (5.3.41); for the element located above P, calculate coefficients $\bar{T}_r$, $\bar{Q}_r$ and $\bar{B}_r$ with Eqs. (5.3.42) to (5.3.44);

Step 3. Merge the coefficients of unknown concentrations and their time derivatives. As a result, we obtain the following equation for node P:

$$
A_{pp}C_p + \sum_{q \neq p} A_{pq}C_q + B_{pp}\frac{dC_p}{dt} + \sum_{q \neq p} B_{pq}\frac{dC_q}{dt} + F_p = 0, \tag{5.3.45}
$$

where $\sum\limits_{q \neq p}$ represents the summation over all neighboring nodes of node P, A_{pq} is determined by coefficients $\underline{T}_r$, $\bar{T}_r$, $\underline{Q}_r$, and $\bar{Q}_r$. B_{pq} is determined by the coefficients $\underline{B}_r$ and $\bar{B}_r$, and F_p depends on the source and sink and on the boundary conditions.

When Eq. (5.3.45) is listed for all nodes where the concentration is un-known, a system of discrete equations are obtained.

The advantages of the three-dimensional MCBM are that the coefficients of the discrete equations are given explicitly, and we do not need to assume that direction z is a principal direction of dispersion.

Some numerical examples of three-dimensional MCBM can be found in Wang et al. (1986).

5.4 The Solution of Finite Element Systems

5.4.1 Features of Finite Element Systems and Direct Solutions

We have seen that, no matter what kind of FEM is used in the discretization of space variables, the advection-dispersion equation will be tranferred into a system of ordinary differential equations:

$$[A]C + [B]\frac{dC}{dt} + F = 0. \tag{5.4.1}$$

After the discretization of the time variable, the system of ODEs will become a system of algebraic equations. Finite element discretization can also be used for the time variable as well as for space variables. However, the results of practical calculations show that using the finite difference approximation of time derivatives is simpler than using the finite element discretization, and the accuracy of the two approaches is almost same. The finite difference approximation for dC/dt appearing in Eq. (5.4.1) is

$$\frac{dC}{dt} = \frac{C_{t+\Delta t} - C_t}{\Delta t}. \tag{5.4.2}$$

The concentration, C, in the first term on the left-hand side of Eq. (5.4.1), can be taken as

$$C = \theta C_{t+\Delta t} + (1 - \theta)C_t, \tag{5.4.3}$$

where θ is a weighting coefficient, $0 \le \theta \le 1$.

By substituting Eqs. (5.4.2) and (5.4.3) into Eq. (5.4.1), and rearranging, we have

$$[T]C_{t+\Delta t} = R, \tag{5.4.4}$$

where

$$[T] = \theta[A] + \frac{[B]}{\Delta t}, \tag{5.4.5}$$

$$R = \left\{ \frac{[B]}{\Delta t} - (1 - \theta)[A] \right\} C_t - F. \tag{5.4.6}$$

Assume that the concentration distribution C_t at time t has been obtained (at $t = 0$ using the initial condition), then, $\mathbf{C}_{t+\Delta t}$ at the next time $(t + \Delta t)$ can be solved by Eq. (5.4.4). Thus, a step-by-step process can be formed. Using the FDM terminology, when $\theta = 0$, the relevant discrete scheme is an explicit one; when $\theta = 1/2$, it is a Crank-Nicolson scheme; and when $\theta = 1$, it is an implicit one.

We know that the coefficient matrix of the finite element system derived from flow equations is sparse, symmetric and positive definite. This type of system of equations can be solved by a special LU decomposition method. In contrast, owing to the existence of the advection term, the coefficient matrix $[\mathbf{T}]$ of Eq. (5.4.4) is definitely asymmetric. Both the computer memory for this matrix and the solution of this system of equations require special treatments. These two aspects are not separable: the arrangement of the memory must be adaptive to the solution method.

A lot of studies have been contributed to direct solutions for large, sparse, and asymmetric equation systems. We know that if the general method of principal element elimination is used, the sparse coefficient matrix will be "filled in." That is to say, non-zero elements will appear in the original positions of the zero elements during the elimination. If we wish to reduce the "filling in," there will be no choice but to write more complex programs.

Gupta and Tanji (1977) put forward both a method and relevant programs, which are applicable to the solution of the finite element systems of water quality. It is known from our experience that direct solutions will be of lower efficiency than iterative solutions when the number of nodes exceeds 500.

5.4.2 Point Iteration Methods

The point iteration method is the simplest among the iteration methods, because it does not depend on whether the coefficient matrix is symmetric or not. Therefore, it can be used for both groundwater flow problems and mass transport problems. Another advantage of this method is that it is relatively easy to program.

In order to store the coefficient matrix $[\mathbf{T}]$, we can create a two-dimensional array (MCPN) to store the numbers of the neighboring nodes of each node. In addition, we can use another array (MA) to store the total number of the neighboring nodes of each node. Coefficients A_{ij} and B_{ij} relating to each node (i) with its surrounding nodes (j) can then be stored into one-dimensional arrays (CFA) and (CFB) by the control of (MA) and (MCPN). Using this method to store coefficient matrices $[\mathbf{A}]$ and $[\mathbf{B}]$ of Eq. (5.4.1), the storage of all zero elements can be eliminated. The coefficient matrix $[T]$ can also be stored in a one-dimensional array with no zero elements. It is formed by multiplying the elements of (CFA) by θ, dividing the elements of (CFB) by Δt, and summing these products element by element.

The process of point iteration includes the following main steps:

1. Determine the order for point iteration, which is generally in nodal order, but can also be carried out in any preset order.
2. To obtain the concentration values of the $(k + 1)$th iteration from the concentration values of kth iteration for all nodes, $C^{(k)}$, we first calculate the following intermediate value for each node i:

$$C_i^* = \left(R_i - \sum_{M_i} T_{ij} C_j^{(k)} - \sum_{K_i} T_{ij} C_j^{(k+1)} \right) \bigg/ T_{ii}, \qquad (5.4.7)$$

where $\{M_i\}$ are those neighboring nodes of node i for which the $(k + 1)$th iteration has not been completed, while $\{K_i\}$ are the neighboring nodes of node i for which the $(k + 1)$th iteration has been completed. The following equation is then used to determine the $(k + 1)$th iteration value for the concentration at node i:

$$C_i^{(k+1)} = C_i^{(k)} + \omega(C_i^* - C_i^{(k)}) = \omega C_i^* + (1 - \omega)C_i^{(k)},$$

$$(i = 1, 2, \ldots, N) \qquad (5.4.8)$$

where ω is the relaxation factor. When $0 < \omega < 1$, the method is called the *successive under relaxation method* (SUR); when $\omega = 1$, it is the *Gauss-Seidel iteration method* (G-S); and when $1 < \omega < 2$, it is the *successive over relaxation method* (SOR).
3. Set up a criterion for convergence (EPS). When the concentration difference between two successive iterations for all nodes is smaller than the value of EPS for every node, the iteration is considered to have converged. The result of iteration is the concentration distribution at the end of this time step.
4. Go to the next time step and repeat the above steps until the total simulation time is reached.

Sun (1981) put forward a revised form for the point iteration method called the *selected-node iteration method*. This kind of method is well suited for solving water quality problems and may save significant computation effort. Its main steps are:

1. Let every node be associated with a characteristic number, $E(i)$. It is defined such that if node i needs to iterate continuously, then $E(i) = 0$; if not, $E(i) = 1$. At the beginning of each step, value 1 is assigned to all boundary nodes where the concentration is given and 0 to other nodes.
2. Use a method of extrapolation to predict the unknown concentration $C_i^{(0)}$ of each node i when its characteristic number is zero, and store these predicted values in an array (C_0).

3. For each node whose characteristic number is zero, use the following formula to calculate its modified value $C_i^{(1)}$

$$C_i^{(1)} = (1 - \omega)C_i^{(0)} + \omega\left(R_i - \sum_{M_i} T_{ij}C_j^{(0)} - \sum_{K_i} T_{ij}C_j^{(1)}\right)\bigg/ T_{ii},$$

$$(i \in VE) \tag{5.4.9}$$

where VE is a set constituted by the nodes whose characteristic numbers are zero. If the absolute value of the difference between $C_i^{(1)}$ and $C_i^{(0)}$ is already smaller than a given error criterion, then the characteristic number of this node is changed to 1, otherwise it remains zero. For all cases, $C_i^{(0)}$ is replaced by $C_i^{(1)}$.

4. If there are still nodes with zero characteristic numbers, then change the characteristic numbers of all their surrounding nodes to zero and return to step 3. This process is repeated until the characteristic numbers of all nodes become 1, whereupon the calculation for this time step is completed.

It is unnecessary to use the iteration formula, Eq. (5.4.9), for all nodes and all time steps. Calculations are only performed for the nodes whose characteristic numbers are equal to zero. After one or two iterations, there may be only a few nodes left where the iteration has not converged. These nodes are generally in the vicinity of the concentration front. Therefore, the selected-node iteration method can automatically track the migration of the front, and thus, significantly reduce the computation effort required. The modified SOR method is adopted in the program given in Appendix B.

Yeh (1985) noticed that when the Peclet number is large, the successive over relaxation method (SOR) may not be convergent. However, the successive under relaxation (SUR) and the Gauss-Seidel methods always give convergent solutions.

5.4.3 Block and Layer Iteration Methods

In many problems, especially the three-dimensional problems, the use of the element iteration method is advantageous. The so-called *successive element iteration method* is designed to update all the nodal values of concentration for each element by solving lower-order equation systems. For example, for each triangular prism element, a sixth-order equation system must be solved. As a result, the concentration values at its six vertices are modified at the same time. If there are m nodes in an element (L), then, we can draw the equations relevant to these m nodes from equation system (5.4.4). By moving the terms which connect with the nodes of other elements on the left-hand side of these m equations to the right-hand side, and using the updated values in the iteration process to replace the unknown values of concentration, we then arrive at a system of m equations as shown below:

$$[\mathbf{T}]\mathbf{C}_L^{(k+1)} = \mathbf{R}_L - [\mathbf{T}_L']\mathbf{C}_L', \tag{5.4.10}$$

where the component of C_L is the nodal concentration of element (L), C'_L the concentrations of the nodes which are adjacent to these nodes but belong to other elements. The coefficient matrix $[T_L]$ in Eq. (5.4.10) is a low-order matrix, so we can adopt the general Gauss elimination to obtain the solution directly.

When using the element iteration method, the submatrices of the coefficient matrix $[T]$ in Eq. (5.4.4) will be stored according to the order of elements. Most of the zero elements of matrix $[T]$ do not need to store. Since this method can modify values of more nodes at the same time, its convergence rate is usually faster than that of the point iteration methods.

In order to speed up the convergence of element iteration, a relaxation factor may be added. Take the solution of Eq. (5.4.10) as the intermediate value, C_L^*, and define the result of the $(k + 1)$th iteration as

$$C_L^{(k+1)} = \omega C_L^* + (1 - \omega)C_L^{(k)}. \tag{5.4.11}$$

When $1 < \omega < 2$, it is the *over relaxation successive element iteration method.*

Element iteration methods are applicable to various kinds of three-dimensional problems, including FEM with isoparametric three-dimensional elements and the mixed method of FEM and FDM.

If we use the partition method suggested in Section 5.3.2, i.e., partitioning the three-dimensional region into many layers of triangular prism elements which are aligned from top to bottom, then a layer-by-layer iteration method can be used. No matter what method is used, either the Galerkin FEM, or the mixed method of FEM and FDM, we only need to collect the equations associated with the mth layer to obtain the following equation:

$$[T_1]_m C_m = R_m - [T_2]_m C_{m+1} - [T_3]_m C_{m-1}, \tag{5.4.12}$$

where C_m is the nodal concentration of the mth layer, and C_{m-1} and C_{m+1} the nodal concentrations of layer $m - 1$ and layer $m + 1$, respectively. If we use the known values of C_{m-1} and C_{m+1} in the iteration, then the modified values of C_m can be solved from Eq. (5.4.12). The coefficient matrix $[T]$ can now be expressed as the following tridiagonal block matrix:

$$[T] = \begin{bmatrix} [T_1]_1 & [T_2]_1 & & & & \\ [T_3]_2 & [T_1]_2 & [T_2]_2 & & [0] & \\ & [T_3]_3 & [T_1]_3 & [T_2]_3 & & \\ \cdots & \cdots & \cdots & \cdots & \cdots & \\ \cdots & \cdots & \cdots & \cdots & \cdots & \\ & [0] & & & [T_2]_{M-1} \\ & & & & [T_3]_M & [T_1]_M \end{bmatrix} \tag{5.4.13}$$

where M is the total number of layers.

Storage space can be significantly saved by storing only the block matrices instead of the entire $[T]$ matrix. The method for solving Eq. (5.4.12) can be used to for each layer to eliminate the zero elements in each block.

Layer iteration can be done in both directions, i.e., from bottom to top and vice versa. First, rewrite Eq. (5.4.12) in the ascending layer order:

$$[\mathbf{T}_1]_m \mathbf{C}_m^{(k+1)} = \mathbf{R}_m - [\mathbf{T}_2]_m \mathbf{C}_{m+1}^{(k)} - [\mathbf{T}_3]_m \mathbf{C}_{m-1}^{(k+1)}, \qquad (5.4.14)$$

and get the solution for each layer ($m = 1, 2, \ldots, M$). Then, rewrite Eq. (5.4.12) as

$$[\mathbf{T}_1]_m \mathbf{C}_m^{(k+2)} = \mathbf{R}_m - [\mathbf{T}_2]_m \mathbf{C}_{(m+1)}^{(k+2)} - [\mathbf{T}_3]_m \mathbf{C}_{m-1}^{(k+1)}, \qquad (5.4.15)$$

and find the solution of each layer successively in the descending layer order ($m = M, M - 1, \ldots, 1$). We must pay attention to the application of boundary conditions in setting up each system of Eqs. (5.4.14) and (5.4.15).

To find the solution for the current time step, the above procedure are repeated until a convergence criterion is satisfied. The next time step can then be calculated in the same way.

Exercises

5.1. Derive the weighted residual method for the following equation:

$$\frac{\partial(\theta C)}{\partial t} = div(\theta \mathbf{D} \, grad \, C) - div(\theta C V) - \lambda \theta C.$$

5.2. What are the advantages of selecting basis functions which satisfy the requirements presented in Section 5.1.2? Are the basis functions uniquely determined by these requirements? Prove that any basis functions satisfying these requirements must be independent of each other.

5.3. Give the details of deriving the finite element equations (5.1.17) when triangle elements and linear basis functions in Eq. (5.1.32) are used.

5.4. Prove that the basis functions given in Table 5.1 for the quadratic element satisfy all requirements presented in Section 5.1.2.

5.5. Suppose that an arbitrary quadrilateral is used as an element, and its four vertices are defined as nodes. What are the basis functions associated with the nodes in this element? Give algorithms for calculating the coefficient matrices $\mathbf{A}$ and $\mathbf{B}$ of the finite element equations.

5.6. Derive the Multiple Cell Balance Method for the following equation

$$\frac{\partial C}{\partial t} = div(\mathbf{D} \, grad \, C) - div(CV) - \lambda C,$$

subject to the first and second types of boundary conditions.

5.7. Using the programs given in Appendix B, reproduce the results presented in Section 5.2.4.

5.8. Design a two-dimensional problem that may be countered in the field and obtain the solution using the program given in Appendix B.

6
Numerical Solutions of Advection-Dominated Problems

6.1 Advection Dominated Problems

6.1.1 Fourier Analysis of Numerical Errors

When the advection term dominates in the advection-dispersion equation, most traditional numerical methods will encounter difficulties. In Chapter 4, we have stated that when the FDM is used for solving advection-dominated problems, two kinds of errors, numerical dispersion and overshoot, will occur. As a result, oscillations will appear around concentration fronts and steep concentration fronts cannot be accurately calculated. In Chapter 5, we pointed out that the Galerkin FEM cannot avoid the above two kinds of errors, either. Consequently, improvement of the accuracy and stability of numerical solutions has become an important subject in current research.

In order to further analyze the origins of numerical errors discussed in Chapter 4, let us again consider the simplest one-dimensional advection-dispersion equation:

$$\frac{\partial C}{\partial t} + V\frac{\partial C}{\partial x} - D\frac{\partial^2 C}{\partial x^2} = 0, \qquad (6.1.1)$$

where the dispersion coefficient D and average velocity V are assumed to be constant. Using Fourier's analysis method, the general solution of Eq. (6.1.1) may be expressed by the following *Fourier's series*:

$$C(x, t) = \sum_{n=-\infty}^{\infty} C_n \exp(i\beta_n t + i\sigma_n x) \qquad (6.1.2)$$

where β_n is the time frequency of the nth component; σ_n the wave number or spatial frequency; $i = \sqrt{-1}$ is the imaginary unit. Since Eq. (6.1.1) is linear, each component of the series in Eq. (6.1.2) may be individually considered according to the principle of superposition. For instance, consider the nth component, and let

$$C \sim C_n \exp(i\beta_n t + i\sigma_n x). \qquad (6.1.3)$$

Substituting Eq. (6.1.3) into Eq. (6.1.1), we have

$$\beta_n + V\sigma_n - iD\sigma_n^2 = 0$$

or

$$\beta_n = \sigma_n(iD\sigma_n - V). \tag{6.1.4}$$

Thus, the relationship between frequency β_n, and wave number σ_n is established. Substituting Eq. (6.1.4) into Eq. (6.1.3), we obtain

$$C \sim C_n \exp[i\sigma_n(x - Vt)] \exp(-D\sigma_n^2 t),$$

where the first exponent on the right-hand side represents the displacement of the nth wave and the second exponent represents the variation of wave amplitude. After a time increment Δt, the wave moves a distance $V\Delta t$, while its amplitude changes with the factor $\exp(-D\sigma_n^2 \Delta t)$. Since

$$C_{t+\Delta t} \sim C_n \exp[i\beta_n(t + \Delta t)] \exp(i\sigma_n x)$$

$$= C_n \exp(i\beta_n t + i\sigma_n x) \exp(i\beta_n \Delta t),$$

we have

$$C_{t+\Delta t} = C_t \exp(i\beta_n \Delta t). \tag{6.1.5}$$

In Eq. (6.1.5), $\exp(i\beta_n \Delta t)$ is called the *analytic enlargement factor* and written as λ_n. From Eq. (6.1.4), we know

$$\lambda_n = \exp(i\beta_n \Delta t) = \exp(-D\sigma_n^2 \Delta t) \exp(-i\sigma_n V\Delta t),$$

thus the magnitude of the analytic enlargement factor is

$$|\lambda_n| = \exp(-D\sigma_n^2 \Delta t), \tag{6.1.6}$$

and its phase angle is

$$\theta_n = \sigma_n V\Delta t. \tag{6.1.7}$$

Similarly, we can define the *numerical enlargement factor* of the nth wave, λ_n', which shows the ratio between the numerical solutions of time $t + \Delta t$ and t. Once we know the discrete scheme of a numerical solution, it is easy to find the expression for λ_n'. For example, the implicit scheme of finite difference of Eq. (6.1.1) is (see Eq. (4.1.14)):

$$-\left[\frac{D\Delta t}{(\Delta x)^2} + \frac{V\Delta t}{2\Delta x}\right]C_{i-1,k+1} + \left[1 + \frac{2D\Delta t}{(\Delta x)^2}\right]C_{i,k+1} - \left[\frac{D\Delta t}{(\Delta x)^2} - \frac{V\Delta t}{2\Delta x}\right]C_{i+1,k+1}$$

$$= C_{i,k} \tag{6.1.8}$$

Using the dimensionless parameters

$$\tilde{D} = \frac{D\Delta t}{(\Delta x)^2}, \quad \tilde{V} = \frac{V\Delta t}{\Delta x},$$

we can rewrite Eq. (6.1.8) as

$$-\left(\tilde{D} + \frac{\tilde{V}}{2}\right)C_{i-1,k+1} + (1 + 2\tilde{D})C_{i,k+1} - \left(\tilde{D} - \frac{\tilde{V}}{2}\right)C_{i+1,k+1} = C_{i,k}. \quad (6.1.9)$$

Assuming that

$$C_{i,k} = C_n \exp(i\beta_n t + i\sigma_n x),$$

we then have

$$C_{i,k+1} = \lambda'_n C_{i,k} = \lambda'_n C_n \exp(i\beta_n t + i\sigma_n x),$$
$$C_{i-1,k+1} = \lambda'_n C_{i-1,k} = \lambda'_n C_n \exp[i\beta_n t + i\sigma_n(x - \Delta x)],$$
$$C_{i+1,k+1} = \lambda'_n C_{i+1,k} = \lambda'_n C_n \exp[i\beta_n t + i\sigma_n(x + \Delta x)].$$

Substituting these equations into (6.1.9), we obtain:

$$-\left(\tilde{D} + \frac{\tilde{V}}{2}\right)\lambda'_n \exp(-i\sigma_n \Delta x) + (1 + 2\tilde{D})\lambda'_n - \left(\tilde{D} - \frac{\tilde{V}}{2}\right)\lambda'_n \exp(i\sigma_n \Delta x) = 1.$$

Hence, the solution for λ'_n is

$$\lambda'_n = \frac{(1 + \tilde{D} - 2\tilde{D}\cos\sigma_n\Delta x) - i\tilde{V}\sin\sigma_n\Delta x}{(1 + 2\tilde{D} - 2\tilde{D}\cos\sigma_n\Delta x)^2 + \tilde{V}^2 \sin^2\sigma_n\Delta x}. \quad (6.1.10)$$

From this equation, it is not difficult to obtain the magnitude $|\lambda'_n|$ and phase angle θ'_n of the numerical enlargement factor λ'_n.

Since the length of the nth wave is $L_n = 2\pi/\sigma_n$, the number of time steps needed by the numerical solution to propagate the whole wavelength is:

$$N_n = \frac{L_n}{V\Delta t} = \frac{L_n}{\tilde{V}\Delta x} = \frac{l_n}{\tilde{V}},$$

where $l_n = L_n/\Delta x$ is the ratio between the wavelength and Δx.

For each time step, the ratio between the numerical and the analytic enlargement factors is λ'_n/λ_n, and its magnitude can be taken as a measure of the amplitude error of the numerical solution. After propagating a whole wavelength, the amplitude ratio is

$$A_n = \left(\frac{|\lambda'_n|}{|\lambda_n|}\right)^{N_n} = \left(\frac{|\lambda'_n|}{\exp(-\sigma_n^2 D\Delta t)}\right)^{N_n}$$
$$= \left(\frac{|\lambda'_n|}{\exp(-4\pi^2\tilde{D}/l_n^2)}\right)^{l_n/\tilde{V}}. \quad (6.1.11)$$

Substituting the expression of $|\lambda'_n|$ into Eq. (6.1.11), we can obtain the relationship between A_n and l_n. If $A_n = 1$, the numerical solutions have no amplitude errors; if $A_n > 1$, the numerical method causes the amplitude to increase; and for $A_n < 1$, the amplitude becomes small, which is called the damping effect.

The *phase angle lag* caused by the numerical solutions while propagating a whole wavelength is defined as

$$\Phi_n = \left(\frac{\theta'_n}{\theta_n}\right) 2\pi - 2\pi = 2\pi\left(\frac{\theta'_n}{\sigma_n V \Delta t}\right) - 2\pi$$

$$= \frac{L_n \theta'_n}{\tilde{V} \Delta x} - 2\pi = \frac{l_n}{\tilde{V}} \theta'_n - 2\pi. \tag{6.1.12}$$

Substituting phase angle θ'_n of the numerical enlargement factor λ'_n into this equation, we can obtain the relationship between Φ_n and l_n. $\Phi_n = 0$ means that the numerical solution has no phase angle error; $\Phi_n < 0$ means the numerical solution has a phase angle lag; and $\Phi_n > 0$ means the numerical solution has an advanced phase angle.

Gray and Pinder (1976) discussed the amplitude error and phase angle lag generated by the linear Galerkin FEM. The discretization equation of Eq. (6.1.1) for an internal node is

$$\frac{1}{6}\left[\frac{C_{i+1,k+1} - C_{i+1,k}}{\Delta t} + 4\frac{C_{i,k+1} - C_{i,k}}{\Delta t} + \frac{C_{i-1,k+1} - C_{i-1,k}}{\Delta t}\right]$$

$$+ V\left[\varepsilon\frac{C_{i+1,k+1} - C_{i-1,k+1}}{2\Delta x} + (1-\varepsilon)\frac{C_{i+1,k} - C_{i-1,k}}{2\Delta x}\right]$$

$$- D\left[\varepsilon\frac{C_{i+1,k+1} - 2C_{i,k+1} + C_{i-1,k+1}}{(\Delta x)^2} + (1-\varepsilon)\frac{C_{i+1,k} - 2C_k + C_{i-1,k}}{(\Delta x)^2}\right] = 0, \tag{6.1.13}$$

where ε is a variable coefficient. When ε is equal to 0, 1 or 1/2, Eq. (6.1.13) becomes the explicit, implicit, and Crank-Nicolson schemes, respectively.

Following the derivation of Eq. (6.1.10), we can obtain the numerical enlargement factor as follows:

$$\lambda'_n = \frac{[2 + \cos(\sigma_n \Delta x)]/3 - (1 - \varepsilon)\{i\tilde{V}\sin(\sigma_n \Delta x) + 2\tilde{D}[1 - \cos(\sigma_n \Delta x)]\}}{[2 + \cos(\sigma_n \Delta x)]/3 + \varepsilon\{i\tilde{V}\sin(\sigma_n \Delta x) + 2\tilde{D}[1 - \cos(\sigma_n \Delta x)]\}}. \tag{6.1.14}$$

Its magnitude $|\lambda'_n|$ and argument θ'_n can be calculated from Eq. (6.1.14). Substituting $|\lambda'_n|$ and θ'_n into Eqs. (6.1.11) and (6.1.12), we can obtain amplitude ratio A_n and phase angle lag Φ_n corresponding to various l_n. Figures 6.1 and 6.2, which are cited from the paper written by Gray and Pinder (1976), show the curves of l_n vs. A_n and l_n vs. Φ_n, respectively. The dimensionless parameters used here are $\tilde{D} = 0.069$ and $\tilde{V} = 0.369$, and the enlargement factor of FDM is determined by Eq. (6.1.10).

As shown by Figure 6.2, neither FEM nor FDM are good at propagating shorter waves, and both have significant phase angle lag. Incorrect propagation of phase angle may cause the numerical solution to oscillate. Figure 6.1 shows that the FEM has some damping effect on shorter waves, thus the

FIGURE 6.1. The ratio of amplitudes for FEM and FDM. ——— = FEM, ------- = FDM.

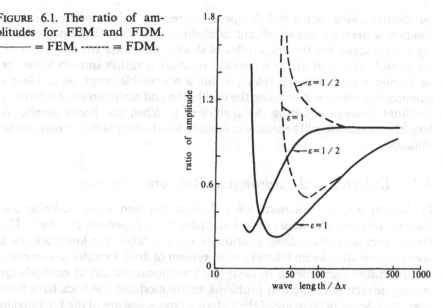

FIGURE 6.2. The phase angle lag for FEM and FDM. ——— = FEM, ------- = FDM.

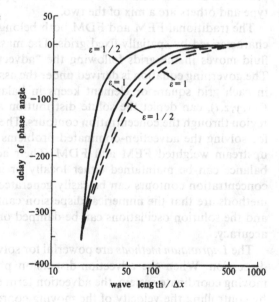

solution of FEM has slightly less oscillation than that of FDM. However, shorter waves play important roles in depicting the shape of a steep front. If shorter waves are damped, the front of the numerical solution will become smooth due to the increased numerical dispersion error. Two extreme cases, $\varepsilon = 0.5$ and $\varepsilon = 1.0$, are shown in Figures 6.1 and 6.2, respectively. For $\varepsilon = 0.5$, when the wavelength increases, the phase angle lag disappears quickly, but shorter waves are not well damped. As a result, the front of the numerical solution is steep but has significant oscillations. For $\varepsilon = 1.0$, the phase angle lag is significant, but the amplitudes of shorter waves are strongly damped. As a result, the front of the numerical solution is rather smooth while the oscillation is weakened. Therefore, within a reasonable range, we can find a compromise between decreasing the oscillation and maintaining the shape of the front through adjusting the coefficient ε. When the Peclet number is larger, the situation will become worse and a satisfactory solution may not be obtained.

6.1.2 Eulerian and Lagrangian Reference Frames

In Section 4.2, we discussed the difference between using *Eulerian and Lagrangian reference frames* for solving advection-dispersion problems. The former uses a spatially-fixed coordinate system, while the latter adopts a coordinate system which follows the movement of fluid. In order to overcome the difficulties encountered in using the traditional numerical methods for solving advection-dominated problems, many modified methods have been presented. Some of them are of the Eulerian type, some are of the Lagrangian type and others are a mix of the two.

The traditional FEM and FDM both belong to the *Eulerian type* and are characterized by spatially-fixed grids. The mass dissolved and mixed in the fluid moves in the grids following the "advection" and "dispersion" rules. The governing equation is derived under the assumption that the solute mass in each grid square or element keeps in balance at all time. Its solution, $C(x, y, z, t)$, can depict the solute distribution at any moment of the whole region through the concentration contours. The Eulerian methods applicable for solving the advection-dominated problems are mainly composed of the upstream weighted FEM and FDM. Their advantages are that the mass balance can be maintained either locally or in the whole region, and the concentration contours can be easily generated. The disadvantages of these methods are that the numerical dispersion cannot be completely eliminated and the solution oscillations can be damped only at the expense of solution accuracy.

The *Lagrangian methods* are powerful for solving the problem of numerical dispersion. When the advection-dispersion phenomena are observed in a moving coordinate system, the advection term may be reduced or eliminated by controlling the velocity of the moving coordinate system. The phenomenon of advection-dispersion observed in the moving coordinate system is always dispersion-dominated, and the effects of advection and dispersion can be computed separately. As a result, the numerical dispersion may be reduced or eliminated. Lagrangian methods involve the moving coordinate method, the deforming grid method, the moving point method, and the Random-Walk method mentioned in Chapter 4. Although these methods can eliminate the numerical dispersion, they cannot guarantee mass balance and usually have difficulties in treating complicated practical problems.

In recent years, *Lagrangian-Eulerian methods*, such as the characteristic-FDM, characteristic-FEM and so on, have been suggested. These methods have the advantages, and, to some degree, the disadvantages, of both the Lagrangian and the Eulerian methods. In the following sections, we will give an introduction to some of the major methods that are currently under development in this field.

6.2 Upstream Weighted Methods

6.2.1 Upstream Weighted Finite Difference Methods

In Eq. (6.1.1), if the first-order and second-order derivatives are replaced by the central difference approximations as follows

$$\frac{\partial C}{\partial x}\bigg|_i \approx \frac{C_{i+1} - C_{i-1}}{2\Delta x}, \quad \frac{\partial^2 C}{\partial x^2}\bigg|_i \approx \frac{C_{i+1} - 2C_i + C_{i-1}}{(\Delta x)^2},$$

then difference equation (6.1.8) at an internal node i can be derived. In Section 4.1.6, we analyzed the relationship between truncation error and numerical dispersion of a difference equation. If we want to obtain a satisfactory solution through this method, it is necessary to control the size of Δx and Δt. Generally, the following two conditions should be satisfied:

1. The *local Peclet number* $Pe < 1$, i.e., $V\Delta x/D < 1$. In order to satisfy this condition, $\Delta x < D/V$ should be adopted.
2. The *local Courant number* $V\Delta t/\Delta x < 1$, which means that the average displacement of the fluid is less than the length of one grid space in one time step. To satisfy this condition, $\Delta t < \Delta x/V$ should be adopted.

For example, suppose that the length of a one-dimensional dispersion region is 12800, $D = 2$ and $V = 0.5$, then $\Delta x = 4$ should be used. The whole region should be divided into 3200 grid spaces, and the time step Δt must not exceed 8. Thus, it requires 1200 time steps to simulate the total time of 96000. This is a highly inefficient method which requires an unacceptable amount of calculation.

In order to search for more efficient methods, let us further analyze the truncation errors and the stability of FDM. The central difference approximation of the second-order derivative may be expressed as

$$\frac{C_{i+1} - 2C_i + C_{i-1}}{(\Delta x)^2} = \frac{\partial^2 C}{\partial x^2}\bigg|_i + \frac{1}{12} C_i^{(4)}(\Delta x)^2 + \text{HOT}, \qquad (6.2.1)$$

where HOT represents terms of a higher order than $(\Delta x)^2$ in the Taylor expansion. Since the last two terms on the right-hand side of Eq. (6.2.1) only involve fourth or higher order derivatives of concentration C, the finite difference approximation is completely accurate when C is a cubic polynomial. The first-order derivative is approximated by the central difference as follows:

$$\frac{C_{i+1} - C_{i-1}}{2\Delta x} = \frac{\partial C}{\partial x}\bigg|_i + \frac{1}{6}C_i^{(3)}(\Delta x)^2 + \text{HOT}. \tag{6.2.2}$$

The truncation error in the above equation involves the third-order derivative of C. Therefore, it is completely accurate when C is a quadratic polynomial.

Although the truncation errors of Eqs. (6.2.1) and (6.2.2) are both proportional to $(\Delta x)^2$, they are not identical from the point of view of the algorithm accuracy.

Now, let us analyze the stability of the algorithm. The canonical equation (6.1.1) may be rewritten as

$$\frac{\partial C}{\partial t} = -V\frac{\partial C}{\partial x} + D\frac{\partial^2 C}{\partial x^2}. \tag{6.2.3}$$

In Section 4.1.2, we stated that von Neumann's concept of stability sometimes fails to guarantee that there will be no oscillation in the solution. Leonard (1979) pointed out that if calculation errors of concentration C are generated at a time level, the finite difference approximation on the right-hand side of Eq. (6.2.3) must have the ability to *self-correct* the solution at every node in subsequent time steps. In other words, a *"negative feedback"* of the errors should exist. Leonard (1979) proposed the following type of stability condition:

$$\frac{\partial(\text{RHS})}{\partial C_i} < 0 \tag{6.2.4}$$

where RHS represents the finite difference approximation of the right-hand side of Eq. (6.2.3). If $\partial(\text{RHS})/\partial C_i > 0$, it is unstable; and if $\partial(\text{RHS})/\partial C_i = 0$, it is in a critical state.

Applying the difference approximations of Eqs. (6.2.1) and (6.2.2) to the right-hand side of Eq. (6.2.3), and taking the partial derivative with respect to C_i, we obtain:

$$\frac{\partial(\text{RHS})}{\partial C_i} = \frac{2D}{(\Delta x)^2}. \tag{6.2.5}$$

When $D > 0$, there is a negative feedback. When D is decreased, the negative feedback decreases and tends to the critical state. Therefore, when D decreases, the stability of the solution is reduced. We have observed this phenomenon before.

In order to increase the effect of the negative feedback, the *upstream FDM* is proposed. Instead of the central difference approximation, we use the following for $\partial C/\partial x$:

1. When $V > 0$, let

$$\frac{C_i - C_{i-1}}{\Delta x} = \frac{\partial C}{\partial x}\bigg|_i - \frac{1}{2}C_i^{(2)}\Delta x + \text{HOT}. \qquad (6.2.6a)$$

2. When $V < 0$, let

$$\frac{C_{i+1} - C_i}{\Delta x} = \frac{\partial C}{\partial x}\bigg|_i + \frac{1}{2}C_i^{(2)}\Delta x + \text{HOT}. \qquad (6.2.6b)$$

Since the finite difference approximations used on left-hand sides of these equations are obtained always by the upstream concentration minus the downstream concentration, Eq. (6.2.6) is called the *upstream finite difference approximation*. If the second-order derivative on the right-hand side of Eq. (6.2.3) is still approximated by the central difference of Eq. (6.2.1), but the first-order derivative is replaced with the upstream finite difference approximation, we have:

$$\frac{\partial(\text{RHS})}{\partial C_i} = -\frac{|V|}{\Delta x} - \frac{2D}{(\Delta x)^2}. \qquad (6.2.7)$$

It shows that this method has a good negative feedback, even for pure advection problems. However, since the truncation errors of Eq. (6.2.6) are proportional to (Δx), the accuracy of the solution is lowered. As a result, the numerical dispersion is increased and thus steep concentration fronts cannot be accurately calculated. In other words, the upstream FDM is unable to solve both the overshoot and numerical dispersion problems at the same time. We can only take a compromise between decreasing oscillations and increasing accuracy. When a weighted mean of the central and upstream finite difference is used to replace the first-order derivative, i.e., let

$$\frac{\partial C}{\partial x}\bigg|_i = \alpha\frac{C_i - C_{i-1}}{\Delta x} + (1-\alpha)\frac{C_{i+1} - C_{i-1}}{2\Delta x}, \quad \text{when } V > 0; \quad (6.2.8a)$$

$$\frac{\partial C}{\partial x}\bigg|_i = \alpha\frac{C_{i+1} - C_i}{\Delta x} + (1-\alpha)\frac{C_{i+1} - C_{i-1}}{2\Delta x}, \quad \text{when } V < 0, \quad (6.2.8b)$$

the corresponding FDM method is called the *Upstream Weighted Finite Difference Method* (UWFDM).

In order to improve the accuracy of FDM solutions, some authors recommend using high-order finite difference approximations. Leonard (1979) proposed that the first-order derivative be approximated by the following *third-order upstream finite differences*:

1. When $V > 0$, use

$$\left.\frac{\partial C}{\partial x}\right|_i = \frac{C_{i+1} - C_{i-1}}{2\Delta x} - \frac{C_{i+1} - 3C_i + 3C_{i-1} - C_{i-2}}{6\Delta x} - \frac{1}{12} C_i^{(4)}(\Delta x)^3 + \text{HOT},$$

(6.2.9a)

2. When $V < 0$, use

$$\left.\frac{\partial C}{\partial x}\right|_i = \frac{C_{i+1} - C_{i-1}}{2\Delta x} - \frac{C_{i+2} - 3C_{i+1} + 3C_i - C_{i-1}}{6\Delta x} - \frac{1}{12} C_i^{(4)}(\Delta x)^3 + \text{HOT}.$$

(6.2.9b)

When C is a cubic polynomial, the two equations are completely accurate because their truncation errors only involve derivatives of fourth-order or higher orders. Consequently, they have the same accuracy as the central finite difference approximation of the second-order derivative in Eq. (6.2.1). Also, since

$$\frac{\partial(\text{RHS})}{\partial C_i} = -\frac{|V|}{2\Delta x} - \frac{2D}{(\Delta x)^2} < 0,$$

(6.2.10)

the method has good stability. It should be pointed out that since C_{i-2} and C_{i+2} appear in Eq. (6.2.9), the coefficient matrix of the finite difference equations is no longer in the tridiagonal form.

The idea of the upstream weighted methods mentioned above is very valuable. It leads us to the advent of many new methods, including various UWFEMs and related methods, which have been widely studied in the past decade.

6.2.2 Upstream Weighted Finite Element Methods

Methods of this kind have various names in the literature. Customarily, they are called *Upwind FEM*, and also called the *Petrov-Galerkin FEM* or the *asymmetric weighted function FEM*. These methods were put forward in the late 1970s (see Heinrich et al., 1977 and Huyakorn and Nilkuha, 1979). Since then, the scope of their application have been extended continuously from one-dimensional to two-dimensional, from low order to high order, and from linear to non-linear equations.

In Section 5.1.2, we discussed general weighted residual methods. The Galerkin method is a special case of weighted residual methods in which the weighting functions are identical with the *basis functions*. To solve advection-dominated problems, we may consider other selections of weighting functions. The key problem is how to select them to obtain appropriate upstream weighted effect.

First of all, let us consider a case of using one-dimensional linear basis functions. For a standard element $(-1, 1)$ in a local coordinate system, the

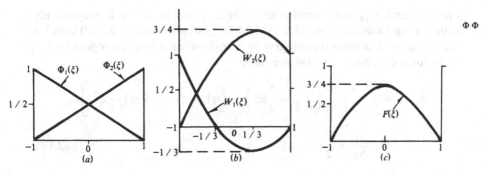

FIGURE 6.3. A standard element $-1 \le \xi \le 1$ in a local coordinate system. (a) basis function; (b) weighting function; (c) modifying function.

linear basis functions corresponding to the ends -1 and 1 are

$$\begin{cases} \phi_1(\xi) = \tfrac{1}{2}(1 - \xi), \\ \phi_2(\xi) = \tfrac{1}{2}(1 + \xi). \end{cases} \tag{6.2.11}$$

The *weighting functions* for the UWFEM or the Petrov-Galerkin method are given as

$$\begin{cases} W_1(\xi) = \phi_1(\xi) - \alpha F(\xi), \\ W_2(\xi) = \phi_2(\xi) + \alpha F(\xi), \end{cases} \tag{6.2.12}$$

where

$$F(\xi) = \tfrac{3}{4}(1 - \xi)(1 + \xi) \tag{6.2.13}$$

is called the *modifying function* and α is a parameter to be determined. The basis functions, weighting functions, and modifying functions are shown in Figure 6.3. In order to illustrate the upstream weighted effect of this method, let us consider the following steady one-dimensional advection-dispersion equation:

$$V\frac{\partial C}{\partial x} - D\frac{\partial^2 C}{\partial x^2} = 0. \tag{6.2.14}$$

In this case, B_{ij} and F_i in Eqs. (5.1.13) to (5.1.15) are equal to zero, and A_{ij} is reduced to

$$A_{ij} = \int_{(R)} \left(D\frac{\partial W_i}{\partial x}\frac{\partial \phi_j}{\partial x} + VW_i\frac{\partial \phi_j}{\partial x} \right) dx. \tag{6.2.15}$$

To derive the discrete finite element equation for node i, only $A_{i,i-1}$, $A_{i,i}$ and $A_{i,i+1}$, need to be calculated. According to the selection of the basis functions, the integral of Eq. (6.2.15) is non-zero only in the subregion contains both nodes i and j. Thus, for $A_{i,i+1}$, we have

$$A_{i,i+1} = \int_{x_i}^{x_{i+1}} \left(D\frac{\partial W_i}{\partial x}\frac{\partial \phi_j}{\partial x} + VW_i\frac{\partial \phi_j}{\partial x} \right) dx, \tag{6.2.16}$$

where x_i and x_{i+1} are coordinates of node i and node $i+1$, respectively. Substituting basis functions (6.2.11) and weighting functions (6.2.12) into Eq. (6.2.16), and calculating the integral by transforming it into the region $(-1, 1)$ in the local coordinate system, we have:

$$A_{i,i+1} = \int_{-1}^{1} \frac{1}{2} \left[\frac{2D}{\Delta x} \left(-\frac{1}{2} + \frac{3}{2}\alpha\xi \right) + \frac{V}{2}(1 - \xi) - \frac{3}{4}V\alpha(1 - \xi^2) \right] d\xi$$

$$= -\frac{D}{\Delta x} + \frac{V}{2}(1 - \alpha). \tag{6.2.17a}$$

Similarly,

$$A_{i,i-1} = -\frac{D}{\Delta x} - \frac{V}{2}(1 + \alpha), \tag{6.2.17b}$$

and

$$A_{i,j} = \frac{2D}{\Delta x} + V\alpha. \tag{6.2.17c}$$

The discrete equation of node i is then obtained as

$$\left[-\frac{D}{\Delta x} - \frac{V}{2}(1 + \alpha) \right] C_{i-1} + \left[\frac{2D}{\Delta x} + V\alpha \right] C_i + \left[-\frac{D}{\Delta x} + \frac{V}{2}(1 - \alpha) \right] C_{i+1} = 0. \tag{6.2.18}$$

It is interesting to note that if the equations of UWFDM, Eqs. (6.2.8) and (6.2.1), are used to replace the advection and dispersion terms in Eq. (6.2.14), the result will be (assuming $V > 0$):

$$\alpha V - \frac{C_i - C_{i-1}}{\Delta x} + (1 - \alpha)V\frac{C_{i+1} - C_{i-1}}{2\Delta x} - D\frac{C_{i-1} - 2C_i + C_{i+1}}{(\Delta x)^2} = 0. \tag{6.2.19}$$

After Eq. (6.2.19) is rearranged, it becomes Eq. (6.2.18). This fact illustrates that if the basis functions and weighting functions are selected according to Eqs. (6.2.11) and (6.2.12), then for the typical equation (6.2.14), the Petrov-Galerkin method is the same as UWFDM. In fact, in Eq. (6.2.12) the undetermined parameter α is an upstream weighting coefficient. When $\alpha = 0$, it is not weighted, and the Petrov-Galerkin method is reduced to the general Galerkin method. Heinrich et al. (1977) pointed out that the optimal value of the weighting coefficient α is

$$\alpha_0 = \coth\left(\frac{Pe}{2} \right) - \frac{2}{Pe}, \tag{6.2.20a}$$

where the local Peclet number $Pe = V\Delta x/D$. This choice of α results in the best balance between oscillation and numerical dispersion errors.

Noorishad et al. (1992) suggested using

$$\alpha = C_r - \frac{2}{P_e} \tag{6.2.20b}$$

as a weighting coefficient when the Crank-Nicolson Galerkin FFM is used to solve the one-dimensional advection-dominated problem. In (6.2.20b), $C_r = V\Delta t/\Delta x$ is the Courant number. When $P_e \cdot C_r \le 2$, upwind weighting is not needed.

In the Petrov-Galerkin method, high-order basis functions can also be used. Heinrich and Zienkilwicz (1977) proposed some cubic asymmetric weighting functions, and used them in combination with the quadratic basis functions.

Now let us move on to *two-dimensional problems*. We only need to determine the corresponding basis and weighting functions. We have already known that the bilinear basis functions corresponding to the four vertices of a standard rectangle element located in the local coordinate system (Figure 6.4) are given by

$$\begin{cases} \phi_1(\xi, \eta) = \phi_1(\xi)\phi_1(\eta), \\ \phi_2(\xi, \eta) = \phi_2(\xi)\phi_1(\eta), \\ \phi_3(\xi, \eta) = \phi_2(\xi)\phi_2(\eta), \\ \phi_4(\xi, \eta) = \phi_1(\xi)\phi_2(\eta), \end{cases} \tag{6.2.21}$$

where ϕ_1 and ϕ_2 on the right-hand side of the above equation are the one-dimensional linear basis functions given in Eq. (6.2.11). Their corresponding weighting functions are

$$\begin{cases} W_1(\xi, \eta) = [\phi_1(\xi) - \alpha_1 F(\xi)][\phi_1(\eta) - \alpha_4 F(\eta)], \\ W_2(\xi, \eta) = [\phi_2(\xi) + \alpha_1 F(\xi)][\phi_1(\eta) - \alpha_2 F(\eta)], \\ W_3(\xi, \eta) = [\phi_2(\xi) + \alpha_3 F(\xi)][\phi_2(\eta) + \alpha_2 F(\eta)], \\ W_4(\xi, \eta) = [\phi_1(\xi) - \alpha_3 F(\xi)][\phi_2(\eta) + \alpha_4 F(\eta)], \end{cases} \tag{6.2.22}$$

where F is the modifying function given in Eq. (6.2.13), and $\alpha_1, \alpha_2, \alpha_3$, and α_4

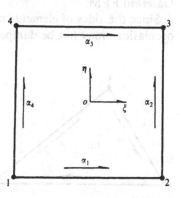

FIGURE 6.4. A standard element in a local two-dimensional coordinate system.

are the upstream weighting coefficients. When the flow direction is illustrated by the arrows in Figure 6.4, the weighting coefficients are taken as positive. It is evident that the two-dimensional weighting functions given by Eq. (6.2.22) are a direct extension of the one-dimensional form given by Eq. (6.2.12). The upstream and downstream relations for the nodes are shown by arrows.

The triangle element is very important both in theory and application. The linear basis functions corresponding to the three nodes i, j, k in a triangle element (Δ) have been given in Eq. (5.1.32) and are repeated here as

$$\phi_l(x, y) = \frac{1}{2\Delta}(a_1 + b_l x + c_l y),$$

$$l = i, j, k; \quad (x, y) \in (\Delta) \tag{6.2.23}$$

where the coefficients a_l, b_l, and c_l are determined by the coordinates of the nodes (see Eq. (5.1.23)). The *asymmetric weighting functions* are defined as

$$\begin{cases} W_i(x, y) = \phi_i + F_i, \\ W_j(x, y) = \phi_j + F_j, \\ W_k(x, y) = \phi_k + F_k. \end{cases} \tag{6.2.24}$$

The expressions of the three modifying functions F_i, F_j, and F_k, which contain the upstream weighting coefficients α_i, α_j, and α_k, are

$$\begin{cases} F_i = -3\alpha_k \phi_i \phi_j + 3\alpha_j \phi_i \phi_k, \\ F_j = -3\alpha_i \phi_j \phi_k + 3\alpha_k \phi_j \phi_i, \\ F_k = -3\alpha_j \phi_k \phi_i + 3\alpha_i \phi_k \phi_j. \end{cases} \tag{6.2.25}$$

Substituting these equations into Eq. (6.2.24), we can obtain the required quadratic weighting functions. When the upstream and downstream relations are shown by the arrows in Figure 6.5, all the upstream weighting coefficients in the equations are taken as positive. Conversely, if the flow directions are reversed, negative values should be adopted. These types of asymmetric weighting functions were proposed by Huyakorn (1977). In this case, the integrant in Eq. (5.1.13) is a quadratic function; therefore, more computation effort is required for obtaining coefficient $A_{i,j}$ than that in the Galerkin FEM.

Since the sides of elements may not coincide with the flow directions, the oscillations may not be damped much but the accuracy of the solution may

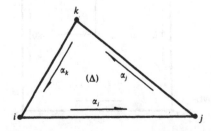

FIGURE 6.5. A triangle element (Δ). The upstream and downstream relations between the nodes are shown by arrows.

be lowered. Some authors called it the *cross wind* phenomenon. In order to eliminate this phenomenon, Brooks and Hughes (1982) proposed a modified form named the *Streamline Upwind/Petrov-Galerkin (SU/PG) method*. In this method, the weighting functions are taken as

$$W_l = \phi_l + k\left(V_x \frac{\partial \phi_l}{\partial x} + V_y \frac{\partial \phi_l}{\partial y}\right)/|V|^2. \tag{6.2.26}$$

where $l = 1, 2, 3, 4$ for rectangle elements, $|V|$ is the absolute value of velocity, and the two-dimensional upstream weighting coefficient k is

$$k = (\alpha V_x \Delta x + \beta V_y \Delta y)/2. \tag{6.2.27}$$

The optimal values for α and β in Eq. (6.2.27) are

$$\alpha = \coth(2Pe_x) - \frac{2}{Pe_x}, \quad \beta = \coth(2Pe_y) - \frac{2}{Pe_y}. \tag{6.2.28}$$

where Pe_x and Pe_y are the local Peclet numbers along the x and y directions, respectively.

Since the effect of crosswind is eliminated, this technique may produce better results than the Petrov-Galerkin method when solving two-dimensional advection dominated problems. Muzukami and Hughes (1985) pointed out that the SU/PG method is not appropriate in all cases. They put forward a new SU/PG method, which can satisfy the mass conservation and the maximum modulus principle simultaneously.

Zienkiewicz (1981) pointed out that although the methods of the asymmetric weighting functions had been extended to two-dimensional, and even three-dimensional problems, their computation is too complex. In addition, when the Peclet number becomes larger, numerical dispersion is inevitable.

6.2.3 The Upstream Weighted Multiple Cell Balance Method

Sun and Yeh (1983) suggested an upstream weighted multiple cell balance method, in which the upstream weight is directly added to the basis functions by introducing a dummy node in each element. Thus, asymmetric weighting functions are not necessary. This method requires less computational effort, and is especially suitable for solving complicated two-dimensional, multiple-layer, and even three-dimensional problems that might be encountered in the field.

Let us go back to Section 5.2, and consider the integral form of the two-dimensional solute transport equation:

$$\int_{(L)} m\left[\left(D_{xx}\frac{\partial C}{\partial x} + D_{xy}\frac{\partial C}{\partial y}\right)dy - \left(D_{xy}\frac{\partial C}{\partial x} + D_{yy}\frac{\partial C}{\partial y}\right)dx\right]$$
$$+ \int_{(L)} mC(V_y\,dx - V_x\,dy) = \int\int_{(D)}\left[\frac{\partial(mC)}{\partial t} + M\right]dx\,dy, \tag{6.2.29}$$

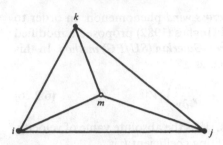

FIGURE 6.6. A triangular element and its subelements, in which center point m is a dummy node.

where all the symbols have been stated in Eq. (5.2.3). The domain D is divided into a triangle net. Assume that (e) is an arbitrary element, whose three vertices are nodes i, j, k with coordinates (x_i, y_i), (x_j, y_j), and (x_k, y_k), respectively.

Now, we define the center of the element as a *dummy node*. Letting the dummy node be denoted by subscript m and have coordinates (x_m, y_m), then we have

$$x_m = \tfrac{1}{3}(x_i + x_j + x_k), \quad y_m = \tfrac{1}{3}(y_i + y_j + y_k). \tag{6.2.30}$$

The concentration C_m of the central point m is defined as the weighted mean of the three vertice concentrations, C_i, C_j, and C_k, that is

$$C_m = \omega_i C_i + \omega_j C_j + \omega_k C_k,$$
$$\omega_i + \omega_j + \omega_k = 1, \tag{6.2.31}$$

where ω_i, ω_j, ω_k are the upstream weights of the nodes i, j, k in element (e), respectively. The values of ω_i, ω_j, and ω_k will be determined later.

The element (e) is divided into three triangular subelements by lines $\overline{mi}$, $\overline{mj}$, and $\overline{mk}$, which are formed by connecting the center point m with three vertices, as shown in Fig. 6.6. In each sub-element, a linear function determined by the concentrations of its three vertices is used to approximately replace the unknown function $C(x, y, t)$. For the whole element (e), $C(x, y, t)$ is then approximately replaced by three planes butted together, as shown in Fig. 6.7(b). Note that there is an additional pillar associated with C_m. The shape of $C(x, y, t)$ depends on the weighting coefficients ω_i, ω_j and ω_k. For sub-element Δijm, we have

$$C(x, y, t) = \phi_{ki} C_i + \phi_{kj} C_j + \phi_{km} C_m, \tag{6.2.32}$$
$$(x, y) \in \Delta ijm$$

where ϕ_{ki}, ϕ_{kj}, and ϕ_{km} are linear functions corresponding to nodes i, j, and m of the sub-element Δijm, respectively. The first subscript in ϕ_{ki} denotes that the considered sub-element is opposite to node k. From Eqs. (5.1.32) and (5.1.33), we know

$$\phi_{kl} = \frac{3}{2\Delta_e}(a_{kl} + b_{kl}x + c_{kl}y), \; l = i, j, m,$$

FIGURE 6.7. Approximate expressions of the unknown concentration. (a) Linear Galerkin FEM or general MCBM; (b) Upstream weighted method.

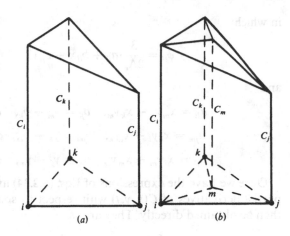

(a) (b)

and

$$a_{ki} = x_j y_m - x_m y_j, \quad b_{ki} = y_j - y_m, \quad c_{ki} = x_m - x_j,$$

$$a_{kj} = x_m y_i - x_i y_m, \quad b_{kj} = y_m - y_i, \quad c_{kj} = x_i - x_m,$$

$$a_{km} = x_i y_j - x_j y_i, \quad b_{km} = y_i - y_j, \quad c_{km} = x_j - x_i. \qquad (6.2.33)$$

Substituting the definition of C_m, Eq. (6.2.31), into Eq. (6.2.32), we have

$$C(x, y, t) = \bar{\phi}_{ki} C_i + \bar{\phi}_{kj} C_j + \bar{\phi}_{kk} C_k, \qquad (6.2.34)$$

$$(x, y) \in \Delta ijm$$

where

$$\bar{\phi}_{ki} = \phi_{ki} + \omega_i \phi_{km}, \qquad (6.2.35a)$$

$$\bar{\phi}_{kj} = \phi_{kj} + \omega_j \phi_{km}, \qquad (6.2.35b)$$

$$\bar{\phi}_{kk} = \omega_k \phi_{km}. \qquad (6.2.35c)$$

As expressed in Eq. (6.2.34), $\bar{\phi}_{ki}$, $\bar{\phi}_{kj}$, and $\bar{\phi}_{kk}$ act as basis functions in the sub-element Δijm and contain weighting coefficients, so they are called *weighted basis functions*.

Similarly, for another sub-element Δimk with node i as its vertex, we have

$$C(x, y, t) = \bar{\phi}_{ji} C_i + \bar{\phi}_{jj} C_j + \bar{\phi}_{jk} C_k,$$

$$(x, y) \in \Delta imk, \qquad (6.2.36)$$

where the weighted basis functions are

$$\bar{\phi}_{ji} = \phi_{ji} + \omega_i \phi_{jm}, \qquad (6.2.37a)$$

$$\bar{\phi}_{jj} = \omega_j \phi_{jm}. \qquad (6.2.37b)$$

$$\bar{\phi}_{jk} = \phi_{jk} + \omega_k \phi_{jm}, \qquad (6.2.37c)$$

in which

$$\phi_{jl} = \frac{3}{2\Delta_e}(a_{jl} + b_{jl}x + c_{jl}y), \quad l = i, k, m \tag{6.2.38}$$

and

$$a_{ji} = x_m y_k - x_k y_m, \quad b_{ji} = y_m - y_k, \quad c_{ji} = x_k - x_m,$$

$$a_{jm} = x_k y_i - x_i y_k, \quad b_{jm} = y_k - y_i, \quad c_{jm} = x_i - x_k,$$

$$a_{jk} = x_i y_m - x_m y_i, \quad b_{jk} = y_i - y_m, \quad c_{jk} = x_m - x_i.$$

Once we have the expressions of Eqs. (6.2.34) and (6.2.36) for $C(x, y, t)$, the partial derivatives of $C(x, y, t)$ with respect to space and time variables can then be obtained directly. They are:

$$\frac{\partial C}{\partial x} = \begin{cases} \dfrac{3}{2\Delta}(\bar{b}_{ki}C_i + \bar{b}_{kj}C_j + \bar{b}_{kk}C_k), & (x, y) \in \Delta ijm \\[2ex] \dfrac{3}{2\Delta}(\bar{b}_{ji}C_i + \bar{b}_{jj}C_j + \bar{b}_{jk}C_k), & (x, y) \in \Delta imk \end{cases} \tag{6.2.39a}$$

$$\frac{\partial C}{\partial y} = \begin{cases} \dfrac{3}{2\Delta}(\bar{c}_{ki}C_i + \bar{c}_{kj}C_j + \bar{c}_{kk}C_k), & (x, y) \in \Delta ijm \\[2ex] \dfrac{3}{2\Delta}(\bar{c}_{ji}C_i + \bar{c}_{jj}C_j + \bar{c}_{jk}C_k), & (x, y) \in \Delta imk \end{cases} \tag{6.2.39b}$$

$$\frac{\partial C}{\partial t} = \begin{cases} \bar{\phi}_{ki}\dfrac{\partial C_i}{\partial t} + \bar{\phi}_{kj}\dfrac{\partial C_j}{\partial t} + \bar{\phi}_{kk}\dfrac{\partial C_k}{\partial t}, & (x, y) \in \Delta ijm \\[2ex] \bar{\phi}_{ji}\dfrac{\partial C_i}{\partial t} + \bar{\phi}_{jj}\dfrac{\partial C_j}{\partial t} + \bar{\phi}_{ik}\dfrac{\partial C_k}{\partial t}, & (x, y) \in \Delta imk \end{cases} \tag{6.2.39c}$$

where

$$\bar{b}_{ki} = b_{ki} + \omega_i b_{km}, \quad \bar{b}_{kj} = b_{kj} + \omega_j b_{km}, \quad \bar{b}_{kk} = \omega_k b_{km},$$

$$\bar{b}_{ji} = b_{ji} + \omega_i b_{jm}, \quad \bar{b}_{jj} = \omega_j b_{jm}, \quad \bar{b}_{jk} = b_{jk} + \omega_k b_{jm},$$

$$\bar{c}_{ki} = c_{ki} + \omega_i c_{km}, \quad \bar{c}_{kj} = c_{kj} + \omega_j c_{km}, \quad \bar{c}_{kk} = \omega_k c_{km},$$

$$\bar{c}_{ji} = c_{ji} + \omega_i c_{jm}, \quad \bar{c}_{jj} = \omega_j c_{jm}, \quad \bar{c}_{jk} = c_{jk} + \omega_k c_{jm}, \tag{6.2.40}$$

Clearly, when $\omega_i = \omega_j = \omega_k = 1/3$, the above equations will be simplified to the previously discussed linear FEM or MCBM without weighting. In this case, Eqs. (6.2.39a) to (6.2.39c) will reduce to Eqs. (5.2.6) to (5.2.8), respectively.

As we did in previous sections, in order to derive the discrete equations for node i, we will consider all the elements with node i as one of their vertices. The centers of each element are then connected with the midpoints of the sides to form a polygon surrounding node i, which is called the *exclusive subdomain* for node i. Then, the equation of local mass balance over the subdomain can be established. The exclusive subdomain of node i is written as (D_i), its boundary as (L_i), and the part occupied by both (D_i) and element (e) is written as (e_i), as shown in Figure 5.10. Equation (6.2.29) will become the mass balance equation for this subdomain by replacing (D) and (L) with (D_i) and (L_i), respectively. The integral along (L_i) can be computed element by element. Calculating the integrals along the line segments $\overline{Am}$ and $\overline{mB}$ in element (e), we have

$$
I_1 = \int_{\overline{Am}} \left(D_{xx}\frac{\partial C}{\partial x} + D_{xy}\frac{\partial C}{\partial y} \right) dy - \left(D_{xy}\frac{\partial C}{\partial x} + D_{yy}\frac{\partial C}{\partial y} \right) dx
$$

$$
= \frac{1}{6}\left[\left(D_{xx}\frac{\partial C}{\partial x} + D_{xy}\frac{\partial C}{\partial y} \right)(b_i - b_j) + \left(D_{xy}\frac{\partial C}{\partial x} + D_{yy}\frac{\partial C}{\partial y} \right)(c_i - c_j) \right].
$$

Substituting the first expressions of (6.2.39a) and (6.2.39b) into this equation, we obtain:

$$
I_1 = -\frac{1}{4\Delta_e}\{[(D_{xx}\bar{b}_{ki} + D_{xy}\bar{c}_{ki})(b_i - b_j) + (D_{xy}\bar{b}_{ki} + D_{yy}\bar{c}_{ki})(c_i - c_j)]C_i
$$

$$
+ [(D_{xx}\bar{b}_{kj} + D_{xy}\bar{c}_{kj})(b_i - b_j) + (D_{xy}\bar{b}_{ki} + D_{yy}\bar{c}_{kj})(c_i - c_j)]C_j
$$

$$
+ [(D_{xx}\bar{b}_{kk} + D_{xy}\bar{c}_{kk})(b_i - b_j) + (D_{xy}\bar{b}_{kk} + D_{yy}\bar{c}_{kk})(c_i - c_j)]C_k\}, \qquad (6.2.41)
$$

where b_l, c_l $(l = i, j, k)$ are determined from Eq. (5.1.33). In a similar way, the second expressions of Eqs. (6.2.39a) and (6.2.39b) are substituted into the equation to yield:

$$
I_2 = \int_{\overline{mB}} \left(D_{xx}\frac{\partial C}{\partial x} + D_{xy}\frac{\partial C}{\partial y} \right) dy - \left(D_{xy}\frac{\partial C}{\partial x} + D_{yy}\frac{\partial C}{\partial y} \right) dx
$$

$$
= -\frac{1}{4\Delta_e}\{[(D_{xx}\bar{b}_{ji} + D_{xy}\bar{c}_{ji})(b_i - b_k) + (D_{xy}\bar{b}_{ji} + D_{yy}\bar{c}_{ji})(c_i - c_k)]C_i
$$

$$
+ [(D_{xx}\bar{b}_{jj} + D_{xy}\bar{c}_{jj})(b_i - b_k) + (D_{xy}\bar{b}_{jj} + D_{yy}\bar{c}_{jj})(c_i - c_k)]C_j
$$

$$
+ [(D_{xx}\bar{b}_{jk} + D_{xy}\bar{c}_{jk})(b_i - b_k) + (D_{xy}\bar{b}_{jk} + D_{yy}\bar{c}_{jk})(c_i - c_k)]C_k\}. \qquad (6.4.42)
$$

I_1 plus I_2 is just the value of the line integral along (L_i) in element (e). Repeating this process for all elements, which has node i as one of their vertices, we finally have

$$
\int_{(L_i)} m\left[\left(D_{xx}\frac{\partial C}{\partial x} + D_{xy}\frac{\partial C}{\partial y} \right) dy - \left(D_{xy}\frac{\partial C}{\partial x} + D_{yy}\frac{\partial C}{\partial y} \right) dx \right]
$$

$$
= \sum_{e_i} (\alpha_{ii}^e C_i + \alpha_{ij}^e C_j + \alpha_{ik}^e C_k), \qquad (6.2.43)
$$

where

$$\alpha_{il}^e = -\frac{m_i}{4\Delta_e}[(D_{xx}\bar{b}_{kl} + D_{xy}\bar{c}_{kl})(b_i - b_j) + (D_{xy}\bar{b}_{kl} + D_{yy}\bar{c}_{kl})(c_i - c_j)]$$

$$+ (D_{xx}\bar{b}_{jl} + D_{xy}\bar{c}_{ji})(b_l - b_k) + (D_{xy}\bar{b}_{jl} + D_{yy}\bar{c}_{jl})(c_i - c_k)].$$

$$(l = i, j, k) \qquad (6.2.44)$$

The second line integral on the left-hand side of Eq. (6.2.29) can be calculated in the same way:

$$\int_{(L_i)} mC(V_y\,dx - V_x\,dy) = \sum_{e_i}(\beta_{ii}^e C_i + \beta_{ij}^e C_j + \beta_{ik}^e C_k), \qquad (6.2.45)$$

where

$$\beta_{ii}^e = \frac{m_i}{4}[V_x(\bar{b}_{jl} + \bar{b}_{kl}) + V_y(\bar{c}_{jl} + \bar{c}_{kl})],$$

$$l = i, j, k, \qquad (6.2.46)$$

In the derivation of Eq. (6.2.46), we assumed that div$(m\mathbf{V}) = 0$. If div$(m\mathbf{V}) \neq 0$, the terms in Eq. (5.2.14) should also be added to β_{ii}^e (Sun and Yeh, 1983).

In order to calculate the right side of Eq. (6.2.29), we can divide it into three parts which are commensurate with Eq. (5.2.17), that is,

$$\iint_{(D_i)}\left[\frac{\partial(mC)}{\partial t} + M\right]dx\,dy = \iint_{(D_i)} m\frac{\partial C}{\partial t}dx\,dy + \iint_{(D_i)} C\frac{\partial m}{\partial t}dx\,dy$$

$$+ \iint_{(D_i)} M\,dx\,dy. \qquad (6.2.47)$$

Using the expressions of C and $\partial C/\partial t$ in different parts of subdomain (D_i), the three integrals on the right-hand side of the above equation can be obtained easily. The result is

$$\iint_{(D_i)} m\frac{\partial C}{\partial t}dx\,dy = \sum_{e_i}\left(B_{ii}^e\frac{\partial C_i}{\partial t} + B_{ij}^e\frac{\partial C_j}{\partial t} + B_{ik}^e\frac{\partial C_k}{\partial t}\right), \qquad (6.2.48)$$

where

$$B_{ii}^e = \frac{m_i\Delta_e}{3}\left(\frac{1}{2} + \frac{\omega_i}{3}\right), B_{ij}^e = \frac{m_i\Delta_e}{3}\left(\frac{1}{12} + \frac{\omega_j}{3}\right), B_{ik}^e = \frac{m_i\Delta_e}{3}\left(\frac{1}{12} + \frac{\omega_k}{3}\right) \qquad (6.2.49)$$

and

$$\iint_{(D_i)} C\frac{\partial m}{\partial t}dx\,dy = \sum_{e_i}(\delta_{ii}^e C_i + \delta_{ij}^e C_j + \delta_{ik}^e C_k), \qquad (6.2.50)$$

where

$$\delta_{ii}^e = \frac{\Delta_e}{3}\left(\frac{1}{2} + \frac{\omega_i}{3}\right)\frac{\partial m_i}{\partial t}, \qquad \delta_{ij}^e = \frac{\Delta_e}{3}\left(\frac{1}{12} + \frac{\omega_j}{3}\right)\frac{\partial m_i}{\partial t},$$

$$\delta_{ik}^e = \frac{\Delta_e}{3}\left(\frac{1}{12} + \frac{\omega_k}{3}\right)\frac{\partial m_i}{\partial t}. \tag{6.2.51}$$

If a source or sink exists at node i, the third integral on the right-hand side of Eq. (6.2.47) can be expressed with R_i. Substituting all the above results into the integration equation (6.2.29), which expresses the local mass balance, where (D) is substituted by (D_i), we can obtain the following equation which relates the concentration of node i and concentrations of its surrounding nodes:

$$\sum_{e_i}(A_{ii}^e C_i + A_{ij}^e C_j + A_{ik}^e C_k) + \sum_{e_i}\left(B_{ii}^e \frac{\partial C_i}{\partial t} + B_{ij}^e \frac{\partial C_j}{\partial t} + B_{ik}^e \frac{\partial C_k}{\partial t}\right) + R_i = 0,$$

$$\tag{6.2.52}$$

where

$$A_{il}^e = -\alpha_{il}^e + \beta_{il}^e + \delta_{il}^e, \qquad l = i, j, k. \tag{6.2.53}$$

Equation (6.2.52) can be rewritten as

$$A_{ii}C_i + \sum_{j \neq i} A_{ij}C_j + B_{ii}\frac{\partial C_i}{\partial t} + \sum_{j \neq i} B_{ij}\frac{\partial C_j}{\partial t} + R_i = 0, \tag{6.2.54}$$

where $\sum_{j \neq i}$ represents the sum over all neighboring nodes of node i and

$$A_{ii} = \sum_{e_i} A_{ii}^e, \quad B_{ii} = \sum_{e_i} B_{ii}^e,$$

$$A_{ij} = \sum_{e_j} A_{ij}^e, \quad B_{ij} = \sum_{e_{ij}} B_{ij}^e, \tag{6.2.55}$$

where $\sum_{e_{ij}}$ represents the sum over all elements with i and j as their common nodes. Establishing Eq. (6.2.54) for the nodes with unknown concentrations and applying the given boundary conditions, we then have the following discrete equations of UWMCBM:

$$[A]C + [B]\frac{dC}{dt} + F = 0, \tag{6.2.56}$$

where $[A]$ and $[B]$ contain upstream weighting coefficients. When these coefficients are all equal to $1/3$, this set of equations will reduce to the MCB equations (5.2.25) without any weighting. The methods introduced in Section 5.4, such as point and block iteration methods and direct solutions may be used to solve the set of equations (6.2.56).

The only problem left is how to determine the weighting coefficients in each element. Assume that V is an average velocity vector in element Δijk, and V_{ij}, V_{jk}, V_{ki} are the projections of V on $\overline{ij}$, $\overline{jk}$ and $\overline{ki}$, respectively. Let

$$\tau_{ij} = \frac{V_{ij}|ij|}{\alpha_L|V|}, \tag{6.2.57}$$

where $|ij|$ is the distance between nodes i and j, $|V|$ is the absolute value of velocity V, and $|\tau_{ij}|$ represents the local Peclet number between nodes i and j. When $\tau_{ij} > 0$, node i is located upstream of node j. Conversely, if $\tau_{ij} < 0$, node j is situated upstream of node i. We may similarly define τ_{ik}. The upstream location of node i, in relation with other two nodes in the element, is represented by $\tau_i = \tau_{ij} + \tau_{ik}$. Factors τ_j and τ_k can be defined in the same way as τ_i. Finally, we stipulate that

$$\omega_i = \frac{1}{3}(1 + \lambda\tau_i) = \frac{1}{3} + \lambda\frac{V_{ij}|ij| - V_{ki}|ik|}{\alpha_L V}, \tag{6.2.58a}$$

$$\omega_j = \frac{1}{3}(1 + \lambda\tau_j) = \frac{1}{3} + \lambda\frac{V_{jk}|jk| - V_{ij}|ij|}{\alpha_L V}, \tag{6.2.58b}$$

$$\omega_k = \frac{1}{3}(1 + \lambda\tau_k) = \frac{1}{3} + \lambda\frac{V_{ki}|ik| - V_{jk}|jk|}{\alpha_L V}. \tag{6.2.58c}$$

Obviously, the condition $\omega_i + \omega_j + \omega_k = 1$ is satisfied in the above three equations, where the ratio is a positive constant to be determined. In the examples we have calculated, as determined by a trial and error procedure, the suitable value of λ is around 0.001.

In order to illustrate the efficiency of this method, let us consider the transport problem in a semi-infinite sand column discussed in Section 5.2.4. Using the elements shown in Figure 5.12 and the data given in that section, i.e., $C_0 = 10\ g/m^3$, $V = 1\ m/d$, $\Delta x = 5\ m$, and $D = 0.05\ m^2/d$, so $Pe = 100$. This is a case of a large Peclet number. If the common FDM, FEM or MCBM are used to solve the problem, we know that oscillations will occur around the fronts of the numerical solutions (see Figure 5.15). Now let us solve this problem by the weighted MCBM. Letting $\lambda = 0.0014$ and using formula (6.2.58), we can calculate the weighting coefficients for two kinds of triangular elements as shown in Figure 6.8. The results are $\omega_i = 0.6$, $\omega_j = \omega_k = 0.2$ and $\omega_i = \omega_k = 0.47$, $\omega_j = 0.06$, respectively.

The numerical solutions are shown in Figure 6.9. In this figure, we can see that oscillations of the numerical solutions are controlled, but the steep front cannot be well depicted. In other words, numerical dispersion is still evident. As Figure 6.10 shows, if λ is too small, the solution will oscillate, but if λ is too big, there will be a larger numerical dispersion. In practical computation, taking $\lambda = 0.001$ often gives a good compromise.

The method given in this section is self-adaptive. According to Eq. (6.2.58), the upstream weights depend on the relative upstream locations of the nodes and local Peclet number. The weights will increase automatically at a loca-

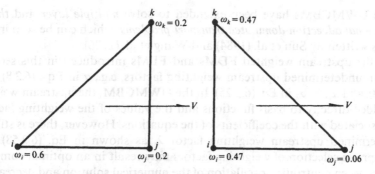

FIGURE 6.8. Weighting coefficients for two types of triangular elements.

FIGURE 6.9. Comparison between the analytic and the numerical solutions of the upstream weighted MCBM. (1) Analytic solution; (2) MCBM ($\lambda = 0.0014$).

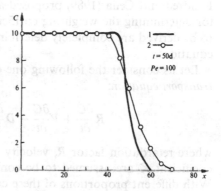

FIGURE 6.10. Effects on the solution with different values of λ. (1) Analytic solution; (2) MCBM ($\lambda = 0.0012$); (3) MCBM ($\lambda = 0.0018$).

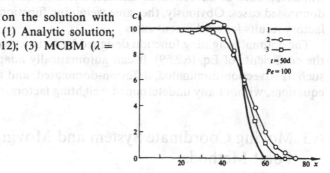

tion where the Peclet number is large, and decrease at a location where the Peclet number is small. This property allows the method to be applied to solve the advection dominated problems in unsteady flow fields.

Oscillations of numerical solutions may occur in practical problems if the concentration fronts are steep or the pollution sources change rapidly with time, even though the velocity of groundwater is low. Therefore, eliminating oscillations of numerical solutions is important both in theory and in practice.

The UWMCBMs have been extended to solve *multiple layer* and *three-dimensional advection-dominated transport problems*, which can be seen in the papers written by Sun et al. (1984) and Wang et al. (1986).

All the upstream weighted FDMs and FEMs introduced in this section contain undetermined upstream weighting factors, e.g., α in Eqs. (6.2.8) and (6.2.18), and α_i, α_j, α_k in Eq. (6.2.25). In the UWMCBM, the upstream weights are added directly to basis functions and the values of the weighting factors are associated with the coefficients of the equations. However, there is still an undetermined upstream weighting factor λ, as shown in Eq. (6.2.58). An appropriate selection of weighting factors may result in an optimal compromise between controlling oscillation of the numerical solution and decreasing numerical dispersion, but in practice it is not easy to find the optimal factors.

Herrera (1985a,b), Celia and Herrera (1987), Celia et al. (1989), and Kindred and Celia (1989) proposed an optimal weighting function method for determining the weighting coefficients. The optimal weighting functions to be derived are completely determined by the coefficients of the governing equations.

Let us consider the following one-dimensional *advective-diffusive-reactive transport equation*:

$$R\frac{\partial C}{\partial t} + V\frac{\partial C}{\partial t} - D\frac{\partial^2 C}{\partial x^2} + KC = Q(x,t), \qquad (6.2.59)$$

where retardation factor R, velocity V, diffusion coefficient D and reaction coefficient K are assumed to be constants, and Q is the source/sink term. With different proportions of these coefficients, various cases of Eq. (6.2.59) will occur, such as advection-dominated, diffusion-dominated and reaction-dominated cases. Obviously, the same weighting function cannot yield satisfactory results for all different cases.

The optimal weighting function designed by Celia et al. (1989) depends on the coefficients of Eq. (6.2.59). It can automatically adapt to various cases, such as advection-dominated, diffusion-dominated, and reaction-dominated equations, without any undetermined weighting factors.

6.3 Moving Coordinate System and Moving Point Methods

6.3.1 Moving Coordinate System Methods

Let us return to the canonical one-dimensional advection-dispersion equation:

$$\frac{\partial C}{\partial t} + V\frac{\partial C}{\partial x} - D\frac{\partial^2 C}{\partial x^2} = 0. \qquad (6.3.1)$$

If the advection-dispersion phenomena are observed in a coordinate system ξ moving with the velocity V, only dispersion will be seen. Using the coordinate transformation

$$\xi = x - Vt, \tag{6.3.2}$$

Eq. (6.3.1) can be transformed into

$$\frac{\partial C^*}{\partial t} - D\frac{\partial^2 C^*}{\partial \xi^2} = 0, \tag{6.3.3}$$

where $C^*(\xi, t)$ is the concentration observed *in the moving coordinate system*. Since there is no advection term in Eq. (6.3.3), any numerical scheme can be used to solve it without numerical difficulty. For instance, we can use the explicit difference method to obtain

$$\frac{C^*_{i,k+1} - C^*_{i,k}}{\Delta t} - D\frac{C^*_{i+1,k} - 2C^*_{i,k} + C^*_{i-1,k}}{(\Delta \xi)^2} = 0, \tag{6.3.4}$$

where $C^*_{i,k}$ and $C^*_{i,k+1}$ are concentrations of node i in the moving coordinate system at time t_k and $t_{k+1} = t_k + \Delta t$, respectively, while $C^*_{i-1,k}$ and $C^*_{i+1,k}$ are concentrations of nodes $i - 1$ and $i + 1$ in the system at time t_k.

If the spatial step is taken as $\Delta \xi = V\Delta t$, then from Eq. (6.3.4), we have:

$$C^*_{i,k+1} = C^*_{i,k} + D(C^*_{i-1,k} - 2C^*_{i,k} + C^*_{i+1,k})/V^2\Delta t. \tag{6.3.5}$$

The concentrations of all nodes in the system can be solved using the explicit scheme. This is the solution associated with the moving coordinate system. Customarily, we wish to obtain the spatial distribution of the concentration at any assigned time. Thus we have to transform, or *project*, the results obtained from the moving coordinate system into a fixed coordinate system. For the simplest case being discussed now, we may take $x = V\Delta t$ as the spatial distance, i.e., take the Courant number $V\Delta t/\Delta x = 1$. Consequently, the nodes in the moving coordinate system always coincide with those in the fixed coordinate system. Suppose that node x_j in the fixed system coincides with node ξ_i in the moving system at time t_{k+1}. We then have the following corresponding relationships:

$$x_{j-1} \leftrightarrow \xi_i, \quad C_{j-1,k} \leftrightarrow C^*_{i,k}, \quad \text{at } t_k,$$

$$x_j \leftrightarrow \xi_i, \quad C_{j,k+1} \leftrightarrow C^*_{i,k+1}, \quad \text{at } t_{k+1},$$

as shown in Figure 6.11. Equation (6.3.5) is therefore transformed into

$$C_{j,k+1} = C_{j-1,k} + D(C_{j-2,k} - 2C_{j-1,k} + C_{j,k})/V^2\Delta t. \tag{6.3.6}$$

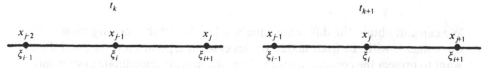

FIGURE 6.11. Corresponding relationship of nodes between the moving and fixed coordinate systems.

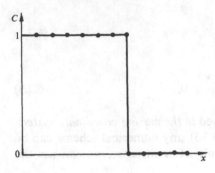

FIGURE 6.12. Comparison between the analytic and the numerical solutions in the moving coordinate system when $D = 0$. —— = analytic solution; ● = numerical solution; $V = 0.369$.

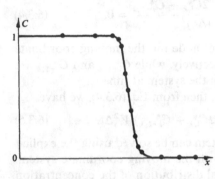

FIGURE 6.13. Comparison between the analytic and numerical solutions in the moving coordinate system when $D \neq 0$. —— = analytic solution; ● = numerical solution; $V = 0.369$; $D = 0.05$.

If this technique is applied to calculate the example mentioned at the end of last section, that is, the advection-dispersion problem in a semi-infinite sand column, then, for the extreme case of $D = 0$, Eq. (6.3.6) can be reduced to

$$C_{j,k+1} = C_{j-1,k}.$$

Thus, a perpendicular front, which exactly coincides with the analytic solution, can be obtained. This case is shown in Figure 6.12. When D is very small, we can also obtain a satisfactory solution from Eq. (6.3.6) as shown in Figure 6.13.

If the velocity changes with time, the coordinate transformation of Eq. (6.3.2) should be modified to

$$\xi = x - \int_0^t V(\tau)\,d\tau. \qquad (6.3.7)$$

We can still obtain the difference equation (6.3.4), but the moving nodes will not coincide with the fixed nodes. In this case, interpolation is required if we want to project the computed results from the moving coordinate system into the fixed coordinate system. This process causes a small amount or numerical dipersion.

The simple one-dimensional problem discussed above shows that Lagrangian methods are powerful in handling advection-dominated problems. Unfortunately, this kind of moving coordinate system method is not easily extended to multidimensional and other complex practical problems.

6.3.2 Element Deformation Methods

The idea of this method is that the finite element nodes can move within a fixed coordinate system. If the nodes are concentrated around the front and can automatically move with the front of migration, the local Peclet number will become smaller and the difficulties of numerical solutions can be overcome.

Assume that the spatial distribution of the nodes at time t_k is $\{X_k\}$ and their concentration distribution is known to be $C^k(X_k)$. We wish to find the distribution of nodes $\{X_{k+1}\}$ at time $t_{k+1} = t_k + \Delta t$ and the corresponding concentration distribution $C^{k+1}(X_{k+1})$. The outline of this method is given as follows:

1. Predict the position of the steep front at time t_{k+1} using the common Galerkin FEM.
2. Rearrange the nodes so as to increase the density near the steep front and form the distribution $\{X_{k+1}\}$.
3. Solve $C^k(X_{k+1})$ from the known $C^k(X_k)$ as the initial values of the solution by means of interpolation.
4. Reform the coefficients of the finite element equations according to the positions of new nodes $\{X_{k+1}\}$ and the given boundary conditions.
5. Solve the finite element equations to obtain $C^{k+1}(X_{k+1})$.

As shown by the steps, the positions of the nodes change with time, so the basis functions and the coefficients of the equations all depend on time. Since the numerical system has to be updated each time step, the computational effort is significantly increased.

O'Neill (1981) studied the *one-dimensional element deformation method* combined with moving coordinates. The position of each moving point x at time t can be expressed as

$$x = x(x_0, t) \tag{6.3.8}$$

where x_0 is the initial position of the moving point. Because $C(x, t) = C(x(x_0, t), t)$, we have

$$\left(\frac{\partial C}{\partial t}\right)_{x_0} = \left(\frac{\partial C}{\partial t}\right)_x + \frac{\partial C}{\partial x}\left(\frac{\partial x}{\partial t}\right)_{x_0}, \tag{6.3.9}$$

where $(\partial x/\partial t)_{x_0} = dx/dt$ represents the velocity of moving point x, and $(\partial C/\partial t)_x$ indicates the change of concentration with time. Then Eq. (6.3.1)

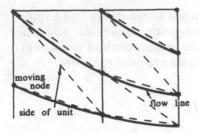

FIGURE 6.14. Two-dimensional Lagrangian element.

becomes

$$\left(\frac{\partial C}{\partial t}\right)_{x_0} + \left(V - \frac{dx}{dt}\right)\frac{\partial C}{\partial x} - D\frac{\partial^2 C}{\partial x^2} = 0. \tag{6.3.10}$$

From this equation, we can see that if the velocity of the moving point dx/dt can be maintained close to velocity V, the advection-dominated equation (6.3.1) will become the dispersion-dominated equation (6.3.10).

If we want to extend the element deformation method or the moving point method to two-dimensional problems, we have to find a way to generate the element deformation. Lynch and O'Neill (1980) proposed a two-dimensional element deformation method combined with the moving boundaries. Thomson et al. (1984) adopted linear triangle elements that can move along the flowlines. The moving nodes are distributed along the flowlines and the strips between every two flowlines are triangulated as shown in Figure 6.14. Yang Jinzhong (1985) designed a similar method for two-dimensional problems using moving nodes. He gave a specific method for increasing or decreasing the elements and modifying the shapes of elements.

6.3.3 *Moving Point Methods*

Moving point methods are another kind of Lagrangian method which can substitute for the element deformation and the moving coordinates methods. Suppose that there are a group of moving points in the flow region and they are "carriers" of the solute concentrations. The displacements of the moving points represent the advective transport. In the advection-dominated case, this is the major portion of transport. The dispersion effect can be realized by appropriate modification of the concentrations of the moving points. We have applied this technique in the method of characteristics and the Random-Walk method presented in Section 4.2 for solving advection-dominated problems.

In the *method of characteristics*, the flow region is divided into a set of finite difference elements. Moving points move in the element system, and the concentrations of both the moving points and the finite difference nodes (the centers of grid squares) can be "projected" onto each other by means of interpolation.

In the *Random-Walk method*, the modification of the concentrations of moving points does not depend on a fixed grid. In other words, it is unnecessary to project the concentration of the moving points onto the nodes at each time step. The modification of the concentration of moving points generated by the dispersion effect is determined by the rule of random walk and so it is not necessary to project the concentration of the fixed nodes onto the moving points either. Thus, the moving points are allowed to move freely in the domain for many steps. Only when we wish to know the distribution of the concentration is it necessary to project the concentrations of the moving points onto the fixed grid. In this way, we can avoid calculation errors resulting from repeated interpolations. The accuracy of Random-Walk methods depends basically on the number of moving points. If the number is insufficient, local mass balance cannot be maintained.

The *techniques of moving points* are flexible and powerful in handling of advection-dominated problems, but when the dispersion effect increases, solution accuracy may be decreased. As pointed out by Neuman (1981), the convergence of these methods has not yet been proven. Although the numerical test made by Farmer (1985) showed that the method of characteristics with quadratic interpolation is always convergent, the detail of the numerical analysis was not given. Besides the Random-Walk methods, Raviat (1985) proposed another moving point method that did not require a fixed grid. Some improved forms of moving point methods will be presented in the next section.

6.4 The Modified Methods of Characteristics

6.4.1 The Single Step Reverse Method

Eulerian methods seem to be quite suitable for solving dispersion-dominated problems, while Lagrangian methods are very powerful for handling advection-dominated problems. The problem is how to combine the two types to get satisfactory results for both large and small Peclet numbers. In this section, we will introduce several Eulerian-Lagrangian methods.

The *single step reverse method* is a modified characteristics method. It was first proposed by Hinstrup (1977), and further analyzed by Neuman and Sorek (1982), Douglass and Russell (1982), and others. Cheng et al. (1984) presented a finite difference version associated with the natural coordinate system, which makes the method more practical.

Consider the following advection-dispersion equation:

$$\frac{\partial C}{\partial t} + \mathbf{V} \cdot \nabla C = \nabla \cdot (\mathbf{D} \nabla C). \tag{6.4.1}$$

Using the total derivative

$$\frac{DC}{Dt} = \frac{\partial C}{\partial t} + \mathbf{V} \cdot \nabla C, \tag{6.4.2}$$

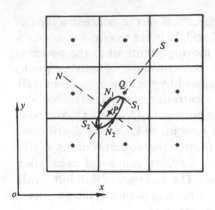

FIGURE 6.15. Relationship between the finite difference grids of the local nature coordinate system and Cartesian coordinates.

Eq. (6.4.1) can be rewritten as

$$\frac{DC}{Dt} = \nabla \cdot (\mathbf{D} \nabla C). \tag{6.4.3}$$

The total derivative DC/Dt represents the rate of concentration change observed by following a moving particle. In this case, only the pure dispersion phenomenon described by the right-hand side of Eq. (6.4.3) is observed. So Eq. (6.4.3) is actually an advection-dispersion equation from the Lagrangian point of view.

For a finite difference node Q (i,j) shown in Figure 6.15, a point P can be found which is located on a streamline passing through node Q. It will arrive at node Q just within one time step. Point P is called the *single step reverse point* of node Q.

According to the meaning of the total derivative DC/Dt, we have the following finite difference approximation:

$$\frac{DC}{Dt} = \frac{C_{i,j}^{k+1} - C^k(P)}{\Delta t}, \tag{6.4.4}$$

where $C_{i,j}^{k+1}$ is the concentration of node Q at $t_{k+1} = t_k + \Delta t$, and $C^k(P)$ is the concentration of point P at t_k. In order to derive the discrete form of the right-hand side of Eq. (6.4.3), we will construct a local natural coordinate system at point P, which is formed by the tangent and orthogonal directions to the streamline; see Fig. 6.15. In this coordinate system, Eq. (6.4.3) becomes

$$\frac{DC}{Dt} = \frac{\partial}{\partial s}\left(D_s \frac{\partial C}{\partial s}\right) + \frac{\partial}{\partial n}\left(D_n \frac{\partial C}{\partial n}\right), \tag{6.4.5}$$

where D_s and D_n are principal dispersion coefficients along the principal dispersion directions, i.e., the longitudinal and transverse dispersion coefficients, respectively.

Using the explicit central difference to discretize the right-hand side of Eq. (6.4.5), we have

$$\frac{\partial}{\partial s}\left(D_s\frac{\partial C}{\partial s}\right) + \frac{\partial}{\partial n}\left(D_n\frac{\partial C}{\partial n}\right) = D_s\frac{C^k(S_1) - 2C^k(P) + C^k(S_2)}{(\Delta s)^2}$$

$$+ D_n\frac{C^k(N_1) - 2C^k(P) + C^k(N_2)}{(\Delta n)^2}, \qquad (6.4.6)$$

where Δs and Δn are grid distances in the natural coordinate system, and S_1, S_2 and N_1, N_2 are the relevant nodes; see Figure 6.15. Substituting Eqs. (6.4.4) and (6.4.6) into Eq. (6.4.5), we obtain

$$C_{i,j}^{k+1} = C^k(P) + \left[\frac{\Delta tD_s}{(\Delta s)^2}\right][C^k(S_1) - 2C^k(P) + C^k(S_2)]$$

$$+ \left[\frac{\Delta tD_n}{(\Delta n)^2}\right][(C^k(N_1) - 2C^k(P) + C^k(N_2)], \qquad (6.4.7)$$

letting

$$\Delta s = (D_s\Delta t)^{1/2}, \quad \Delta n = (D_n\Delta t)^{1/2}, \qquad (6.4.8)$$

we have

$$\frac{\Delta tD_s}{(\Delta s)^2} = 1, \quad \text{and} \quad \frac{\Delta tD_n}{(\Delta n)^2} = 1.$$

Thus, Eq. (6.4.7) is simplified to

$$C_{i,j}^{k+1} = C^k(P) + [C^k(S_1) + C^k(S_2) + C^k(N_1) + C^k(N_2) - 4C^k(P)], \quad (6.4.9)$$

where the terms on the right-hand side can be determined from interpolation of known concentrations of all nodes at time t_k. Although Eq. (6.4.9), which is associated with node (i,j), has a very simple form, it correctly expresses the tensor property of dispersion. To assume that the principal directions of dispersion always coincide with the coordinate directions is impractical. However, if all components of the dispersion tensor are taken into consideration, the equation will be too tedious. Therefore, Eq. (6.4.9), which adopts the local natural coordinate system, is an interesting discrete form. It is a typical Eulerian-Lagrangian algorithm. On its left-hand side are the unknown concentrations of the fixed node. The first term on its right-hand side represents the advection effect determined by the concentration of moving point P. The other terms within the square brackets represent the dispersion effect. When the dispersion coefficient is zero, only the advection term is left in Eq. (6.4.7). Thus, there is no numerical dispersion in this case.

If the concentrations of all nodes, $C_{i,j}^k$, at time t_k are known, we may use the single step reverse method to solve the concentrations of all nodes $C_{i,j}^{k+1}$ at time $t_{k+1} = t_k + \Delta t$. The major steps of the single step reverse method are given as follows:

1. Find the single step reverse point of each node at t_k. For instance, the coordinates of node Q are assumed to be (x_Q, y_Q), so the coordinates of its single step reverse point P can be calculated by the following approximations:

$$x_P^k = x_Q - V_x^k \cdot \Delta t, \quad y_P^k = y_Q - V_y^k \cdot \Delta t. \tag{6.4.10}$$

2. Apply Eq. (6.4.8) to determine Δs and Δn, then find points S_1, S_2, N_1 and N_2 along the direction of line PQ and its perpendicular direction, respectively.
3. Use an interpolation method to obtain $C^k(P)$, $C^k(S_1)$, $C^k(S_2)$, $C^k(N_1)$ and $C^k(N_2)$.
4. Substitute these values into Eq. (6.4.9) to obtain $C_{i,j}^{k+1}$.

Evidently, Δt must be sufficiently small so that the coordinates of point P calculated from Eq. (6.4.10) can be guaranteed to be accurate enough, and the convergence of the explicit scheme (6.4.9) can also be ensured. The key problem here is the choice of an interpolation formula for step 3 which can maintain the accuracy of the solution. It is easy and feasible to apply the bilinear interpolation with the neighboring four nodes taken as the basis points. For instance, in Figure 6.15, the concentration of point P, i.e., C_p^k, can be determined by the bilinear interpolation based on $C_{i-1,j}^k$, $C_{i,j-1}^k$, $C_{i+1,j}^k$, and $C_{i,j+1}^k$. However, Cheng et al. (1984) pointed out that when the bilinear interpolation formula is used, the accuracy of their method is equivalent to the upstream FDM. Therefore, they suggested using the nine neighboring nodes to generate a *higher-order finite element interpolation*.

In the single step reverse method, it is not necessary to locate and trace a group of moving points. This is the main difference between the single step reverse and the method of characteristics. As a result, the calculation process of the single step reverse method is relatively simple. However, it still uses moving points, such as P, S_1, S_2, N_1, N_2 and so on, but these moving points are redefined in each time step.

Although we have applied the finite difference approximation to the dispersion part on the right-hand side of Eq. (6.4.5), the finite element approximation could be used instead. Neuman and Sorek (1982), Douglass and Russell (1982) and others discussed the single step reverse techniques in combination with FEM. Donea (1984) combined the linear Galerkin FEM and a high-accuracy interpolation to form a single step reverse method.

6.4.2 The Hybrid Single Step Reverse–Moving Point Method

Although moving points are already involved in the single step reverse method, their number may be insufficient for depicting a steep concentration front. With few moving points, numerical dispersion will still exist near the front. Neuman (1984) and Farmer (1985) suggested combining the single step reverse method with a *self-adaptive moving point technique*. Their basic idea is that the single step reverse method alone is used in the portion where the front is smooth, while moving points are added automatically around the steep front. When the front becomes smooth, the moving points will disappear automatically. The purpose of this method is to take advantage of the simplicity of the single step reverse method and the effectiveness of the moving point method in defining the shape of steep front. This makes the presented hybrid method applicable for both large and small Peclet number problems.

In general, for each transient state, the steep front only appears in a small portion of the flow region. Let us first design a method to automatically control the distribution of moving points so that they can be concentrated around the steep front.

At time t_k, define

$$S_{i,j}^k = (C_{i+1,j}^k - C_{i-1,j}^k)^2 + (C_{i,j+1}^k - C_{i,j-1}^k)^2, \qquad (6.4.11)$$

and prescribe a critical value S^*. When $S_{i,j}^k < S^*$, the front in element (i,j) is smooth, otherwise, it is steep. Furthermore, let the allowed minimum and maximum numbers of moving points in the element with a smooth front be m_b, M_b, respectively. Similarly, define m_s and M_s for the element with a sharp front. The number of moving points, $v_{i,j}^k$, can be controlled automatically by the following rules:

1. $m_b \leq v_{i,j}^k \leq M_b$, when $S_{i,j}^k < S^*$;
2. $m_s \leq v_{i,j}^k \leq M_s$, when $S_{i,j}^k > S^*$.

If the number of moving points in an element exceeds the requirements of the above rules, some moving points should be eliminated; otherwise, if they are insufficient, new points need to be generated. This process can be controlled automatically by the computer program.

In using the hybrid method, we first form the initial concentrations of all nodes $\{C_{i,j}^0\}$ according to the initial conditions, then compute $\{S_{i,j}^0\}$. If $S_{i,j}^0 < S^*$, m_b moving points should be placed in the element; otherwise, m_s moving points should be placed. The concentrations of the moving points $\{C_v^0\}$ are determined based on the initial conditions, where the subscript v represents the moving point. If concentrations $\{C_{i,j}^k\}$, moving points $\{v^k\}$, and the concentrations of these points $\{C_v^k\}$ have been obtained for time t_k, we can use the following steps to compute $\{C_{i,j}^{k+1}\}$, $\{v^{k+1}\}$, and $\{C_v^{k+1}\}$:

1. Apply the single step reverse method to solve for the concentrations of all nodes at t_{k+1}, which are written as $\{\overline{C}_{i,j}^{k+1}\}$. The dispersion part, i.e., the term within the square brackets on the right-hand side of Eq. (6.4.9), is written as $\delta C_{i,j}^k$;
2. Use interpolation to estimate the effect of dispersion on the concentrations of the moving points, $\{\delta C_v^k\}$, and modify the moving point concentrations as follows

$$C_v^{k+1} = C_v^k + \delta C_v^k; \qquad (6.4.12)$$

3. Project the concentrations of the moving points onto the fixed nodes. This can be done by using the simple arithmetic mean, distance weighted mean or an interpolation method. The concentrations of the nodes thus obtained are written as $\{\tilde{C}_{i,j}^{k+1}\}$;
4. Define $C_{i,j}^{k+1}$ to be the weighted mean between $\bar{C}_{i,j}^{k+1}$ and $\tilde{C}_{i,j}^{k+1}$, i.e., let

$$C_{i,j}^{k+1} = \lambda\bar{C}_{i,j}^{k+1} + (1 - \lambda)\tilde{C}_{i,j}^{k+1}, \tag{6.4.13}$$

where $0 < \lambda < 1$. The weighting parameter λ is taken to be close to zero in elements occupied by a steep front because these elements contain more moving points and $\tilde{C}_{i,j}^{k+1}$ is rather accurate. For elements located in a smooth front, λ is chosen to be close to one because there are fewer moving points and $\bar{C}_{i,j}^{k+1}$ is more reliable;
5. Increase or reduce the number of moving points in each element according to the criteria given above.

6.4.3 The Hybrid Moving Point—Characteristics Finite Element Method

For the standard Galerkin finite element procedure, we assume the approximate solution

$$C(x, y, t) \approx \sum_{j=1}^{N} C_j(t)\phi_j(x, y), \tag{6.4.14}$$

and require that it satisfy

$$\iint_{(R)} \left[\frac{DC}{Dt} - \nabla \cdot (\mathbf{D}\nabla C)\right]\phi_i \, dR = 0,$$

$$(i = 1, 2, \ldots, N). \tag{6.4.15}$$

Eliminating the second order derivative term by using Green's formula, we have

$$\iint_{(R)} \frac{DC}{Dt}\phi_i \, dR + \iint_{(R)} \mathbf{D}\nabla C \cdot \nabla\phi_i \, dR - \int_{(\Gamma)} g_2\phi_i \, d\Gamma = 0,$$

$$(i = 1, 2, \ldots, N). \tag{6.4.16}$$

In the above equation, boundary condition, $-\mathbf{D}\nabla C \cdot \mathbf{n} = g_2$, has been used along boundary (Γ).

The basis function $\phi_i(x, y)$ are non-zero only in the elements involving node i. For instance, if triangle elements and linear basis functions are adopted, $\phi_i(x, y)$ are non-zero only in those triangle elements which contain node i as their common vertex. Node i is located almost at the center of the subdomain. So approximately, we have

$$\iint_{(R)} \frac{DC}{Dt}\phi_i \, dR = \frac{DC_i}{Dt} \iint_{(R)} \phi_i \, dR, \tag{6.4.17}$$

where DC_i/Dt is the total derivative at node i. Similar to Eq. (6.4.4), we have the following difference approximation at time t_k

$$\frac{DC_i}{Dt} \approx \frac{C_i^{k+1} - C_{p_i}^k}{\Delta t}, \tag{6.4.18}$$

where P_i is a single step reverse point of node i at time t_k. In other words, the moving point located at point P_i at time t_k will move to node i at time t_{k+1}. Substituting Eq. (6.4.18) into Eq. (6.4.17), we obtain:

$$\iint_{(R)} \frac{DC}{Dt} \phi_i \, dR = (b_i/\Delta t)(C_i^{k+1} - C_{pi}^k), \tag{6.4.19}$$

where

$$b_i = \iint_{(R)} \phi_i \, dR. \tag{6.4.20}$$

Using Eq. (6.4.14), we get

$$\iint_{(R)} \mathbf{D}\nabla C \cdot \nabla \phi_i \, dR = \sum_{j=1}^{N} a_{ij} C_j, \tag{6.4.21}$$

where

$$a_{ij} = \iint_{(R)} \mathbf{D}\nabla \phi_i \cdot \nabla \phi_j \, dR. \tag{6.4.22}$$

Substituting Eqs. (6.4.19), (6.4.21), and (6.4.23) into Eq. (6.4.16), we have, at time t_{k+1}:

$$\sum_{j=1}^{N} a_{ij} C_j^{k+1} + \left(\frac{b_i}{\Delta t}\right) C_i^{k+1} = \left(\frac{b_i}{\Delta t}\right) C_{p_i}^k + F_i,$$

$$(i = 1, 2, \dots, N) \tag{6.4.23}$$

where

$$F_i = \int_{(\Gamma)} g_2 \phi_i \, d\Gamma. \tag{6.4.24}$$

Using the notations of vectors and matrices, Eq. (6.4.23) can be expressed as:

$$\left([\mathbf{A}] + \frac{[\mathbf{B}]}{\Delta t}\right) \mathbf{C}^{k+1} = \left(\frac{[\mathbf{B}]}{\Delta t}\right) \mathbf{C}_p^k + \mathbf{F}, \tag{6.4.25}$$

where $[\mathbf{A}]$ is a symmetric matrix with its elements a_{ij} defined by Eq. (6.4.22), and $[\mathbf{B}]$ is a diagonal matrix whose elements are determined by Eq. (6.4.20). Therefore, the coefficient matrix of Eq. (6.4.25), $([\mathbf{A}] + [\mathbf{B}]/\Delta t)$, is also a symmetric matrix. The elements of vector F are determined from Eq. (6.4.23). If the equation contains a source/sink term, it may also be included in vector F, just as it is in the standard FEM. $\mathbf{C}_P^k$ on the right-hand side of Eq. (6.4.25) represents the concentration vector of the single step reverse point at time t_k, which may be determined from the known concentration vector $\mathbf{C}^k$ by means of interpolation. $\mathbf{C}^{k+1}$ can then be computed using Eq. (6.4.25). This is the *single step reverse FEM*, which is also called the *modified characteristics FEM*.

For each node i, $C_i^{k+1} - C_{P_i}^k$ represents the difference between the total concentration and the concentration due to pure advection, i.e., the concentration variation caused by dispersion. It is written as $\overset{o}{C}_i(t_{k+1})$, and we have

$$\overset{o}{C}_i(t_{k+1}) = C_i^{k+1} - C_{pi}^k. \tag{6.4.26}$$

Since C_i^{k+1} has been solved by Eq. (6.4.25), $\overset{o}{C}_i(t_{k+1})$ is also known. Using finite element interpolation,

$$\overset{o}{C}_i(x, y, t_{k+1}) = \sum_{i=1}^{N} \overset{o}{C}_i(t_{k+1})\phi_i(x, y), \tag{6.4.27}$$

we can obtain the concentration change generated by the dispersion effect for an arbitrary point (x, y) at time t_{k+1}.

In order to increase the accuracy in simulating the steep front, we may place some moving points in the domain to trace the development of the front. It is assumed that there are v moving points, and their coordinates are (x_m, y_m), $m = 1, 2, \ldots, v$. If their concentrations C_m^k $(m = 1, 2, \ldots, v)$ at time t_k are known, their concentrations at time t_{k+1} should be the sum of advection and dispersion effects:

$$C_m^{k+1} = C_m^k + \overset{o}{C}_i(x_m, y_m, t_{k+1})$$

$$= C_m^k + \sum_{i=1}^{N} \overset{o}{C}_i(t_{k+1})\phi_i(x_m, y_m). \tag{6.4.28}$$

Neuman (1984) proposed a smooth criterion for controlling the number of moving points in the elements. If the concentrations of all moving points calculated from Eq. (6.4.28) in an element are between the maximum and minimum concentrations of the element, then the smooth criterion is considered to be satisfied in the element. If an element satisfies the smooth criterion for several time steps in succession, the moving points within the element may be eliminated to increase the efficiency of the solution process. If the criterion is not satisfied, we can update the nodal concentrations by projecting the moving points onto the nodes and introducing new moving points. The concentrations of moving points are defined by equations similar to Eq. (6.4.13). Furthermore, we can use

$$C_m^{k+1} = \sum_{i=1}^{N} C_i^{k+1}\phi_i(x_m, y_m) \tag{6.4.29}$$

to define the concentrations of the nodes.

This method has advantages of both the Eulerian and the Lagrangian methods. It is applicable for any Peclet number, from 0 to ∞, and even if the

Courant number exceeds 1, satisfactory results will still be obtained. Since the moving points are placed only near the steep front, the computational efficiency is rather high. Another advantage of this method is that the coefficient matrix in the finite element equation remains symmetric. This also contributes to the efficiency of the method. Some applications of this method were given by Neuman (1984), Cady and Neuman (1987), and Huang Kangle (1985). Instead of using the FEM, Bentley and Pinder (1992) used the least squares collocation method to combine with the Eulerian-Lagrangian method.

Yeh (1990) proposed an Eulerian-Lagrangian method, in which the technique of self-adaptive element partition is used instead of moving points. In each element there are K hidden nodes. When the concentration front in an element becomes steep, the hidden nodes will emerge as common nodes. The element is then subdivided into many finer elements in order to reduce the local Peclet number and increase the accuracy of the solution. Conversely, when the concentration front in the element becomes smooth, these nodes will return to the hidden state for the sake of reducing the amount of computation. Yeh (1990) called this method the *Lagrangian-Eulerian method with zoomable hidden fine-grid* (LEZOOM). Its basic steps are similar to those of the hybrid moving points FEM (FDM), but the nodes which can hide or emerge are used to substitute for the moving points that may be added or removed. This approach is more convenient in programming and more accurate for depicting the front than the moving points method.

A new approach, the *Eulerian-Lagrangian localized adjoint method* (ELLAM), for solving the advection-dispersion equation was developed by Celia et al. (1990) and Russell (1990). ELLAM can consistently treat boundary fluxes and maintain the mass balance. Thus, the problem of large mass balance error associated with characteristic methods may be avoided. Recently, Healy and Russell (1993) combined ELLAM with an integrated finite difference discretization. Both global and local mass balances are guaranteed.

When the flow velocity changes with time and space, the key to increasing the accuracy of Lagrangian-Eulerian methods is to improve the calculation of the velocity field. Generally, the groundwater flow problem is solved first for obtaining the head distribution, and then Darcy's law is employed to compute the distribution of velocities. During this process, the solution accuracy is greatly affected by the numerical differentiation. Segol et al. (1975), Ewing et al. (1983), and Russell and Wheeler (1983) suggested the use of the *mixed FEM*, in which Darcy's velocities are taken as the solutions of the first-order partial differential equations to be solved simultaneously with the distribution of the water heads. This improves the accuracy in computing the velocity field. We shall give an introduction to this method in combination with the discussion of salt intrusion problems in Section 8.2. Chiang et al. (1989) presented a numerical method for solving the problems of mass transport in groundwater, in which the modified characteristics method (MMOC) is combined with the mixed finite element (MFE) method.

Regarding other numerical solutions to advection-dominated problems, the work of some experts should be mentioned: the finite analytic solution by Hwang et al. (1985), the boundary element method by Taigbenu and Liggett (1986), the advection controlling method by Sun and Liang (1988), and the Laplace transform Galerkin method by Sudicky (1989). Li et al. (1992) extended the finite analytic method to solve the transient advection-dispersion problem through the Laplace transformation. When advection dominates dispersion, this method can produce accurate results for a wide range of Peclet numbers. Zeitoun and Pinder (1993) used an optimal control least squares method to solve coupled flow-transport problems. At each time step, the solution of the discretized differential system is transformed into an optimal control problem. A one-dimensional example shows that this method can produce more accurate results than FEM when the Peclet number is large.

Exercises

6.1. Calculate the numerical enlargement factor when the Crank-Nicolson finite difference scheme is used for solving Eq. (6.1.1).

6.2. Derive the third-order upstream difference formula in Eq. (6.2.9).

6.3. Use the example given in Section 5.2.4 to compare the results obtained by FDM, FEM, or MCBM, with and without upstream weights.

6.4. Extend Eq. (6.3.10) to the two-dimensional case.

6.5. Solve the one-dimensional advection-dispersion problem presented in Section 5.2.4 with the single step reverse method.

6.6. Write a flow chart for the hybrid moving point and characteristics FEM given in Section 6.4.3.

7
Mathematical Models of Groundwater Quality

7.1 The Classification of Groundwater Quality Models

7.1.1 Hydrodynamic Dispersion Models

We have introduced the mechanism of hydrodynamic dispersion, derived the hydrodynamic dispersion equations, and presented various methods for solving this kind of equation in previous chapters. Let us now consider how to construct and solve the general *hydrodynamic dispersion models*, or the *advection-dispersion models*.

Hydrodynamic dispersion equations contain some parameters such as the hydrodynamic dispersion coefficients, mean flow velocity, fluid density and source/sink terms. We should first determine their values, and then solve for the concentration distribution. Generally, the variation of solute concentration may affect the density and viscosity of the fluid, and conversely, the changes of fluid density and viscosity may cause the state of the flow field to change. In other words, the concentration distribution and the velocity distribution are interconnected. The concentration distribution is dependent on the velocity distribution, and vice versa. They are both unknown functions. Therefore, a single equation of hydrodynamic dispersion is not enough in the general case. To solve the problem of groundwater quality in a saturated zone, we need the following system of non-linear partial differential equations:

the *hydrodynamic dispersion equation*,

$$\frac{\partial C}{\partial t} = div\left[\mathbf{D}\rho \, grad\left(\frac{C}{\rho}\right)\right] - div(C\mathbf{V}) + I; \qquad (7.1.1)$$

the *continuity equation*,

$$\frac{\partial \rho}{\partial t} + div(\rho\mathbf{V}) = 0; \qquad (7.1.2)$$

the *kinetic equations*,

$$V_i = -\frac{k_{ij}}{\mu n}\left(\frac{\partial p}{\partial x_j} + \rho g \frac{\partial z}{\partial x_i}\right); \quad (i,j = 1,2,3) \qquad (7.1.3)$$

187

and the *state equations*,

$$\rho = \rho(C, p), \quad \mu = \mu(C, p), \tag{7.1.4}$$

where C denotes the solute concentration; p the pressure; ρ and μ the density and viscosity of the fluid; $\mathbf{D}$ the hydrodynamic dispersion coefficient; k_{ij} the component of hydraulic conductivity tensor; n the effective porosity; V_1, V_2, V_3 the three components of mean velocity $\mathbf{V}$; and I the source/sink term. In Eq. (7.1.3), Einstein's summation convention is used. For an incompressible fluid of low concentration, we can use the following first-order approximation of Eq. (7.1.4):

$$\rho = \rho_0 + \alpha(C - C_0), \quad \mu = \mu_0 + \beta(C - C_0), \tag{7.1.5}$$

where C_0 is a reference concentration, ρ_0 and μ_0 the density and viscosity at C_0, and α and β are constants.

Equations (7.1.1) to (7.1.4) contain seven equations and seven unknowns, which are C, p, ρ, μ, V_1, V_2, and V_3. With appropriate initial and boundary conditions, those unknown functions can be uniquely determined. This set of equations are called the *hydrodynamic dispersion system* or *generalized advection-dispersion model*. For this model, the porous medium can be considered to be heterogeneous, anisotropic and having an arbitrary geometry. Furthermore, the fluid may be inhomogeneous, the density and viscosity may change with the solute concentration, and so forth. However, the flow velocity should not exceed the effective range of Darcy's law and the fluid temperature should be approximately constant.

From both practical and computational considerations, it is very important to distinguish between the cases of a *homogeneous fluid* (ρ and μ are constants) and a *heterogeneous fluid*. If the solute concentration is extremely low, the fluid may be regarded as homogeneous and the solute as an ideal tracer. Therefore, the two cases are also called the *tracer case* and the *general case*, respectively. The solution procedures for the two cases are greatly different. It is simple for the tracer case and complex for the general case.

The Tracer Case

In this case, ρ and μ are constants. Thus, the dispersion equation, continuity equation, and kinetic equations have no effect on the state equations. The solutions of these equations can be separated into two separate subproblems: First, obtaining the velocity distribution from the continuity equation and the kinetic equations, and secondly, substituting the velocity distribution into the advection-dispersion equation to obtain the concentration distribution. The flow chart in Figure 7.1 shows the process of computation.

The General Case

In this case, ρ and μ are determined by the state equations. The changes of the concentration may cause the values of ρ and μ to change. Through the

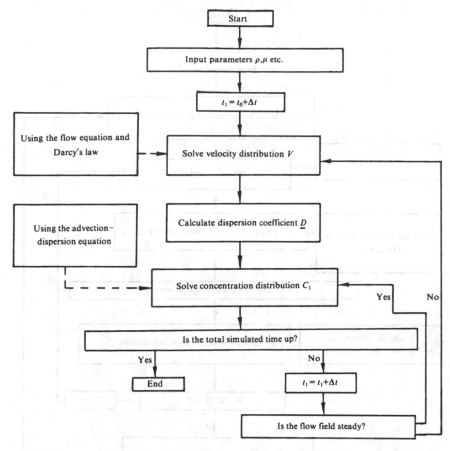

FIGURE 7.1. Flow chart for solving advection-dispersion model in the tracer case.

continuity and kinetic equations, the mean velocity distribution will depend on concentration distribution. Thus, the equations of the hydrodynamic dispersion system must be solved simultaneously. The iteration method for this case is described as follows: assume that the solution at time t has been obtained, and we want to find the solution at the next time, $t + \Delta t$. We may estimate the concentration distribution at $t + \Delta t$ by an extrapolation method and use the state equations to calculate the relevant values of ρ and μ. These values of ρ and μ are then inserted into the continuity and kinetic equations to obtain the velocity $\mathbf{V}$. Next, we calculate the coefficients of the advection-dispersion equation using $\mathbf{V}$ and solve the advection-dispersion equation to update the concentration distribution at $t + \Delta t$. Using the new concentration distribution, we can modify the values of ρ and μ. This process is repeated until the concentration distribution does not change within a certain accu-

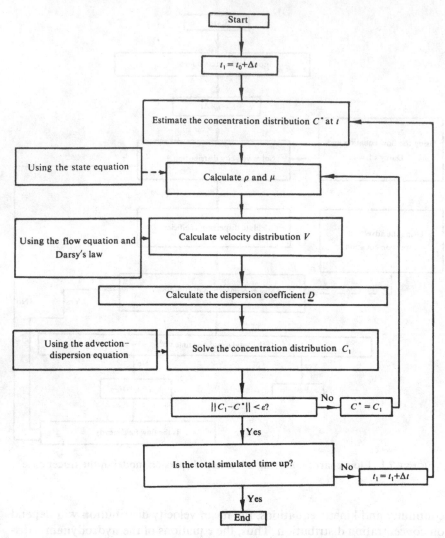

FIGURE 7.2. Flow chart for solving advection-dispersion model in the general case.

racy. We then have the solution at $t + \Delta t$. The flow chart in Figure 7.2 shows the process of computation. Obviously, more computation effort is required in the general case than that in the tracer case. Fortunately, most practical problems of groundwater pollution fall into the tracer case category. One exception is the salt water intrusion problem, which should be taken as the general case, because the salinity will significantly change the density and viscosity of water.

7.1.2 Coupled Equations of Groundwater Flow and Mass Transport

In both the tracer and general cases, the advection-dispersion equation is coupled with the flow equation through the continuity and kinetic equations. In this section, we will give some specific forms of these equations which are often encountered in groundwater pollution problems.

When studying the groundwater flow in a saturated zone, we always use the water head, $h = z + p/\rho g$, as a dependent variable. For an isotropic porous medium, the kinetic equation (7.1.3), i.e., Darcy's law, can be expressed as:

$$V_i = -\frac{K}{n}\frac{\partial h}{\partial x_i}, \quad (i = 1, 2, 3), \tag{7.1.6}$$

where $K = k\rho g/\mu$ is the hydraulic conductivity of the isotropic porous medium. It depends on the density and viscosity of the fluid, as well as, the permeability, k, of the porous medium. For a homogeneous fluid, the partial differential equation of groundwater flow can be obtained by combining Darcy's law and the continuity equation:

$$S_s\frac{\partial h}{\partial t} = div(K \; grad \; h) + W, \tag{7.1.7}$$

where S_s is the specific storativity, and W is the source/sink term. In this case, a model may be constructed by coupling the advection-dispersion equation (7.1.1) with the flow equation (7.1.7) by the aid of Darcy's law, rather than using the general system of hydrodynamic dispersion Eqs. (7.1.1) to (7.1.4). Several common combinations are listed below. All equations are written in scalar forms in the Cartesian coordinates.

Three-Dimensional Dispersion in a Three-Dimensional Flow Field

The flow equation:

$$S_s\frac{\partial h}{\partial t} = \frac{\partial}{\partial x}\left(K\frac{\partial h}{\partial x}\right) + \frac{\partial}{\partial y}\left(K\frac{\partial h}{\partial y}\right) + \frac{\partial}{\partial z}\left(K\frac{\partial h}{\partial z}\right) + W; \tag{7.1.8}$$

the dispersion equation:

$$\begin{aligned}
\frac{\partial C}{\partial t} &= \frac{\partial}{\partial x}\left(D_{xx}\frac{\partial C}{\partial x} + D_{xy}\frac{\partial C}{\partial y} + D_{xz}\frac{\partial C}{\partial z} - CV_x\right) \\
&+ \frac{\partial}{\partial y}\left(D_{xy}\frac{\partial C}{\partial x} + D_{yy}\frac{\partial C}{\partial y} + D_{yz}\frac{\partial C}{\partial z} - CV_y\right) \\
&+ \frac{\partial}{\partial z}\left(D_{xz}\frac{\partial C}{\partial x} + D_{yz}\frac{\partial C}{\partial y} + D_{zz}\frac{\partial C}{\partial z} - CV_z\right) + I. \tag{7.1.9}
\end{aligned}$$

Two-Dimensional Dispersion in a Two-Dimensional Horizontal Flow Field

The flow equation:

$$S\frac{\partial h}{\partial t} = \frac{\partial}{\partial x}\left(Km\frac{\partial h}{\partial x}\right) + \frac{\partial}{\partial y}\left(Km\frac{\partial h}{\partial y}\right) + W'; \tag{7.1.10}$$

the dispersion equation:

$$\frac{\partial(mC)}{\partial t} = \frac{\partial}{\partial x}\left[m\left(D_{xx}\frac{\partial C}{\partial x} + D_{xy}\frac{\partial C}{\partial y} - CV_x\right)\right]$$

$$+ \frac{\partial}{\partial y}\left[m\left(D_{xy}\frac{\partial C}{\partial x} + D_{yy}\frac{\partial C}{\partial y} - CV_y\right)\right] + I', \tag{7.1.11}$$

where m is the saturated thickness of the aquifer; Km is the transmissivity for a confined aquifer; and S is the storativity. For the phreatic aquifer which satisfies the Dupuit assumptions, S should be changed to the effective porosity, n. The dimensions of the two-dimensional source/sink term W' are $[L/T]$, and the dimensions of I' are $[M/L^2T]$. If W' denotes water extraction, then the relevant $I' = -CW'/n$; while if W' denotes water injection, the relevant $I' = C_0W'/n$, in which C_0 is the tracer concentration contained in the injected water.

Two-Dimensional Dispersion in a Two-Dimensional Vertical Flow Field

The flow equation:

$$\frac{\partial\theta}{\partial t} = \frac{\partial}{\partial x}\left(D(\theta)\frac{\partial\theta}{\partial x}\right) + \frac{\partial}{\partial z}\left(D(\theta)\frac{\partial\theta}{\partial z}\right) + \frac{\partial K}{\partial z} + W; \tag{7.1.12}$$

the dispersion equation:

$$\frac{\partial(\theta C)}{\partial t} = \frac{\partial}{\partial x}\left[\theta\left(D_{xx}\frac{\partial C}{\partial x} + D_{xz}\frac{\partial C}{\partial z} - CV_x\right)\right]$$

$$+ \frac{\partial}{\partial z}\left[\theta\left(D_{xz}\frac{\partial C}{\partial x} + D_{zz}\frac{\partial C}{\partial z} - CV_z\right)\right] + I, \tag{7.1.13}$$

where $D(\theta)$ is the diffusion coefficient of water in soil, and θ is the moisture content. If we use the pressure head, ψ, as the dependent variable, and translate the flow equation into

$$(\zeta + \beta S_s)\frac{\partial\psi}{\partial t} = \frac{\partial}{\partial x}\left(K\frac{\partial\psi}{\partial x}\right) + \frac{\partial}{\partial z}\left(K\frac{\partial\psi}{\partial z}\right) + \frac{\partial K}{\partial z} + W, \tag{7.1.14}$$

then the two-dimensional dispersion problem in a saturated-unsaturated flow field can be solved. In Eq. (7.1.14), $\zeta = \partial\theta/\partial\psi$; $\beta = 1$ is taken in the saturated zone and $\beta = 0$ in the unsaturated zone. In the unsaturated zone, the value of K is dependent on ψ and is smaller than that in the saturated zone. This problem will be discussed in detail in the next chapter.

Besides the above cases, some other hydrodynamic dispersion models, such as the water quality in three-dimensional unsaturated zones, the water quality in multilayer leaking aquifers, and so on, can also be listed without difficulty.

From the charts shown in Figures 7.1 and 7.2, we can see that the solution of hydrodynamic dispersion models consists of two parts—solving the flow equation and the advection-dispersion equation separately or simultaneously. We have introduced various numerical methods for handling these two kinds of equations. As a result, now we are able to use these numerical methods to solve the advection-dispersion models in various cases.

7.1.3 Pure Advection Models

Hydrodynamic dispersion always creates a transition zone between tracer marked water and unmarked water. However, if the width of the transition zone is relatively small in comparison with the study area, the *pure advection model* may be adopted without considering the effect of dispersion. This case may be encountered when the contamination is caused by agriculture, seawater intrusion, artificial recharge, and so on. As a result, the difficulty in determining the dispersion coefficient is avoided.

There are two kinds of pure advection models. The first one uses a flow equation only, and the second one requires coupling of a flow equation with a quality equation, but without considering the dispersion. They are described below.

1. Solute Transport Prediction with Flow Equation Only

With the transition zone neglected, it is assumed that there is an abrupt interface between two different bodies of water with different qualities. The *abrupt interface* is also called a front, which moves uninterruptedly with the development of pollution or the transportation of the solute. The problem is how to determine the location of this moving front. Let us introduce a numerical method, which is very simple in concept, for solving this problem.

Suppose the location of the front at time t is known and is described by a set of points on the front. The velocities of these points at moment t can be calculated by the solution of the flow equation and Darcy's law. Assume that point i is located at $[x_i(t), y_i(t), z_i(t)]$ at time t, and the components of velocity are $V_{i,x}(t)$, $V_{i,y}(t)$, $V_{i,z}(t)$. Then the location of this point at $t + \Delta t$ can be predicted by

$$\begin{cases} x_i(t + \Delta t) \approx x_i(t) + V_{i,x}(t) \cdot \Delta t, \\ y_i(t + \Delta t) \approx y_i(t) + V_{i,y}(t) \cdot \Delta t, \\ z_i(t + \Delta t) \approx z_i(t) + V_{i,z}(t) \cdot \Delta t. \end{cases} \qquad (7.1.15)$$

When all the locations of these points at $t + \Delta t$ are obtained, the approximate location of the front at $t + \Delta t$ can be depicted. In fact, this procedure has been used in the *method of characteristics* (MOC).

It should be noted that the location obtained by this method is only the mean location of the front, and not the real spreading area of the tracer. The real spreading area is always larger than the area defined by the front. Consequently, the travel time of the pollutants from the source to a certain place calculated by the location of the front only represents the *mean travel time*. The estimate of the travel time is very useful in practice. For instance, Cherry et al. (1973) studied the movement of a radioactive substance using this method to see if the substance could pass through an aquifer and enter a river from its disposal site. They obtained a velocity field based on the field studies and the finite element analysis. There is a fault zone between the disposal site and the river. Their calculation results show that the passing time is about 100 years. Even if the uncertainty of the estimated transmissivity is taken into consideration, the travel time is still larger than 20 years. Because of adsorption, the practical transport velocity of the radioactive substance is only about 1/10 of that of the groundwater flow. Therefore, the real travel time should be at least 200 years. During this period, the concentration of the radioactive substance (^{90}Sr and ^{137}Cs) should have been reduced to a very low level by natural decay.

With respect to the study of travel time, the research by Kirkham and Stotres (1978) and Cushman and Kirkham (1978) should be mentioned. They studied the travel time of a solute from a single or multi-layer aquifer to a well, where the flow field is defined by an analytic solution. Nelson (1978) also adopted the concept of travel time in the study of regional groundwater pollution problems. The transient flow field that he used was obtained by the finite difference method.

2. The Method of Coupling the Flow and Water Quality Equations

To explain this method, let us consider the groundwater pollution problem in a two-dimensional aquifer. Partition the flow region into several elements, the shapes of which are usually rectangles or other polygons. Figure 7.3 shows a polygonal element i, where j is an neighboring element of it. To obtain the water mass balance and solute mass balance equations with respect to element i, the following balance factors are usually involved:

1. The flow rate Q_{ij} between element i and its adjacent element j. It is positive for inflow and negative for outflow. The solute concentration of Q_{ij} is denoted by C_{ij};
2. The flow rate N_i from the unsaturated zone. The solute concentration of N_i is denoted by C_{N_i};
3. The flow rate R_i of artificial recharge. The solute concentration of R_i is denoted by C_{R_i};
4. The drainage rate P_i from wells or springs in the element, the solute concentration of which is just the concentration of the element, C_i.

FIGURE 7.3. Water balance and solute mass balance in element i.

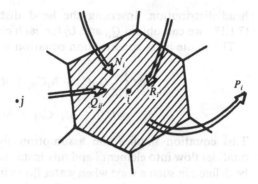

With the above factors considered, the water mass balance equation for element i is

$$\Delta t \left\{ \sum_{(j)} Q_{ij} + N_i + R_i - P_i \right\} = U_i(t + \Delta t) - U_i(t), \qquad (7.1.16)$$

where $\sum_{(j)}$ represents the summation of all the elements adjacent to element i, U_i the water contained in element i, and $U_i(t + \Delta t) - U_i(t)$ denotes the change of water volume in element i within Δt. If a side of the element is a section of the inflow (or outflow) boundary of the region, then we should add (or deduct) the water passing through the boundary in the balance equation (7.1.16).

Using the area A_i, porosity n_i and the height of aquifer bottom b_i of element i, and using the hydraulic conductivity K_{ij}, aquifer thickness m_{ij}, the width of water carrying section l_{ij}, and the distance r_{ij} between nodes i and j, all flux terms in Eq. (7.1.16) can be represented by the hydraulic head. Let h_i and h_j be the water heads of nodes i and j, respectively, we then have

$$Q_{ij} = \frac{K_{ij} m_{ij} l_{ij}}{r_{ij}} (h_j - h_i), \qquad (7.1.17)$$

$$U_i = n_i A_i (h_i - b_i). \qquad (7.1.18)$$

Inserting these equations into Eq. (7.1.16), we arrive at

$$\sum_{(j)} \lambda_{ij}(h_j - h_i) + N_i + R_i - P_i = n_i A_i \frac{h_i(t + \Delta t) - h_i(t)}{\Delta t}, \qquad (7.1.19)$$

where

$$\lambda_{ij} = \frac{K_{ij} m_{ij} l_{ij}}{r_{ij}}.$$

Readers may recognize that Eq. (7.1.19) is actually a finite difference equation of ground water flow associated with the polygonal element. Solving the balance equations for all elements simultaneously, we can obtain the water head distribution. Inserting the head distribution into Eqs. (7.1.17) and (7.1.18), we can obtain Q_{ij} and U_i for each element.

The solute mass conservation equation within element i is

$$\Delta t \left\{ \sum_{(j)} Q_{ij} C_{ij} + N_i C_{N_i} + R_i C_{R_i} - P_i C_i \right\}$$
$$= U_i(t + \Delta t) \cdot C_i(t + \Delta t) - U_i(t) \cdot C_i(t). \tag{7.1.20}$$

This equation implies the assumption that water bodies with different qualities flow into element i and mix instantaneously. In this equation, C_{ij} can be defined in such a way: when water flows from j to i, that is, $Q_{ij} > 0$, $C_{ij} = C_j$ is used; conversely, when water flows from i to j, i.e., $Q_{ij} < 0$, then $C_{ij} = C_i$ should be used. We may also take C_{ij} as the weighted mean of C_i and C_j:

$$C_{ij} = (1 - \delta)C_i + \delta C_j. \tag{7.1.21}$$

The value of $\delta = 0.75$ is suggested for $Q_{ij} > 0$, and $\delta = 0.25$ for $Q_{ij} < 0$. This factor acts as an "upstream weight." Substituting Eq. (7.1.21) into Eq. (7.1.20), we obtain a finite difference equation for the concentration. For elements located on the boundary, we must consider boundary conditions when establishing the balance equation (7.1.20). If the values of concentration on the left-hand side of Eq. (7.1.20) are taken at time t, the solution scheme is explicit and we can directly calculate the unknown concentration $C_i(t + \Delta t)$ on the right-hand side. If the values of concentration on the left-hand side are taken at time $t + \Delta t$, the solution scheme is implicit. In this case, we must solve a set of equations containing all the elements to obtain the concentration distribution at $t + \Delta t$. Regardless of whether the explicit or the implicit scheme is used, numerical dispersion cannot be avoided, and a "transition zone" will be generated. The "upstream weight" in Eq. (7.1.21) may reduce the oscillations of the numerical solution.

Pure advection models of water quality should also be divided into the tracer case and the general case. For the tracer case, Q_{ij} and U_i can be obtained by solving Eqs. (7.1.17) to (7.1.19) for each time step. Then, inserting them into Eq. (7.1.20), we can obtain the concentration distribution. The flow equation and the quality equation are solved separately. For the general case, the flow and quality equations must be solved iteratively in combination with the state equations (7.1.5).

7.1.4 Lumped Parameter Models

The advection-dispersion models and pure advection models mentioned above are all distributed parameter models. Although in the pure advection model, the problem of determining dispersion coefficients can be avoided, input of the values of hydraulic conductivity and porosity, input of infiltration and artificial recharge factors for different locations, input of the locations of pollution sources and their intensities, and input of boundary conditions are still required. In practice, it is very difficult to obtain all of these data.

If we only concern the average level of pollution in an aquifer and want to predict its variation with time, the so-called *"black-box model"* or a single element model can be used. In this kind of model, the concentration only depends on time, not on space. Therefore, it is often called the *lumped parameter model* of water quality.

Let us imagine an aquifer as a black-box where pollutants flow in or out. By analyzing the observed input-output data, we may find an input-output relationship of the box, although the structure of the box is unknown. Once the input-output relationship is found, we can use it to predict the aquifer response (model output) corresponding to any excitation to the aquifer. For instance, rainfall may bring contaminants from the ground surface into the aquifer (input) and then drains them into a river (output). If we only wish to know the relationship between input and output, rather than the pollution level in different places of the aquifer, we may construct a black-box model for the aquifer, as shown in Figure 7.4.

In the black-box method, the contaminant inputted is regarded as a signal, $e(t)$, and the function of the aquifer is regarded as an operator, A, which is called the *transfer function*. Signal $e(t)$ is transferred into an output $S(t)$ via A. This relationship can be expressed by *convolution* as

$$S(t) = \int_0^t A(t - \tau)e(\tau)\,d\tau. \tag{7.1.22}$$

To apply this model, we must calculate first the transfer function A based on the observed data of input $e(t)$ and output $S(t)$. In other words, we have to solve the inverse problem of the black-box model. This kind of algorithm is called *inverse convolution*. Fried (1975) gave an introduction to the Emsellem method of inverse convolution, which is based on successive approximations. Once the transfer function A is obtained, one can apply Eq. (7.1.22) to infer model outputs corresponding to different model inputs.

There is another starting point of constructing lumped parameter models, in which the whole flow region is considered as a single element. The water balance and solute mass balance equations for the single element are:

$$\Delta t\{N + R - P - Q\} = U(t + \Delta t) - U(t) \tag{7.1.23}$$

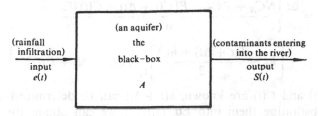

FIGURE 7.4. The black-box model.

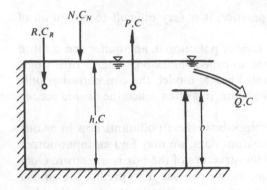

FIGURE 7.5. The model with only one element.

and

$$\Delta t\{NC_L + RC_R - PC - QC\} = U(t + \Delta t)\cdot C(t + \Delta t) - U(t)\cdot C(t), \quad (7.1.24)$$

respectively. Meanings of all symbols in Eqs. (7.1.23) and (7.1.24) are the same as those in Eqs. (7.1.16) and (7.1.20). The subscript i has been omitted because there is only one element now. Since the whole region is taken as one element, concentration C depends only on time. In Eq. (7.1.23), Q represents the rate of outflow from the region. Let this rate be proportional to the thickness of the drainage layer, i.e.,

$$Q = \alpha(h - h_0), \quad (7.1.25)$$

where α is called the drainage coefficient; h is the elevation of water table. When $h \leq h_0$, there will be no outflow at all, as shown in (Figure 7.5). With $U = nAh$, where n is the effective porosity and A the area of the aquifer, Eq. (7.1.23) can be rewritten as

$$h(t + \Delta t) - h(t) = \frac{\Delta t}{A}\left\{\frac{N + R - P}{n} + \left(h_0 - \frac{h(t + \Delta t) - h(t)}{2}\right)\frac{1}{t_h}\right\}, \quad (7.1.26)$$

where $t_h = n/\alpha$ is called the *response time of the system*. Eq. (7.1.24) can then be written as

$$h(t + \Delta t)\cdot C(t + \Delta t) - h(t)\cdot C(t)$$

$$= \frac{\Delta t}{A}\left\{\frac{NC_N + RC_R - P[C(t + \Delta t) + C(t)]/2}{n}\right.$$

$$\left. + \left[h_0 - \frac{h(t + \Delta t) + h(t)}{2}\right]\frac{1}{t_h}\cdot[C(t + \Delta t) + C(t)]/2\right\}. \quad (7.1.27)$$

When $h(t)$ and $C(t)$ are known, $h(t + \Delta t)$ can be determined using Eq. (7.1.26). Substituting them into Eq. (7.1.27), we can obtain the (average) concentration $C(t + \Delta t)$.

For practical problems, the artificial recharge rate R, solute concentration C_R contained in the recharging water, and pumping rate P are usually known. The remaining three parameters in Eqs. (7.1.26) and (7.1.27) are the natural infiltration rate N, the solute concentration of infiltration water C_N and the reaction time t_R. These three parameters can be obtained by calibration of the model with historical observation data.

Gelhar and Wilson (1974) reported using the one element model to study the impact of highway-frost-proof salt on groundwater quality. Mercado (1976) introduced the application of the lumped parameter model in the management of groundwater quality, which was quoted in a book by Bear (1979).

7.1.5 Criteria of Model Selection

We have already introduced the lumped parameter model and the distributed parameter model of water quality. The latter can be further divided into the pure advection type and the advection-dispersion type. Which type should we select in a practical application? We now present some general principles on model selection.

To select a model, we must make sure what our purpose is. If we want to study the pollutant concentration distribution in an aquifer and to simulate the advance of the concentration front with the model, then the concentration variation with space is a factor which must be taken into consideration. In this case, we have no alternative but to adopt the distributed parameter model. If the problem requires knowing only the variation of the mean solute concentration over time and not the solute distribution in an aquifer, then we can choose the lumped parameter model. It should be noted that the distributed parameter model gives time-space changes of the solute concentration, and therefore, depicts the real situation in a more authentic and more meticulous way. On the other hand, the lumped parameter model is relatively rough. The concentration obtained by a lumped parameter model does not represent the solute concentration at a certain part of an aquifer or in a certain well.

The second consideration for model selection is the quantity and quality of data available. To build up a distributed parameter model requires many distributed parameters, for example, porosity, hydraulic conductivity, dispersion coefficients, and so forth. Moreover, we should also know the boundary conditions of water flow and water quality, and the distribution of source/ sink terms. The acquisition of these data depends on field observation and parameter identification (see the next paragraph for details). If there is only a small amount of reliable information, it will not be worthwhile to establish a complex model because we do not have sufficient and reliable data for the model calibration. Without reliable calibration data, even a complex model will not be accurate. In that case, it would be better to use a simple mass balance model with only one or a few elements.

The third consideration in model selection is the computation effort. A complex distributed parameter model requires solving a large system of equations and thus needs much more computational effort than a lumped parameter model does. Of course, with the modern computer technology, the problem of calculation will no longer be the main factor for consideration.

Based on the above considerations, we often use the distributed parameter model to study groundwater quality problems. If there is sufficient data for model calibration, the distributed parameter model can produce an exact distribution of solute concentration. Next, let us turn to the issue of what kind of distributed parameter model should be used.

The first consideration is still the purpose of using the model. The advection-dispersion model may depict the transition zone accurately because it involves the effects of hydrodynamic dispersion and it conforms better to reality than the pure advection model does. For problems which require high accurate solutions, e.g., to determine the concentration of a poisonous substance in a well, we should not neglect the effect of the transition zone. In this case, we have to use the advection-dispersion model. However, if the ratio of the width of transition zone to the whole region is quite small, the effect of the transition zone on the whole concentration distribution is negligible, and thus the pure advection model may be used.

The second factor to be considered is still the data available. To build an advection-dispersion model, we need extra data to determine the values of dispersion coefficients. A transition zone calculated with incorrect dispersion coefficients will not be reliable.

The third consideration is the computational effort. Compared with the methods of characteristics and Random-Walk, the pure advection model requires a lesser amount of calculation, because it only needs to solve the advection part of water flow equation rather than both the advection and dispersion parts. However, the accuracy of the pure advection model is relatively low. When using the finite difference method, the computational effort required to solve the pure advection model, which is a flow equation coupled with an advection equation, is almost the same as that for the advection-dispersion model. The numerical solution of a pure advection model will also suffer from "numerical dispersion" and "overshoot."

The fourth factor we should consider is the degree of difficulty in solving the inverse problem. Using observed data to calibrate the conceptual model is a necessary step for building any applicable model. Observation of the concentration is the main basis for calibrating a water quality model. When a pure advection model is used, it is difficult to fit the observed concentration values only by adjusting the advection parameters. On the other hand, when an advection-dispersion model is used, the model output may fit the concentration observation quite well by adjusting both advection and dispersion parameters.

With the above considerations, we conclude that generally we should choose advection-dispersion models to simulate the groundwater quality. Another reason for using advection-dispersion models is that we have already had various effective numerical techniques for solving the advection-dispersion equation.

7.2 Model Calibration and Parameter Estimation

7.2.1 Parameter Identification of Advection-Dispersion Equations

Constructing a distributed parameter model for a specified system includes finding a set of governing equations and determining the subsidiary conditions. A correct, reliable mathematical model must be a reproduction of the real system. That is to say, the input-output relation of the mathematical model must be fully coincident with or very close to the excitation-response relation of the real system. Unfortunately, it is very difficult to achieve this requirement in groundwater modeling. First, the governing equations selected may not be suitable; secondly, we do not know the correct value of each parameter entering the equations; and thirdly, the information on boundary conditions and/or source/sink terms is usually not enough. For these reasons, after constructing a model based on a preliminary understanding of the real system, we have to calibrate the model with observation data obtained from pumping and tracer tests or from the historical records. The factors relevant to the model, which include the coefficients of the equations, the subsidiary conditions and the source/sink terms, can all be the objects of calibration. The water flow subproblem and the water quality subproblem can be calibrated separately, but sometimes joint calibration is necessary.

In the case of a homogeneous fluid, the flow part and the dispersion part may be calibrated separately. First, identify the values of hydraulic conductivity, storativity, effective porosity, and other parameters related to flow based on the head observations, and thereby figure out the mean velocity field. Then, depending on the concentration observations, identify the values of dispersivity, retardation factor, and other parameters relevant to dispersion. Parameter identification of groundwater flow equations has been widely studied in the past twenty years (Nelson, 1968; Emsellem and de Marsily, 1971; Neuman, 1973; Chavent et al., 1975; Neuman and Yokowitz, 1979; Sun, 1981; Yeh and Yoon, 1981; Cooley, 1982; Kitamidis and Vomvoris, 1983; Dagan, 1985; Carrera and Neuman, 1986a, b; Yeh, 1986; Sun, 1994). Although the basic concepts and methods used for studying the parameter identification of groundwater flow problems can be borrowed to study the parameter identification of advection-dispersion equations, there are some special difficulties associated with the latter. Uncertainties in the flow parameters (hydraulic conductivity and effective porosity) may propagate to the identified dispersion parameters through the uncertainty in velocity. It is also difficult to differentiate the effects of advection transport and macroscopic dispersion transport. In this section, we will discuss the parameter identification problem of advection-dispersion equations by assuming that the velocity field is given. In the next section, a general coupled groundwater flow and mass transport inverse problem will be considered, in which flow and dispersion parameters are determined simultaneously based on both head and concentration observations.

Assuming that:

1. We have a set of observations of concentration

$$\mathbf{C}^* = \{C_1^*, C_2^*, \ldots, C_L^*\}, \tag{7.2.1}$$

which are the sampling concentrations of certain observation wells at certain times, and which generally include measurement errors.

2. The unknown distribution parameters (e.g., the longitudinal or transverse dispersivity) can be approximately expressed by a vector of finite dimension:

$$\mathbf{p} = \{p_1, p_2, \ldots, p_m\}. \tag{7.2.2}$$

According to the prior information (such as the upper-bound or lower-bound estimations) of unknown parameters, we can define an admissible set P_{ad} of $\mathbf{p}$.

3. There is a simulation model M, which can output the time-space distribution of concentration for each given parameter $\mathbf{p}$, including, of course, a set of calculated concentrations corresponding to the observed concentrations (7.2.1):

$$\mathbf{C}(\mathbf{p}) = \{C_1(\mathbf{p}), C_2(\mathbf{p}), \ldots, C_L(\mathbf{p})\}. \tag{7.2.3}$$

Errors will inevitably exist because model M cannot depict the real physical process exactly. If M is a discrete finite difference or finite element model, its output also contains calculation errors.

The purpose of parameter identification is to find a $\mathbf{p}^* \in P_{ad}$, such that the model output $\mathbf{C}(\mathbf{p}^*)$ is "closest" to the observations of $\mathbf{C}^*$. The simplest way to find $\mathbf{p}^*$ is the trial-error method. First of all, make a guess of an initial value $\mathbf{p}_0$, input it into model M, and get a set of outputs $\mathbf{C}(\mathbf{p}_0)$. Then, analyze the difference between the outputs and observed concentrations in Eq. (7.2.1). An experienced modeler knows how to modify the parameter to reduce the difference between $\mathbf{C}^*$ and $\mathbf{C}(\mathbf{p})$ and obtain a modified parameter $\mathbf{p}_1$. These steps are repeated until no more improvement can be made. The advantage of this process is that it is easy to operate and it does not need any programs except the simulation model. Moreover, the physical intuition of the hydrogeologist can be fully utilized to ensure the reasonableness of the identified parameters. However, this method is inefficient in computation because of the continued involvement of the modeler. In addition, it cannot guarantee convergence to the optimal parameters.

In order to realize the "automatization" of computation, we have to determine a criterion for measuring the "distance" between the observations of Eq. (7.2.1) and the model outputs of Eq. (7.2.3). The most commonly adopted one is the criterion of least squares, that is

$$I(\mathbf{p}) = (\mathbf{C}^* - \mathbf{C}(\mathbf{p}))^T [\mathbf{W}] (\mathbf{C}^* - \mathbf{C}(\mathbf{p})), \tag{7.2.4}$$

where $[\mathbf{W}]$ is a weighted matrix of $L \times L$ and superscript T denotes the transposition. For any two parameters $\mathbf{p}_1$ and $\mathbf{p}_2$, if $I(\mathbf{p}_2) < I(\mathbf{p}_1)$, we then conclude that the output related to parameter $\mathbf{p}_2$ is closer to the observations than that of $\mathbf{p}_1$, and $\mathbf{p}_2$ is better than $\mathbf{p}_1$. The problem mentioned above now becomes the following optimization problem: to obtain $\mathbf{p}^* \in P_{ad}$, such that

$$I(\mathbf{p}^*) = \min I(\mathbf{p}), \qquad \mathbf{p} \in P_{ad}. \tag{7.2.5}$$

The method of Gauss-Newton with constraints is often used to solve the minimization problem (7.2.5), the iteration series of which is (Luenberger, 1984):

$$\mathbf{p}_{r+1} = \mathbf{p}_r + ([\mathbf{J}]_r^T [\mathbf{W}] [\mathbf{J}]_r)^{-1} [\mathbf{J}]_r^T [\mathbf{W}] (\mathbf{C}^* - \mathbf{C}(\mathbf{p}_r)), \tag{7.2.6}$$

where $\mathbf{p}_r$, $\mathbf{p}_{r+1}$ are the values of the unknown parameter obtained at the rth and $(r + 1)$th iterations, respectively. Jacobian

$$[\mathbf{J}]_r = \left[\frac{\partial \mathbf{C}(\mathbf{p})}{\partial \mathbf{p}} \right]_{\mathbf{p} = \mathbf{p}_r} \tag{7.2.7}$$

is also called the *sensitivity matrix*, and its elements are called sensitivity coefficients. We can use the Rosen projection method (Yoon and Yeh, 1976) to ensure that $\mathbf{p}_{r+1}$ is within P_{ad}. In fact, any non-linear programming method can be used to solve the problem of Eq. (7.2.5). For example, Umari et al. (1979) used a quasi-linearization technique, Strecker and Chu (1986) used a quadratic programming method, and Wagner and Gorelick (1986) used a quasi-Newtonian algorithm.

The weighted matrix $[\mathbf{W}]$ in the objective function (7.2.4) is generally a diagonal matrix. Its element w_{ii} can be given artificially or determined by the Maximum Likelihood Estimation method (Carrera and Neuman, 1986a). Wagner and Gorelick (1986) suggested taking $w_{ii} = 1/C_i^2(\mathbf{p})$, where $\mathbf{p}$ is the current value of the parameter determined in the iteration process.

Now we have given a simple introduction for determining the unknown parameters in the advection-dispersion equation based on concentration observations. The inverse techniques can be used to interpret the observed data obtained from the field or from laboratory tracer tests for determining the parameters relevant to dispersion. A lot of practical work has been done in this respect, as can be seen in the following sections. We must point out that the problem of parameter identification is an inverse problem of simulation. Its solution is usually not unique and it is also very sensitive to observation errors. If the observed data are insufficient in quantity and quality, or attention is not paid to the effect of the errors of model structure, the parameters obtained may be very different from their true values.

7.2.2 Field Experiments for Determining Dispersivities

We have mentioned the effect of experimental scale on the identified values of dispersivities in Section 2.5.1. In what follows, we will introduce various laboratory and field experiments for determining the longitudinal dispersivity α_L and transverse dispersivity α_T. These experiments are classified according to their scales.

The *laboratory scale* is usually in the range of 10^0 m. In a classical column experiment, the water with tracer concentration C_0 is continuously injected into a column containing sand or other materials of porous media from one end to displace the original water in the column without tracer. The concentration and volume of effluent water from other end of the column is measured. Since the experimental conditions may be simulated approximately by Problem 1 or Problem 2 in Section 3.2.1, the observation results can be interpreted by the following analytical solution:

$$\frac{C(x,t)}{C_0} = \frac{1}{2}\operatorname{erfc}\left[\frac{x - Vt}{2\sqrt{D_L t}}\right] = \frac{1}{\sqrt{2\pi}}\int_{(x-Vt)/\sqrt{2D_L t}}^{\infty} \exp(-\eta^2)\,d\eta. \qquad (7.2.8)$$

For a given time t, this equation may be represented as $1 - N[(x - m)/\sigma]$, where N is a normal distributed function. Its mathematical expectation m and variance σ are

$$m = Vt \quad \text{and} \quad \sigma = \sqrt{2D_L t}, \qquad (7.2.9)$$

respectively. From the definition of normal distributed function, we have

$$N(1) \approx 0.84, \quad N(-1) \approx 0.16. \qquad (7.2.10)$$

For a fixed time t, the distance between point $x_{0.84}$ and point $x_{0.16}$ may be defined as the width e of the transition zone, where the relative concentration $C/C_0 = 0.84$ at point $x_{0.84}$ and $C/C_0 = 0.16$ at point $x_{0.16}$, respectively. From Eq. (7.2.10), we have $e = 2\sigma$. Based on the observed data, we can obtain an illustration of the relative concentration versus x at a certain time t, as shown in Figure 7.6. From Figure 7.6, we can find the locations of points $x_{0.84}$ and $x_{0.16}$. The distance between them is just the width e of the transition zone. Thus, we have

$$\sigma = \frac{e}{2} = \frac{1}{2}(x_{0.16} - x_{0.84}).$$

Using Eq. (7.2.9), we then obtain

$$D_L = \frac{1}{8t}(x_{0.16} - x_{0.84})^2. \qquad (7.2.11)$$

FIGURE 7.6. The concentration distribution curve shown by error functions.

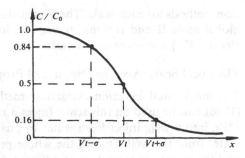

This formula can also be translated into one related to a given point x. Let $t_{0.16}$ and $t_{0.84}$ denote the time required for the relative concentration at a point x to reach 0.16 and 0.84 respectively, we then have

$$D_L = \frac{1}{8}\left[\frac{x - Vt_{0.16}}{(t_{0.16})^{1/2}} - \frac{x - Vt_{0.84}}{(t_{0.84})^{1/2}}\right]^2. \tag{7.2.12}$$

The point $x_{0.5}$ at $C/C_0 = 0.5$ moves with mean velocity V, so that $x_{0.5} = Vt_{0.5}$. Generally, the width of the transition zone is very small in comparison to the length of the sand column. Therefore, $(t_{0.16})^{1/2}$ and $(t_{0.84})^{1/2}$ in (7.2.12) can be substituted approximately by $(t_{0.5})^{1/2}$. Thus

$$D_L = \frac{V^2}{8t_{0.5}}(t_{0.84} - t_{0.16})^2. \tag{7.2.13}$$

The values of $t_{0.16}$, $t_{0.5}$ and $t_{0.84}$ can be measured at the end of the sand column. More conveniently, we can measure the volume of the effluent fluid at the end of the sand column, instead of measuring the time. The effluent volume for per unit section at the end of the sand column is $U = Vt$. Thus, Eq. (7.2.13) can be rewritten as

$$D_L = \frac{V}{8U_{0.5}}(U_{0.84} - U_{0.16})^2, \tag{7.2.14}$$

where $U_{0.16}$, $U_{0.5}$, and $U_{0.84}$ are the volumes of the effluent water from the end of the column when the relative concentrations reach 0.16, 0.5 and 0.84, respectively. Using this method, the value of longitudinal dispersivity generally ranges from 0.01 cm to 2 cm, depending on the grain-size distribution.

When a laboratory experiment cannot be simulated by an analytical model, we can use a numerical model to simulate the experiment and use the least square criterion to determine the unknown dispersivity parameters.

For field dispersion tests, we can either impose a flow field or directly use the natural flow field. According to average propagation distance, Fried (1975) divided test sites into four scales and suggested relevant experimentation methods for each scale. The four scales are: the local scale, global scale I, global scale II and regional scale. The following presentation is based on Fried (1975) and some later reports.

The Local Scale (Average Distance of Propagation Is 2 to 4 m)

The single well injection-extraction method, first suggested by Mercado (1966), can be used on this scale. Inject a radioactive tracer, typically ^{131}I or ^{34}Br, into a well; inject fresh water to push for some time, and then extract water from the well. During the whole process, the changes of the velocity and the tracer concentration at different depths will be recorded by a probe. This process can be easily described by a mathematical model.

Since the experiment is limited to the area near a well, the natural flow velocity in the aquifer can be neglected in comparison to the velocities imposed by injection and extraction. In this case, a radial flow is formed around the well and the dispersion can be simplified to a radial dispersion. Because the flow velocity is large near the well, the effect of molecular diffusion can be ignored. Moreover, the density and viscosity of the fluid do not vary after the radioactive tracer is added. In consideration of the above points, we can formulate a mathematical model as follows:

$$\frac{1}{r}\frac{\partial}{\partial r}\left(r\alpha_L V\frac{\partial C}{\partial r}\right) - \frac{1}{r}\frac{\partial}{\partial r}(rVC) - \lambda C = \frac{\partial C}{\partial t}, \qquad (7.2.15)$$

$$\frac{1}{r}\frac{\partial h}{\partial r} + \frac{\partial^2 h}{\partial r^2} = \frac{S}{T}\frac{\partial h}{\partial t}, \qquad (7.2.16)$$

$$V = -\frac{K}{n}\frac{\partial h}{\partial r}. \qquad (7.2.17)$$

(a) The boundary conditions during injection are:

$$\begin{cases} C(r_0, t) = C_0, & 0 \le t < t_0, \\ C(r_0, t) = 0, & t \ge t_0, \\ C(\infty, t) = 0, & t \ge 0, \\ h(r_0, t) = h_w, & t \ge 0, \\ h(\infty, t) = h_0, & t \ge 0, \end{cases} \qquad (7.2.18)$$

where r_0 is the radius of the well; the injected water contains tracer before t_0, and is fresh after t_0; h_w is the fixed high water head kept in the well during injection; and h_0 is the head in the aquifer before injection.

(b) The initial conditions during injection are:

$$\begin{cases} C(r, 0) = 0, & r > r_0, \\ h(r, 0) = h_0, & r > r_0. \end{cases} \qquad (7.2.19)$$

(c) The boundary conditions during extraction are:

$$\begin{cases} \left.\dfrac{\partial C}{\partial r}\right|_{(r_0,t)} = 0, \\ C(\infty, t) = 0, \quad t > t_1, \\ h(r_0, t) = h'_w, \\ h(\infty, t) = h_0, \end{cases} \qquad (7.2.20)$$

where the first formula shows the concentrations inside and outside of the well sidewall are approximately equal during extraction; h'_w is the fixed low water head kept in the well during extraction; and t_1 is the starting time of extraction.

(d) The initial conditions during extraction are:

$$\begin{cases} C(r, t_1) = C_1(r, t_1), \\ h(r, t_1) = h_1(r, t_1) \end{cases} \qquad r \geq r_0, \qquad (7.2.21)$$

where $C_1(r, t_1)$ and $h_1(r, t_1)$ are, respectively, the concentration distribution and head distribution at the time when injection stops and extraction starts. They are calculated from the solutions of the injection model.

For the tracer case, the flow equation and dispersion equation can be solved individually. In each time step, we can solve head h from Eq. (7.2.16) and substitute it into Eq. (7.2.17) to obtain mean velocity V. Then V is substituted into Eq. (7.2.15) to obtain concentration C. Either FDM or FEM can be used. As the radial flow velocity around the well is relatively large, we must use a small time step and a dense grid in order to reduce the numerical dispersion and solution oscillations. We should also adopt the numerical solution techniques for advection-dominated problems when necessary.

Using the numerical method to simulate the process of extraction, we can obtain a curve of concentration versus time in the well. This curve is shown in Figure 7.7. To find the best estimation of α_L, we can change the value of α_L in the model to best fit the observations. This procedure can be completed automatically by an technique of single variable optimization.

Global Scale I (Average Distance of Propagation Is 4 to 20 m)

On this scale, the single well injection-extraction back method cannot control the velocity field, so dual-well tests or multi-well tests are usually used. In the

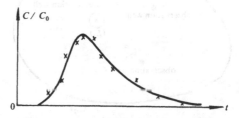

FIGURE 7.7. The curve showing concentration changes with time during extraction.

dual-well extraction-injection test, two wells are drilled about 10 meters apart. Water is pumped from one well and injected into the other at the same rate, so as to form a steady flow field. After steady state conditions are achieved, a tracer is added into the injection well, either continuously or in a pulse. If the aquifer is infinite, homogeneous and isotropic, the effect of the natural flow velocity near the well can be neglected, then it is easy to formulate a mathematical model for this system. The velocity distribution is the superposition of two radial flow velocity fields generated by a source and a sink, respectively.

By sampling from the extraction well, we can obtain a curve of the tracer concentration versus time. Modifying the longitudinal dispersivity, α_L, so that the model output agrees with the observed curve, we then have the value of α_L. If there are observation wells outside the line between the two wells, we can also obtain the transverse dispersivity α_T by fitting the observed concentrations in the observation wells.

Hoopes and Harleman (1967) gave an analytical expression for the tracer concentration distribution in a dual-well test in an integral form, and pointed out that the dispersion in the extraction well mainly shows in the early period of pumping when the relative concentration is rather low. Grove (1971) provided a program for calculating the dispersivity based on the interpretation of a dual-well test. Pickens and Grisak (1981) presented a report of their dual-well tests including the testing conditions, types of tracers, facilities used and the interpretation of their results.

Sometimes multi-well tests are also used on this scale, i.e., injecting a radioactive tracer into a well and recording the concentration changes versus time in the neighboring wells, as shown in Figure 7.8. According to the relationship between the injection velocity and natural velocity, there are three cases:

(a) The natural velocity is dominant and the velocity imposed by injection can be neglected, e.g., after releasing a tracer, the natural flow in the aquifer washes the tracer away.

(b) The natural velocity is of the same order of magnitude as the imposed velocity in the injection well, and the composite effect of them must be taken

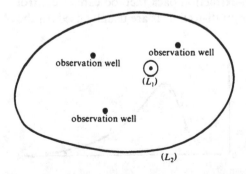

FIGURE 7.8. A multi-well experiment in the field. ● = observation well.

into consideration. This case occurs when the natural velocity is so low that a significant amount of water containing tracer must be added to speed up the test.

(c) The velocity imposed by injection is dominant and the natural velocity can be neglected. This is the case when fresh water is injected to form a relatively steady radial flow between the injection well and the observation wells.

In the last case, when the flow has not reached a steady state, the single well injection model mentioned before should be used. If the flow approaches a steady state, the following simpler model may be used:

$$
\begin{cases}
\dfrac{\partial C}{\partial t} = \dfrac{\alpha_L A}{r} \dfrac{\partial^2 C}{\partial r^2} - \dfrac{A}{r} \dfrac{\partial C}{\partial r}, \\[2mm]
C(r,0) = 0; \quad r > 0, \\[2mm]
C(r_0, t) = C_0, \quad 0 \le t < t_0, \\[2mm]
C(r_0, t) = 0, \quad t \ge t_0, \\[2mm]
C(\infty, t) = 0,
\end{cases}
\tag{7.2.22}
$$

where $A = Q/2\pi Bn$; Q is a constant injection rate; B the thickness of the aquifer; n the effective porosity; and t_0 shows the instant before which the tracer is injected and after which fresh water is injected.

This model can be solved by either FDM or FEM, the solution of which depends on dispersivity α_L. The value of α_L can be identified by fitting the model output with the observed concentration values in the well.

Let us turn to consider the cases (a) and (b). In fact, it is difficult to form a radial flow on this scale, because the radial velocity $V_r = A/r$ decreases rapidly with increasing of distance, r, between the wells.

The real velocity is the superposition of the natural velocity and radial velocity, that is

$$
V_x = V_{n,x} + V_r \cos \alpha, \quad V_y = V_{n,y} + V_r \sin \alpha,
\tag{7.2.23}
$$

where $V_{n,x}$, $V_{n,y}$ are the two components of the natural velocity V_n; and α is the angle between the x axis and the radial vector.

In this case, we have to construct a numerical model to simulate the experiment. Both longitudinal dispersivity α_L and transverse dispersivity α_T can be determined by fitting the output of the numerical model with the concentration records obtained from observation wells.

Global Scale II (Average Distance of Propagation Is 20 to 100 m)

The method of multi-well tests mentioned above can also be used for this scale. However, radial flow cannot be formed by injection because of the long distance of propagation. Therefore, tracer transportation mainly relies on the natural flow, which prolongs the test procedure. Moreover, the flow velocity

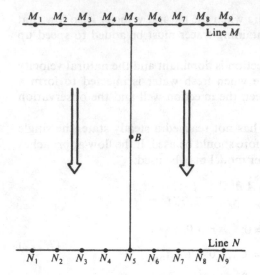

M_1 M_2 M_3 M_4 M_5 M_6 M_7 M_8 M_9

Line M

B

Line N

N_1 N_2 N_3 N_4 N_5 N_6 N_7 N_8 N_9

FIGURE 7.9. Extraction-injection tests in a unidirectional flow field.

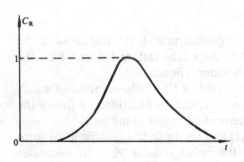

C_R

1

0

t

FIGURE 7.10. Relative concentration versus time in an extraction well.

cannot be well controlled in such a large area, and consequently, the accuracy of the whole test may be low. Sauty (1980) designed a field test with wells in two parallel lines that can form an artificial one-dimensional flow and two-dimensional dispersion. This method may produce accurate experimental results, and the interpretation of the experimental results is relatively easy.

Two rows of wells are drilled at a certain distance apart (not exceeding 100 m in general) in the test field, e.g., the wells on Lines M and Line N, as shown in Figure 7.9. Water is injected into the wells on Line M and pumped out from those on Line N. A certain time later, a steady and uniform one-dimensional flow will form between the two lines. The tracer with mass m is then injected into the whole thickness of the aquifer from a middle well B, and hence, the concentration changes of the tracer can be observed in the extraction wells on Line N. We can get a typical curve shown in Figure 7.10, where C_R is the relative concentration, i.e., the ratio of observed concentration C to its peak value C_{max}. The mathematical model that can describe this test has been given in Problem 3 in Section 3.2.2. Thus, the analytical solution for this problem is:

$$C(x, y, t) = \frac{m/n}{4\pi V \sqrt{\alpha_L \alpha_T}} \exp\left[-\frac{(x - Vt)^2}{4\alpha_L Vt} - \frac{y^2}{4\alpha_T Vt}\right], \quad (7.2.24)$$

where V is the steady one-dimensional flow velocity. Here, we let the direction of the flow coincide with axis x and injection well B be at the origin of the coordinates.

Let (X, Y) be the coordinates of an observation well and $C(X, Y, t)$ be the concentration of tracer in that well. Substituting the following dimensionless variables:

$$t_R = Vt/\alpha_L, \quad (7.2.25)$$

$$a = (X^2/\alpha_L^2 + Y^2/\alpha_L \alpha_T)^{1/2}, \quad (7.2.26)$$

and

$$C_R = C/C_{\max} \quad (7.2.27)$$

into Eq. (7.2.24), we then obtain the concentration of point (X, Y) as:

$$C_R(t_R) = \frac{K}{t_R} \exp\left[-\frac{a^2 + t_R^2}{4t_R}\right], \quad (7.2.28)$$

where

$$K = t_{R,\max} \exp[(a^2 + t_{R,\max}^2)/4t_{R,\max}],$$
$$t_{R,\max} = \sqrt{4 + a^2} - 2. \quad (7.2.29)$$

Equation (7.2.28) only depends on a dimensionless parameter a, so a group of typical curves can be plotted against different values of a, see Figure 7.11. A logarithmic scale is used in the figure for the abscissa t_R.

By making a comparison between the observed curve of N_5 in Figure 7.10 and the typical curves in Figure 7.11, we can identify the best value of a, marked as a_5. Let the coordinates of N_5 be $(X_5, 0)$, then from Eq. (7.2.26) we

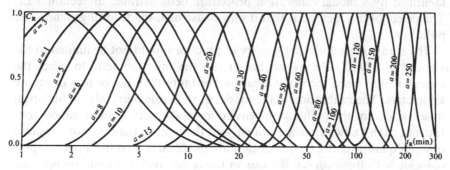

FIGURE 7.11. The typical curves of relative concentration $C_R(t_R)$ with parameter a.

have $a_5 = X_5/\alpha_L$, hence

$$\alpha_L = X_5/a_5. \tag{7.2.30}$$

By correlating the observed curve of N_4 with the typical curves, we can identify another value of a, marked as a_4. Let the coordinates of N_4 be (X_4, Y_4), then from Eq. (7.2.26) we have

$$a_4 = [X_4^2/\alpha_L^2 + Y_4^2/\alpha_L \alpha_T]^{1/2},$$

and consequently

$$\alpha_T = Y_4^2/\alpha_L \left(a_4^2 - \frac{X_4^2}{\alpha_L^2} \right). \tag{7.2.31}$$

The Regional Scale (Average Distance of Propagation Is over 100 m)

On this scale, if the above-mentioned tests are conducted, many wells must be drilled and a large amount of tracer should be injected. As a result, the test period must be very long and the test cost must be very high. When the distance of propagation ranges from 100 to 300 m, a possible alternative is to use the geophysical technique. The application of geophysical prospecting to hydrogeological research can provide the estimates of flow velocity and flow direction, as well as hydrogeological parameters such as porosity and transmissivity. The increase of salinity in water causes its resistance to decrease, so the salinity changes in groundwater can be observed from resistance measurements. During the experiment, an electrolyte solution, for example, saline water, is injected into a well, and transported by the natural flow. Then, electrical resistance changes are measured at some points in the region being considered. The resistance changes may be interpreted as the changes of flow velocity field and concentration field by the geophysical approach. A mathematical model can then be established based on site conditions and the procedure of test. A geophysical survey conducted before the injection may provide necessary initial and boundary conditions for the model. With the mathematical model, as well as observations of the flow velocity field and concentration field, dispersivities α_L and α_T can be identified by a model calibration procedure demonstrated in Section 7.2.1. The defect of this method is that it may cause man-made local groundwater pollution.

The study of groundwater quality in practice should not be limited to the vicinity of a well. We are unable to use the injection tests to affect the overall region when the study area reaches more than ten or tens of square kilometers. The calibration of the model in this scale must rely on historical observations of the pollution or environmental tracer. Concentration contours at each known time are drawn, and then, using the common program of water quality model, the parameters are modified so that the model outputs coincide with these curves. Because of the larger area, the non-homogeneity of porous media should be taken into account. Different values of α_L and α_T

may be taken in different subareas, so there may be a great number of unknown parameters to be determined.

There are a number of real world examples of building water quality models based on existing contamination data and environmental tracers. For instance, the study of the pollution of a munitions factory in Rocky Mount, Arsenal, Denver, Colorado, USA, by Konikow (1977). Another example is the research on the groundwater contamination of Xi'an City, China by the Hydrogeological Team of Shaanxi Province and Shangdong University (1985). Herweijer et al. (1985) reported a case of calibrating a water quality model from historical observations of the environmental isotope tritium since 1951. They used the Random-Walk model introduced in Section 4.2.6 to simulate the transport of tritium. The paper gives a detailed introduction of the hydrogeological conditions, observations and the results of model calibration.

In order to deeply examine the dispersion process and provide validation data for the stochastic theories, several large-scale field dispersion experiments have been conducted in recent years, of which five comprehensive tracer tests should be mentioned: Borden Air Force Base (Sudicky, 1986), Cape Cod (Garabedian et al., 1991), Twin Lake (Killey and Moltyaner, 1988; Moltyaner, 1993), Columbus Air Force Base (Boggs et al., 1992), and Jutland Peninsula, Denmark (Jensen et al. 1993). The characteristics of these tests include:

- The experimental scales were as large as several hundreds of meters, and the test periods as long as one or more years.
- The observation systems were three-dimensional distributed.
- The tracers used were nonreactive.
- The variabilities in hydraulic conductivities and/or velocities were accurately measured.
- The concentration observations were interpreted by three-dimensional stochastic and deterministic numerical models.

These experiments have provided important insight into the dispersion process in natural heterogeneous formations, and have also provided useful data bases for validation of both deterministic and stochastic theories, as well as various numerical models. Readers can find the details of these tests from references mentioned above. Some results of these tests will be reviewed in the following sections.

7.2.3 The Relationship Between Values of Dispersivity and Scales of Experiment

Many field tracer tests have been done which covered various scales as mentioned previously. Some values of dispersivity obtained on different experiment scales, as quoted from the literature, are listed in Tables 7.1 to 7.4. It has been found that the dispersivity values obtained by using mathematical models to interpret tracer injection tests are not constant, but depend on the

TABLE 7.1. The local scale.

Country	References	Experiment method	Longitudinal dispersivity α_L (m)
Canada	Pickens et al. (1977)	Single-well test	$0.034 \sim 0.1$
France	Fried (1975)	Single-well test	$0.1 \sim 0.5$
Canada	Sudicky and Cherry (1972)	Natural gradient	$0.01 \sim 0.22$
France	Fried et al. (1972)	Single-well (pulse injection)	$0.1 \sim 0.6$
Canada	Pickens and Grisak (1981)	Single-well test	$0.3 \sim 0.7$

TABLE 7.2. Global scale I.

Country	References	Experiment method	Longitudinal dispersivity α_L (m)
France	Sauty (1978)	Double-well test	$0.3 \sim 6.9$
UK	Ivanovich and Smith (1978)	Double-well test	$1.0 \sim 3.1$
France	Fried (1975)	Single-well test	$5.0 \sim 8.0$
USA	Robson (1974)	Double-well test	15.2

TABLE 7.3. Global scale II.

Country	References	Experiment method	α_L (m)	α_T (m)
USA	Wilson (1971)	Double-well test	15.2	
France	Sauty (1980)	Double well-line test	15.5	10.4
USA	Grove and Beetem (1971)	Double-well test	38.10	

TABLE 7.4. The regional scale.

Country	References	Distance of nodes (m)	α_L (m)	α_T (m)
USA	Konikow (1977)	305	30.5	30.5
France	Fried (1975)	600	12	4
USA	Gupta et al. (1975)	Changeable	$80 \sim 200$	$8 \sim 20$
USA	Pinder (1973)	Changeable	21.3	4.5
USA	Grove (1977)	640	91	91
China	Jining City (1983)	220	$5 \sim 9$	$0.6 \sim 0.9$
China	Xi'an City (1985)	740	131	48

scale of the test. In other words, the identified values of dispersivity depend on the distance or time of tracer propagation (scale effect). This fact is not consistent with the original physical meaning of dispersivity. Many researchers attempted to explain this phenomenon from both theory and field experiment. One explanation is that since the porous media in the field are all inhomogeneous and anisotropic, the larger the experiment scale is, the more opportunities there will be to meet the local heterogeneity. The macroscopic inhomogeneity and anisotropy can cause the real transport of the solute to be much larger than the result calculated based on the mean velocity. Along this way, the statistical theory of mass transport in porous media has been developed (Gelhar and Axness, 1983; Dagan, 1984; Dagan, 1989). The second explanation is that the observations may be interpreted by an incorrect model. There are two terms in the transport equation: advection and dispersion. An incorrect model may regard the effect of advection as that of dispersion.

Hutton and Lightfoot (1984) pointed out that when the dimension of a mathematical model is lower than that of the corresponding physical model of a test, unrealistic values of dispersivity may be obtained. For example, if there are one-dimensional horizontal flows in a horizontal layered stratum with different velocities, then the corresponding advection-dispersion equation will be:

$$\frac{\partial C}{\partial t} + V(y)\frac{\partial C}{\partial x} = D_{xx}(y)\frac{\partial^2 C}{\partial x^2} + \frac{\partial}{\partial y}\left[D_{yy}(y)\frac{\partial C}{\partial y}\right], \qquad (7.2.32)$$

where V is the horizontal flow velocity; D_{xx} and D_{yy} are the longitudinal and transverse dispersion coefficients, both of which depend on the coordinate y. This model is illustrated in Figure 7.12.

In this case, if the following one-dimensional advection-dispersion equation

$$\frac{\partial C}{\partial t} + V\frac{\partial C}{\partial x} = D\frac{\partial^2 C}{\partial x^2} \qquad (7.2.33)$$

is used to fit the observed concentrations, the identified mean velocity and

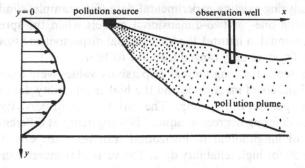

FIGURE 7.12. The pollutant transportation in a horizontal layered aquifer.

dispersion coefficients will not reflect their true values in Eq. (7.2.32). Hutton and Lightfoot proved that if Eq. (7.2.32) is averaged along the vertical direction and transformed into the form of Eq. (7.2.33), the "apparent velocity" V and "apparent dispersion coefficient" D will both depend on time, and an additional "apparent source/sink term" appears. The authors thought that this analysis could explain why the dispersion coefficients depend on the propagation distance when using Eq. (7.2.33) to fit the test data. Dominico and Robbins (1984) got similar results from numerical analysis. Bear and Veriijt (1987) obtained the two-dimensional advection-dispersion equation through averaging the three-dimensional advection-dispersion equation along the vertical direction. From their formulation we can see that the application of two-dimensional models may enlarge the dispersivities when the physical problem is three-dimensional.

Killey and Moltyaner (1988), Moltyaner and Killey (1988a, b) reported the tracer tests conducted in the Twin Lake aquifer of Canada by Chalk River Nuclear Laboratories. They used a three-dimensional model to fit the observed concentration distribution. The "scale effect" was not found when the propagation distance reached 40 m. The longitudinal and transverse dispersivities obtained are almost identical with the results obtained in the laboratory scale. Jensen et al. (1993) reported that the "best fitting" dispersivity parameters for the large-scale tests recently conducted in Denmark are: $\alpha_L = 0.45$ m and $\alpha_T = 0.001$ m in the horizontal direction, and $\alpha_L = 0.05$ m and $\alpha_T = 0.0005$ m in the vertical direction.

It is thus clear that, when making an interpretation of the test data, we must carefully analyze the conditions of the test, find the structure of the aquifer, and observe the plumes in detail. Then, we can select an appropriate model to fit the observations. The model should not be arbitrarily simplified for convenience. Since the three-dimensional numerical model can take all the practical conditions into account, it is most suitable for interpreting the results of field experiments.

In Chapter 2, we have mentioned the review paper of Gelhar et al. (1992), in which 59 different field sites are summarized. The paper presented the following conclusions:

- After reanalyzing existing experimental data (for example, eliminating the effect of using one- or two-dimensional models when the spreading was three-dimensional in nature), the longitudinal dispersivities associated with high reliability data only ranged from 0.1 to 10 m.
- On a given scale, the longitudinal dispersivity values were found to range over 2 to 3 orders of magnitude and the higher reliability data tend to fall in the lower portion of this range. The variations in dispersivity reflect the influence of differing degrees of aquifer heterogeneity at different sites.
- The ratio of longitudinal to horizontal transverse dispersivities ranged from 10 to 50 for high reliability data. The vertical transverse dispersivities are typically an order of magnitude smaller than horizontal transverse dispersivities.

- When the data are classified according to porous versus fractured media, there does not appear to be any significant difference between these aquifer types.

7.2.4 Determination of Mean Flow Velocities

The methods of determining the average flow velocities in an advection-dispersion model can be divided into two categories: *indirect determination* and *direct determination*. The indirect determination involves the use of field data from pumping tests to calibrate the flow model and thus obtain the hydraulic conductivity K and effective porosity n. Then K and n are substituted into Darcy's law to get the average velocity. As K, n and the hydraulic gradient values may contain large errors, the velocities obtained by using this method are usually not reliable. This is a difficult problem in groundwater quality modeling.

In order to calibrate a model, it is better to take direct measurement of the flow velocity in the field, because the transport model is very sensitive to errors in velocity. In the direct method, values of velocity are determined through measuring the movement of artificial tracers or environmental isotopes. The artificial tracers include salts, radioactive isotopes, coloring agents and so on. The single-well tests done in the past was to put salt into a well, measure the changes of the electric resistance from the ground surface and then determine the velocity and direction of the flow according to the changes of electric resistance at different distances. The accuracy of this kind of test is poor. In the multi-well method, tracer is injected into one well and its arrival time is observed in the others. After eliminating the effect of dispersion, the groundwater velocity can be calculated by considering the distance between the injection well and an observation well and the mean arrival time of the tracer. This kind of test, however, is very difficult to successfully conduct because of the low natural velocity, unclear flow direction and the mixture of the water from adjacent aquifers.

To determine the velocity near a well, one can apply a method called "point dilution." The tracer is injected at a certain depth in the well is left to be diluted by the natural flow. It is known that the rate of dilution depends on the velocity of the groundwater flow. From the records of tracer concentration changes in the well we can obtain the values of the velocity. Moreover, we can also determine the velocities at different depths in a well, and find the layer having a relatively larger velocity. The large velocity layer plays an important role in the propagation of pollutants.

For a detailed demonstration of the use of the direct method to measure the flow velocities, readers may refer to Cherry et al. (1975) and Fried (1975). Ciccioli et al. (1980) put forward a method which used various tracers to accurately determine the velocity of groundwater. The discussion of this method is beyond the scope of this book.

It is quite evident that the various field tests can only give some local and steady values of the flow velocity. The velocity distribution of the whole region can only be obtained through calibration of either a flow model or a coupled flow and quality model.

7.2.5 Identification of Retardation Factor and Chemical Reaction Parameters

When the actions of adsorption, ion exchange and chemical reaction are included into a water quality model, some other undetermined parameters may appear in the model.

For linear equilibrium absorption, we have (see Eq. (2.6.27)):

$$F = \beta C, \tag{7.2.34}$$

where F is the tracer concentration in solid, C the tracer concentration in liquid and β a coefficient. Let

$$R_d = 1 + \frac{1 - n}{n} \beta, \tag{7.2.35}$$

where R_d is the retardation factor, n the effective porosity, then we can rewrite the hydrodynamic dispersion equation as (see Eq. (2.6.29)):

$$\frac{\partial C}{\partial t} = \frac{\partial}{\partial x_i} \left(\frac{D_{ij}}{R_d} \frac{\partial C}{\partial x_j} \right) - \frac{V_i}{R_d} \frac{\partial C}{\partial x_i}. \tag{7.2.36}$$

Consequently, we can obtain the retarded longitudinal and transverse dispersivities α_L/R_d and α_T/R_d, as well as the retarded mean velocity V/R_d through fitting the field test data. When we carry out calibration of the model with the observed concentration distribution, the effect of adsorption is already taken into account.

If the effective porosity n is known, we can first obtain the distribution coefficient β through experiments, then determine the retardation factor by Eq. (7.2.35). It is evident that β depends not only on the nature of the porous matrix, but also on that of the solute. Griffin et al. (1976) pointed out that the chlorine ion, sodium ion and organic compositions soluble in water are hardly adsorbed when they pass through a clay layer; the ions of potassium, ammonia, magnesium, silicon and iron are moderately adsorbed; and the ions of lead, cadmium, mercury, zinc are greatly adsorbed. Many experiments have been done for radioactive isotopes. In unsolidified geological materials, the distribution coefficient β of radioactive isotopes can reach $10^3 \sim 10^5$ mL/g, see Johnston and Gillham (1980). For instance, the transport velocity of strontium-90 is only 3% of the true velocity of groundwater due to adsorption. Therefore, when we simulate the transport processes of different chemical compositions, the problem of determining the retardation factor must be taken into consideration.

For a multicomponent solute, the chemical reactions between its components and ion exchanges with the solids should also be considered. In these cases, there will be more parameters contained in the governing equation. Each component has its own conservation equation, and they are connected by a group of isothermal relationships, that is,

$$q_i = f_i(C_1, C_2, \ldots, C_n), \quad (i = 1, 2, \ldots, n) \tag{7.2.37}$$

where q_i and C_i are the concentrations of ion i in solid phase and liquid phase, respectively. To study this aspect, readers may refer to Charbeneau (1981) and Jennings et al. (1982).

7.2.6 Identification of Pollutant Sources

A problem often encountered in practice is the determination of the source/sink term in a water quality equation from concentration observations, which is commonly called the *identification of pollutant sources*. It usually includes two aspects: identification of the locations and the intensities of pollutant sources.

The identification of source/sink terms, of course, is a special case of the general inverse problem. Therefore, we can use the methods mentioned in Section 7.2.1 to solve it. Since the distribution of concentration may be very sensitive to the variation of the source/sink term, generally, the source/sink term in a water quality equation is easier to be identified than other parameters.

In some problems the pollutant sources are known. For example, in the case of recharging with sewage water, the concentration and the volume of recharging sewage can be measured in practice. For another example, when making a tracer test, the quantity of the tracer to be injected is preset and known. In some problems, however, the pollutant source will be a parameter to be determined. For example, when sewage is dispersed through field spreading, how much do the pollutants actually enter? Where is the breakpoint of the buried sewage pipe? Does the sewage discharge of a factory exceed the preset limit? All of these, and many others, are problems of identification of pollutant sources. In an extensive sense, the hydrochemical prospecting for ores can also be included in this type of inverse problem.

Gorelick et al. (1983) analyzed the method of identification of pollutant sources through two case studies. The first case was to find the break location of a sewage pipe and calculate the leakage using groundwater concentration observations. The second case was the identification of an instantaneous pollutant source. Assuming that there are some wells at the upstream in which sewage may be injected, and several observation wells at the downstream which may record the changes of concentration with time. Using these data, Gorelick (1983) calculated the injected quantity of sewage in each injection well for each year.

7.2.7 Computer Aided Design for Field Experiments

It is more difficult to design a tracer experiment for determining the parameters relevant to dispersion (average velocity, dispersivity, distribution coefficient, etc.) than to design a pumping test for determining parameters relevant to water flow (hydraulic conductivity, porosity, etc.). Some problems that may be encountered in the tracer tests include: the peak value of concentration fails to be caught in observation wells, the testing period lasts too long, the observed values of concentration are not accurate, and so on.

The use of the mathematical model in the early stage of designing a tracer test will help to make the field experiment as reasonable as possible, and will enable one to predict the probable consequences. As a result, the expected goal of the test can be achieved. This is the basis for applying *computer aided design* (CAD) to hydrogeology. We shall explain, through a concrete design of a tracer experiment, the application of the CAD approach to hydrogeology as follows.

Suppose that we want to design a tracer test by the dual-well injection-extraction method to determine the dispersivity of an aquifer. The following problems may be raised:

1. What is the suitable rate of injection (or extraction)?
2. What is the suitable distance between the two wells?
3. What depth should the well be drilled to? Should the vertical flow be taken into account if the well is partially penetrated?
4. How long should the experiment last?

In practice, we may have some prior information before the design. Some useful data in geology and hydrogeology may be obtained through field investigations. The problem is how to transfer this information into a quantitative basis for experimental design. In the case given by Mercer and Faust (1980), the aquifer being considered is formed by limestone with a thickness of 200 m to 500 m. From experience, its permeability coefficient can be assessed as 10^{-4} to 10^{-6} m/s, porosity 0.08 to 0.32, the ratio of horizontal and vertical permeabilities (K_r/K_z) ranged from 1 to 100, and the longitudinal dispersivity 5 to 40 m. Suppose we already have a general program for water quality simulation, then a sensitivity analysis can be carried out by the computer for every design parameter. That is, let a parameter vary within the possible range, while the others are fixed, then output the results of the corresponding changes (i.e., the concentration curve of a pumping well). It is just like doing a number of "experiments," but completed on a computer. The cost is low and the time is saved. During this process, the designer can quantitatively understand the effects of different parameters on the observed concentration, and can compare the results generated from different schemes.

There are some examples of sensitivity analyses in Figures 7.13 to 7.16. Figure 7.13 shows the concentration-time curves changing with longitudinal dispersivity α_L. The other parameters are fixed as follows: the transverse

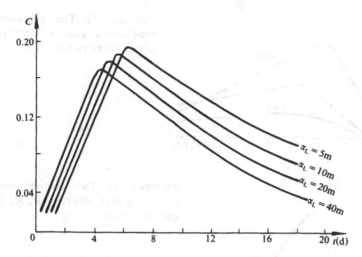

FIGURE 7.13. The concentration-time curves against different α_L in the extraction well.

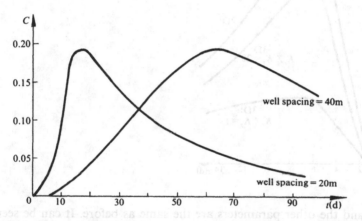

FIGURE 7.14. The concentration-time curves against different well distances in the extraction well.

dispersivity = $\alpha_L/10$, the duration of tracer injected = 4.63 d, porosity = 0.2, hydraulic conductivity = 1.0×10^{-5} m/s, well spacing = 40 m, depth of well penetration in the aquifer = 20 m, and injection (or pumping) rate = 1.61×10^{-2} m^3/s. It may be seen from this figure that the time for the concentration peak to appear in the well may have a difference of 3 to 4 days depending on the value of α_L, but the value of the peaks changes very little.

Figure 7.14 shows the effect of different well distances on the concentration-time curves. It is assumed that the durations of tracer injected are 10.4 d and 46.3 d, the injecting rates 1.80×10^{-3} m^3/s and 1.61×10^{-3} m^3/s, well distances are 20 m and 40 m, respectively, longitudinal dispersivity $\alpha_L =$

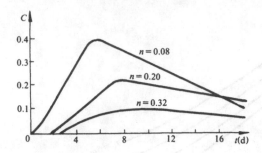

FIGURE 7.15. The concentration-time curves against different porosities in the extraction well.

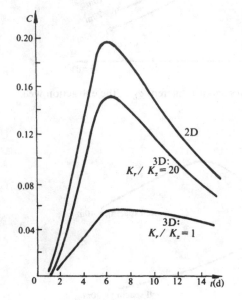

FIGURE 7.16. The concentration-time curves against different K_r/K_z in the extraction well.

10 m, and the other parameters are the same as before. It can be seen that the well distance has a great effect on the occurrence of concentration peaks in the pumping well. When the well distance increases from 20 m to 40 m, the arrival time of the concentration peak increases from 15 days to 60 days or more, however, there is almost no change in the value of the peak.

Figure 7.15 gives the effect of different values of porosity on the concentration-time curves in the pumping well, where the longitudinal dispersivity α_L is assumed to be 10 m, and the rest of parameters are the same as before. From this figure, we can see that the porosity has an obvious effect on the value of peak concentration in the extraction well. When the value of porosity increases, the peak value of front becomes smaller and lasts longer. Also, the time of occurrence is delayed.

The last set of sensitivity curves, Figure 7.16, shows the effects of the horizontal and vertical anisotropies, where the longitudinal dispersivity is taken as 10 m and the other parameters are the same as those in Figure 7.13. From the figure it is seen that, if K_r/K_z is over 50, the results of a three-dimensional model are almost the same as those of a two-dimensional model. If K_r/K_z is relatively small, the results calculated from these two models differ greatly. This means that when we fit the model to the observed curves, we should not simplify the model without taking into consideration the value of K_r/K_z.

A reasonable design may be found after analyzing a great number of sensitivity curves. The best way is to drill one well first, do some single well tests to reduce the ranges of the unknown parameters, and then use the results of sensitivity analyses to determine a suitable distance for the second well. As a result, it is more possible to obtain a successful design than drilling two wells at the same time. Obviously, the CAD method can and should be used to obtain more complex experimental designs.

7.3 Coupled Inverse Problems of Groundwater Flow and Mass Transport

7.3.1 The Definition of Coupled Inverse Problems

According to Section 7.1.1, the groundwater flow and mass transport in the saturated zone of an isotropic aquifer are governed by the following equations:

$$S_s \frac{\partial h}{\partial t} - \frac{\partial}{\partial x_i}\left(K \frac{\partial h}{\partial x_i}\right) + Q = 0, \tag{7.3.1}$$

$$\frac{\partial(\theta C)}{\partial t} - \frac{\partial}{\partial x_i}\left(D_{ij}\theta \frac{\partial C}{\partial x_j}\right) + \frac{\partial}{\partial x_i}(V_i\theta C) + \lambda\theta C + M = 0, \tag{7.3.2}$$

where water head h and concentration C satisfy the following initial conditions:

$$h = f_0, \quad C = g_0; \quad t = 0, \quad \mathbf{x} \in (\Omega) \tag{7.3.3}$$

and boundary conditions:

$$h = f_1, \quad C = g_1; \quad \mathbf{x} \in (S_1), \quad t \geq 0 \tag{7.3.4}$$

$$-K\frac{\partial h}{\partial x_i}n_i = f_2, \quad \left(-D_{ij}\theta\frac{\partial C}{\partial x_j} + V_i\theta C\right)n_i = g_2; \quad \mathbf{x} \in (S_2), \quad t \geq 0. \tag{7.3.5}$$

In Eqs. (7.3.1) to (7.3.5), Q is the source/sink term of the flow equation; M the source/sink term of the quality equation; f_0, g_0, f_1, g_1, f_2, g_2 all known functions; (Ω) the flow domain; and (S_1) and (S_2) its boundary surfaces. The other symbols are the same as before. The components of mean velocity V

and dispersion coefficient **D** are

$$V_i = -\frac{K}{\theta}\frac{\partial h}{\partial x_i}, \tag{7.3.6}$$

$$D_{ij} = \alpha_T V \delta_{ij} + (\alpha_L - \alpha_T)\frac{V_i V_j}{V}. \tag{7.3.7}$$

Here, we assume that molecular diffusion is negligible. The state variables of the coupled problem Eqs. (7.3.1) to (7.3.7) are h and C, and the parameters are S_S, K, θ, α_L, α_T, and λ. Substituting Eqs. (7.3.6) and (7.3.7) into Eq. (7.3.2), it can be seen that the concentration C depends directly on K, θ, α_L, α_T, and λ, and indirectly on S_S via water head h. We have discussed the numerical simulation of this coupled problem in Section 7.1.1. Now, we will consider its inverse problem:

1. Assume that K observations of water head $\mathbf{h}^*$, and L observations of concentrations $\mathbf{C}^*$, at given times and given locations are known:

$$\mathbf{h}^* = (h_1^*, h_2^*, \ldots, h_K^*)^T, \tag{7.3.8}$$

$$\mathbf{C}^* = (C_1^*, C_2^*, \ldots, C_L^*)^T. \tag{7.3.9}$$

Note that both of them may contain *observation errors*.
2. The relationship between state variables and parameters may be represented by model M, which is a numerical model generated by solving the discretized versions of Eqs. (7.3.1) to (7.3.7). The unknown parameters in model M have been parameterized and expressed as an m-dimensional vector:

$$\mathbf{p} = (p_1, p_2, \ldots, p_m). \tag{7.3.10}$$

Model M always contains *modeling errors* and *calculation errors*. The former are caused by the differences between the simplified mathematical model and the true physical model, and also by the differences in structure resulted from parameterization. The latter are mainly numerical errors generated by discretization.
3. Assume that some prior information of unknown parameters has been acquired, which is generally the estimated range of the parameter values. Let us define the admissible set of the parameters as

$$P_{ad} = \{\mathbf{p}|\underline{p}_i \le p_i \le \bar{p}_i, i = 1, 2, \ldots, m\}, \tag{7.3.11}$$

where $\underline{p}_i$ and $\bar{p}_i$, are the lower and upper bounds of p_i, respectively. The *coupled inverse problem* is the determination of the unknown parameter vector $\mathbf{p}$, based on the above data. The inverse problem stated here is different from what was discussed before. In Section 7.2.1, the unknown parameters are identified by using the water flow and quality equations separately. Now, the hydraulic conductivity K is identified not only by water head observations but also by concentration observations. Moreover, in the identification of dispersivities, the velocity field is not required to be given.

Sun and Yeh (1990a), in view of the multi-objective optimization and generalized least squares criteria, formulated the coupled inverse problem into the following non-linear programming problem:

$$\min I(\mathbf{p}) = w_c \sum_{l=1}^{L} \frac{[C_l(\mathbf{p}) - C_l^*]^2}{\sigma_c^2} + w_h \sum_{k=1}^{K} \frac{[h_k(\mathbf{p}) - h_k^*]^2}{\sigma_h^2}, \quad \mathbf{p} \in P_{ad} \quad (7.3.12)$$

where σ_c and σ_h are standard deviations of concentration observation errors and head observation errors respectively; $C_l(\mathbf{p})$ and $h_k(\mathbf{p})$ the model outputs corresponding to the observations C_l^* and h_k^*; and w_c and w_h the weighting coefficients determined by the principles of multi-objective optimization.

If the simplest gradient method is used for solving problem (7.3.12), we have the following iteration series:

$$\mathbf{p}_{r+1} = \mathbf{p}_r + \lambda_r \mathbf{d}_r, \quad (7.3.13)$$

where λ_r is the step factor. The negative gradient direction of objective function $I(\mathbf{p})$ at $\mathbf{p}_r$ may be taken as the search direction $\mathbf{d}_r$, i.e., let

$$\mathbf{d}_r = -\nabla I(\mathbf{p}_r) = -\left(\frac{\partial I}{\partial p_1}, \frac{\partial I}{\partial p_2}, \dots, \frac{\partial I}{\partial p_m}\right)^T \bigg|_{\mathbf{p}_r}. \quad (7.3.14)$$

We can calculate the partial derivatives in Eq. (7.3.14) by their finite difference approximations. If the forward finite difference algorithm is used, we need to solve the simulation problem $m + 1$ times, while if the central finite difference algorithm is used, we have to solve the simulation problem $2m + 1$ times. Meanwhile, in order to determine the step factor λ_r in Eq. (7.3.13), more simulation runs are required. Therefore, the total computation effort is very huge.

The other way for solving the optimized problem (7.3.12) is the Gauss-Newton method. Rewrite the objective function $I(\mathbf{p})$ into the following form:

$$I(\mathbf{p}) = [C(\mathbf{p}) - C^*]^T W_c [C(\mathbf{p}) - C^*] + [h(\mathbf{p}) - h^*]^T W_h [h(\mathbf{p}) - h^*], \quad (7.3.15)$$

where W_c and W_h are diagonal matrices, the elements on the diagonal are w_c/σ_c^2 and w_h/σ_h^2, respectively. Using the Taylor's expansion to linearize $C(\mathbf{p})$ and $h(\mathbf{p})$ around $\mathbf{p}_r$ and using the necessary condition of minimum, we can obtain the improved value of $\mathbf{p}_r$ as:

$$\Delta \mathbf{p}_r = (J_c^T W_c J_c + J_h^T W_h J_h)^{-1} [J_c W_c (C(\mathbf{p}_r) - C^*) + J_h W_h (h(\mathbf{p}_r) - h^*)], \quad (7.3.16)$$

where the sensitivity matrices

$$J_c = \left[\frac{\partial C(\mathbf{p})}{\partial \mathbf{p}}\right] \quad \text{and} \quad J_h = \left[\frac{\partial h(\mathbf{p})}{\partial \mathbf{p}}\right] \quad (7.3.17)$$

are $L \times m$ and $K \times m$ matrices and evaluated at $\mathbf{p}_r$. Their elements may include the partial derivatives $\partial h/\partial k$, $\partial h/\partial S_s$, $\partial C/\partial K$, $\partial C/\partial \theta$, $\partial C/\partial \alpha_L$, $\partial C/\partial \alpha_T$,

$\partial C/\partial \lambda$, etc. To calculate these derivatives, the simplest way is still the use of finite difference approximations. To improve the convergence of the Gauss-Newton method, the following iteration series is often used:

$$\mathbf{p}_{r+1} = \mathbf{p}_r + \lambda_r \Delta \mathbf{p}_r, \tag{7.3.18}$$

where $\Delta \mathbf{p}_r$ is determined from Eq. (7.3.16) and λ_r by one-dimensional optimization. In addition, by controlling the value of λ_r or using Rosen's projection method, the solutions can satisfy the given constraints.

More optimization methods for solving inverse problems in groundwater modeling can be found in Sun (1981), Yeh (1986), Carrera (1987), and Sun (1994). The coupled inverse problem of flow and mass transport was considered by Carrera (1987) and Wagner and Gorelick (1987), where the unknown transmissivities and dispersivities are determined simultaneously. Ewing et al. (1987) described a coupled inverse problem in connection with the simulation of secondary recovery processes in oil reservoirs. Woodbury et al. (1987), and Woodbury and Smith (1988), used temperature measurement to improve the estimation of hydraulic conductivities, in which the inverse problem of flow-heat transport was solved. Mishra and Parker (1989) considered the inverse problem for coupled unsaturated flow and mass transportation. A complete description of coupled inverse problems in groundwater modeling was given by Sun and Yeh (1990a, b). Xiang et al. (1993) used L_1-norm criterion to solve coupled inverse problems of groundwater flow and mass transport. A program that can solve coupled inverse problems of groundwater flow and mass transport by a modified Gauss-Newton method is given in Sun (1994).

7.3.2 Variational Sensitivity Analysis

Another way to calculate the gradient of Eq. (7.3.14) and the sensitivity matrices of Eq. (7.3.17) is the use of *variational calculus*. Consider the objective function in a general form:

$$I(h, C, p) = \int_0^T \int_{(\Omega)} f(h, C, p; \mathbf{x}, t) \, d\Omega \, dt, \tag{7.3.19}$$

where f is a function to be selected, and $[0, T] \times (\Omega)$ the time-space domain being considered. For determination, let us temporarily assume that the unknown parameter p is the hydraulic conductivity $K(\mathbf{x})$. When there is a slight change (or variation) δK in K, it will cause changes (variations) δh and δC in the head h and concentration C, respectively. The PDEs and corresponding initial and boundary conditions for δh and δC can be obtained by taking the variations of the original coupled problem defined by

Eqs. (7.3.1) to (7.3.7) as follows:

$$S_s \frac{\partial \delta h}{\partial t} - \frac{\partial}{\partial x_i}\left(K \frac{\partial \delta h}{\partial x_i}\right) - \frac{\partial}{\partial x_i}\left(\delta K \frac{\partial h}{\partial x_i}\right) = 0, \tag{7.3.20}$$

$$\frac{\partial(\theta \delta C)}{\partial t} - \frac{\partial}{\partial x_i}\left(\theta D_{ij} \frac{\partial \delta C}{\partial x_j}\right) + \frac{\partial}{\partial x_i}(\theta V_i \delta C) + \lambda \theta \delta C$$

$$- \frac{\partial}{\partial x_i}\left(\theta \frac{\partial D_{ij}}{\partial V_s} \delta V_s \frac{\partial C}{\partial x_j}\right) + \frac{\partial}{\partial x_i}(\theta \delta V_i C) = 0, \tag{7.3.21}$$

$$\delta h|_{t=0} = 0, \quad \delta C|_{t=0} = 0, \tag{7.3.22}$$

$$\delta h|_{(S_1)} = 0, \quad \delta C|_{(S_1)} = 0, \tag{7.3.23}$$

$$\left(-K \frac{\partial \delta h}{\partial x_i} - \delta K \frac{\partial h}{\partial x_i}\right) n_i \bigg|_{(S_2)} = 0, \tag{7.3.24}$$

$$\left(-\theta D_{ij} \frac{\partial \delta C}{\partial x_j} - \theta \frac{\partial D_{ij}}{\partial V_s} \delta V_s \frac{\partial C}{\partial x_j} + \theta C \delta V_i + \theta V_i \delta C\right) n_i \bigg|_{(S_2)} = 0, \tag{7.3.25}$$

where Eqs. (7.3.20) and (7.3.21) are the governing equations for δh and δC, respectively. Equations. (7.3.22) to (7.3.25) are their initial and boundary conditions.

The variation of objective function (7.3.19) can be expressed as

$$\delta I = \int_0^T \int_{(\Omega)} \left(\frac{\partial f}{\partial h} \delta h + \frac{\partial f}{\partial C} \delta C + \frac{\partial f}{\partial K} \delta K\right) d\Omega \, dt. \tag{7.3.26}$$

In order to eliminate the terms connected with the unknown variations δh and δC from the above formula, we introduce two new unknown variables ϕ_1 and ϕ_2, which are called *adjoint state variables* of h and C. We also assume that ϕ_1 and ϕ_2 have continuous second-order partial derivatives in the time-space domain of interest.

Multiplying Eq. (7.3.20) by ϕ_1 and integrating it over $[0, T] \times (\Omega)$, we obtain

$$\int_0^T \int_{(\Omega)} \left\{S_s \frac{\partial \delta h}{\partial t} \phi_1 - \frac{\partial}{\partial x_i}\left(K \frac{\partial \delta h}{\partial x_i}\right) \phi_1 - \frac{\partial}{\partial x_i}\left(\delta K \frac{\partial h}{\partial x_i}\right) \phi_1\right\} d\Omega \, dt = 0. \tag{7.3.27}$$

Using *Green's theorem* to transfer the partial derivative operation from δh to ϕ_1, the above equation can be rewritten as

$$\int_{(\Omega)} S_s(\phi_1 \delta h)\bigg|_0^T d\Omega + \int_0^T \int_{(\Omega)} \left\{\left[-S_s \frac{\partial \phi_1}{\partial t} - \frac{\partial}{\partial x_i}\left(K \frac{\partial \phi_1}{\partial x_i}\right)\right] \delta h\right.$$

$$\left. + \frac{\partial \phi_1}{\partial x_i} \frac{\partial h}{\partial x_i} \delta K\right\} d\Omega \, dt + \int_0^T \int_{(S)} \delta h K \frac{\partial \phi_1}{\partial x_i} n_i \, dS \, dt$$

$$- \int_0^T \int_{(S)} \phi_1 \left[K \frac{\partial \delta h}{\partial x_i} + \delta K \frac{\partial h}{\partial x_i}\right] n_i \, dS \, dt = 0. \tag{7.3.28}$$

Using the initial and boundary conditions of δh in Eqs. (7.3.22) to (7.3.24), and in addition, defining the final condition for adjoint state ϕ_1 and its boundary condition on (S_1) as follows:

$$\phi_1|_{t=T} = 0, \tag{7.3.29}$$

$$\phi_1|_{(S_1)} = 0, \tag{7.3.30}$$

then the first and the last integrals on the left-hand side of Eq. (7.3.28) will all vanish, we then have

$$\int_0^T \int_{(\Omega)} \left\{ \left[S_s \frac{\partial \phi_1}{\partial t} + \frac{\partial}{\partial x_i} \left(K \frac{\partial \phi_1}{\partial x_i} \right) \right] \delta h - \left[\frac{\partial h}{\partial x_i} \frac{\partial \phi_1}{\partial x_i} \right] \delta K \right\} d\Omega \, dt$$

$$- \int_0^T \int_{(S)} K \delta h \frac{\partial \phi_1}{\partial x_i} n_i \, dS \, dt = 0. \tag{7.3.31}$$

Similarly, multiplying Eq. (7.3.21) by ϕ_2, integrating over $[0, T] \times (\Omega)$, and using Green's theorem to transfer the calculation of partial derivatives to ϕ_2, we arrive at

$$\int_{(\Omega)} \theta(\phi_2 \delta C) \Big|_0^T d\Omega + \int_0^T \int_{(\Omega)} \left[-\theta \frac{\partial \phi_2}{\partial t} - \frac{\partial}{\partial x_i} \left(\theta D_{ij} \frac{\partial \phi_2}{\partial x_j} \right) \right.$$

$$\left. - V_i \theta \frac{\partial \phi_2}{\partial x_i} + \lambda \theta \phi_2 \right] \delta C \, d\Omega \, dt + \int_0^T \int_{(\Omega)} \left[\theta \frac{\partial D_{ij}}{\partial V_s} \delta V_s \frac{\partial C}{\partial x_j} \frac{\partial \phi_2}{\partial x_i} \right.$$

$$\left. - \theta C \frac{\partial \phi_2}{\partial x_i} \delta V_i \right] d\Omega \, dt + \int_0^T \int_{(S)} \theta \delta C D_{ij} \frac{\partial \phi_2}{\partial x_j} n_i \, dS \, dt$$

$$+ \int_0^T \int_{(S)} \phi_2 \left[-\theta D_{ij} \frac{\partial \delta C}{\partial x_j} - \theta \frac{\partial D_{ij}}{\partial V_s} \delta V_s \frac{\partial C}{\partial x_j} + \theta C \delta V_i + \theta V_i \delta C \right] dS \, dt$$

$$= 0. \tag{7.3.32}$$

Using the initial and boundary conditions for δC in Eqs. (7.3.22), (7.3.23) and (7.3.25), and defining the final and boundary conditions for the adjoint state ϕ_2.

$$\phi_2|_{t=T} = 0, \tag{7.3.33}$$

$$\phi_2|_{(S_1)} = 0, \tag{7.3.34}$$

$$\theta D_{ij} \frac{\partial \phi_2}{\partial x_j} n_i \Big|_{(S_2)} = 0, \tag{7.3.35}$$

the first integral and the last two integrals along the boundary on the left-hand side of (7.3.32) will vanish. From Darcy's law (7.3.6), we can obtain the variation of velocity, that is,

$$\delta V_i = -\frac{K}{\theta} \frac{\partial \delta h}{\partial x_i} - \frac{\delta K}{\theta} \frac{\partial h}{\partial x_i}. \tag{7.3.36}$$

Thus, the third integral on the left-hand side of (7.3.32) may be rewritten as

$$\int_0^T \int_{(\Omega)} \left[\theta \frac{\partial D_{ij}}{\partial V_s} \delta V_s \frac{\partial C}{\partial x_j} \frac{\partial \phi_2}{\partial x_i} - \theta C \frac{\partial \phi_2}{\partial x_i} \delta V_i \right] d\Omega \, dt$$

$$= \int_0^T \int_{(\Omega)} \left\{ \left[\frac{\partial}{\partial x_i} \left(K \frac{\partial D_{ij}}{\partial V_s} \frac{\partial C}{\partial x_j} \frac{\partial \phi_2}{\partial x_i} \right) - \frac{\partial}{\partial x_i} \left(K C \frac{\partial \phi_2}{\partial x_i} \right) \right] \delta h \right.$$

$$\left. - \left[\left(\frac{\partial D_{ij}}{\partial V_s} \frac{\partial C}{\partial x_j} \frac{\partial h}{\partial x_s} \frac{\partial \phi_2}{\partial x_i} - C \frac{\partial \phi_2}{\partial x_i} \frac{\partial h}{\partial x_i} \right) \delta K \right] \right\} d\Omega \, dt$$

$$- \int_0^T \int_{(S)} \left[K \frac{\partial D_{ij}}{\partial V_s} \frac{\partial C}{\partial x_j} \frac{\partial \phi_2}{\partial x_i} - K C \frac{\partial \phi_2}{\partial x_i} \right] \delta h n_i \, dS \, dt. \qquad (7.3.37)$$

Substituting the above formula into Eq. (7.3.32), and adding it to Eq. (7.3.31), we then have:

$$\int_0^T \int_{(\Omega)} \left\{ \left[\theta \frac{\partial \phi_2}{\partial t} + \frac{\partial}{\partial x_i} \left(\theta D_{ij} \frac{\partial \phi_2}{\partial x_j} \right) + V_i \theta \frac{\partial \phi_2}{\partial x_i} - \lambda \theta \phi_2 \right] \delta C \right.$$

$$+ \left[S_s \frac{\partial \phi_1}{\partial t} + \frac{\partial}{\partial x_i} \left(K \frac{\partial \phi_1}{\partial x_i} \right) - \frac{\partial}{\partial x_i} \left(K \frac{\partial D_{ij}}{\partial V_s} \frac{\partial C}{\partial x_j} \frac{\partial \phi_2}{\partial x_i} \right) + \frac{\partial}{\partial x_i} \left(K C \frac{\partial \phi_2}{\partial x_i} \right) \right] \delta h$$

$$\left. + \left[- \frac{\partial h}{\partial x_i} \frac{\partial \phi_1}{\partial x_i} - C \frac{\partial h}{\partial x_i} \frac{\partial \phi_2}{\partial x_i} + \frac{\partial D_{ij}}{\partial V_s} \frac{\partial C}{\partial x_j} \frac{\partial h}{\partial x_s} \frac{\partial \phi_2}{\partial x_i} \right] \delta K \right\} d\Omega \, dt = 0. \qquad (7.3.38)$$

In the above equation, the following boundary surface integral is originally contained:

$$\int_0^T \int_{(S_2)} \delta h \left[- K \frac{\partial \phi_1}{\partial x_i} + K \frac{\partial D_{ij}}{\partial V_s} \frac{\partial C}{\partial x_j} \frac{\partial \phi_2}{\partial x_i} - K C \frac{\partial \phi_2}{\partial x_1} \right] n_i \, dS \, dt. \qquad (7.3.39)$$

However, if we let ϕ_1 satisfy the following condition on (S_2):

$$\frac{\partial \phi_1}{\partial x_i} n_i \bigg|_{(S_2)} = \left(\frac{\partial D_{ij}}{\partial V_s} \frac{\partial C}{\partial x_j} \frac{\partial \phi_2}{\partial x_i} - C \frac{\partial \phi_2}{\partial x_i} \right) n_i \bigg|_{(S_2)}, \qquad (7.3.40)$$

then the integral in Eq. (7.3.39) will certainly be equal to zero. After adding Eq. (7.3.38) to the right-hand side of Eq. (7.3.26), we can see that if we select ϕ_1 and ϕ_2 to satisfy the following equations

$$S_s \frac{\partial \phi_1}{\partial t} + \frac{\partial}{\partial x_i} \left(K \frac{\partial \phi_1}{\partial x_i} \right) = \frac{\partial f}{\partial h} + \frac{\partial}{\partial x_i} \left(K \frac{\partial D_{ij}}{\partial V_s} \frac{\partial C}{\partial x_j} \frac{\partial \phi_2}{\partial x_i} \right) - \frac{\partial}{\partial x_i} \left(K C \frac{\partial \phi_2}{\partial x_i} \right), \qquad (7.3.41)$$

$$\theta \frac{\partial \phi_2}{\partial t} + \frac{\partial}{\partial x_i} \left(\theta D_{ij} \frac{\partial \phi_2}{\partial x_j} \right) + V_i \theta \frac{\partial \phi_2}{\partial x_i} - \lambda \theta \phi_2 = \frac{\partial f}{\partial C}, \qquad (7.3.42)$$

then the terms connected with δh and δC in Eq. (7.3.26) will be eliminated and we finally obtain

$$\delta I = \int_0^T \int_{(\Omega)} \left[\frac{\partial f}{\partial K} + \frac{\partial h}{\partial x_i} \frac{\partial \phi_1}{\partial x_i} - \frac{\partial D_{ij}}{\partial V_s} \frac{\partial C}{\partial x_j} \frac{\partial h}{\partial x_s} \frac{\partial \phi_2}{\partial x_i} + C \frac{\partial h}{\partial x_i} \frac{\partial \phi_2}{\partial x_i} \right] \delta K \, d\Omega \, dt.$$

$$(7.3.43)$$

Summarily, there are two problems here: the original coupled problem defined by Eqs. (7.3.1) to (7.3.5) and its *adjoint problem*. The governing equations of the latter are Eqs. (7.3.41) and (7.3.42), where the final and boundary conditions for ϕ_1 are defined by Eqs. (7.3.29), (7.3.30), and (7.3.40), and those for ϕ_2, by Eqs. (7.3.33), (7.3.34), and (7.3.35). The distributions of h and C can be determined by solving original coupled problem, and the distributions of ϕ_1 and ϕ_2 can be determined by solving the adjoint problem. As a result, integrand (7.3.43) can be calculated.

If the variable substitution, $\tau = T - t$, is made in the adjoint problem, Eqs. (7.3.41) and (7.3.42) will become identical to the original coupled problem in form. The final condition will turn into the initial condition. Hence, we need only to slightly modify the subprogram for solving h and C, which can then be used for solving ϕ_1 and ϕ_2. The detailed solution procedure and numerical examples may be found from Sun and Yeh (1990a). Next, let us demonstrate the usage of Eq. (7.3.43), the variation of the objective function. Assume that K is parameterized by the zonation method, that is, the whole domain (Ω) is divided into m subdomains (Ω_1), (Ω_2), ..., (Ω_m). K is a constant in each subdomain. The unknown parameter vector now becomes $\mathbf{K} = (K_1, K_2, \ldots, K_m)^T$. To calculate the gradient (7.3.14), the function to be selected in Eq. (7.3.19) can be taken as

$$f(h, C, \mathbf{K}; \mathbf{x}, t) = w_C \sum_{l=1}^{L} \frac{[C_l(\mathbf{K}) - C_l^*]^2}{\sigma_C^2} \delta(\mathbf{x} - \mathbf{x}_l)\delta(t - t_l)$$
$$+ w_h \sum_{k=1}^{K} \frac{[h_k(\mathbf{K}) - h_k^*]^2}{\sigma_h^2} \delta(\mathbf{x} - \mathbf{x}_k)\delta(t - t_k), \qquad (7.3.44)$$

where δ is the Dirac-δ function, $(\mathbf{x}_l, t_l)$ the location and time of the lth concentration observation, $(\mathbf{x}_k, t_k)$ the location and time of the kth observation of hydraulic head. Then, the general objective function (7.3.19) will become the objective function (7.3.12). Therefore, from Eq. (7.3.43) we have

$$\frac{\partial I}{\partial K_n} = \int_0^T \int_{(\Omega_n)} \left[\frac{\partial h}{\partial x_i}\frac{\partial \phi_1}{\partial x_i} - \frac{\partial D_{ij}}{\partial V_s}\frac{\partial C}{\partial x_j}\frac{\partial h}{\partial x_s}\frac{\partial \phi_2}{\partial x_i} + C\frac{\partial h}{\partial x_i}\frac{\partial \phi_2}{\partial x_i} \right] d\Omega\, dt.$$
$$(n = 1, 2, \ldots, m) \qquad\qquad (7.3.45)$$

Note that we need only solve the original problem and its adjoint problem once to obtain the m partial derivatives, regardless of the number m. Thus, this kind of algorithm will be useful if $m > 2$.

If the function to be selected in Eq. (7.3.19) is

$$f(h, C, \mathbf{K}; \mathbf{x}, t) = h(\mathbf{x}, t)\delta(\mathbf{x} - \mathbf{x}_k)\delta(t - t_k), \qquad (7.3.46)$$

then objective function (7.3.19) will become $h_k(\mathbf{x}, t)$. Therefore, from Eq. (7.3.43), we can obtain the sensitivity coefficient

$$\frac{\partial h_k}{\partial K_n} = \int_0^T \int_{(\Omega_n)} \left[\frac{\partial h}{\partial x_i}\frac{\partial \phi_1}{\partial x_i} - \frac{\partial D_{ij}}{\partial V_s}\frac{\partial C}{\partial x_j}\frac{\partial h}{\partial x_s}\frac{\partial \phi_2}{\partial x_i} + C\frac{\partial h}{\partial x_i}\frac{\partial \phi_2}{\partial x_i} \right] d\Omega\, dt. \qquad (7.3.47)$$

Similarly, when the function in Eq. (7.3.19) is taken as

$$f(h, C, \mathbf{K}; \mathbf{x}, t) = C(\mathbf{x}, t)\delta(\mathbf{x} - \mathbf{x}_l)\delta(t - t_i), \tag{7.3.48}$$

we will obtain

$$\frac{\partial C_l}{\partial K_n} = \int_0^T \int_{(\Omega_n)} \left[\frac{\partial h}{\partial x_i} \frac{\partial \phi_1}{\partial x_i} - \frac{\partial D_{ij}}{\partial V_s} \frac{\partial C}{\partial x_j} \frac{\partial h}{\partial x_s} \frac{\partial \phi_2}{\partial x_i} + C \frac{\partial h}{\partial x_i} \frac{\partial \phi_2}{\partial x_i} \right] d\Omega \, dt. \tag{7.3.49}$$

For calculating sensitivity matrix (7.3.17), the number of times that the original problem and its adjoint problem are solved is equal to the number of the observation data. Thus, this algorithm is advantageous only when $m > L + K$. This case, of course, is unlikely to occur when solving the inverse problem. However, for the case of evaluating the effect of parameter uncertainties on the model predictions, m may be equal to the number of nodes, while the number of "observations" is only equal to the number of a few locations where the concentrations are to be predicted. In this case, we can take the advantage of Eq. (7.3.49) to save the computational effort.

Using a similar method, we can derive the following formulas, as shown in Eq. (7.3.45), but with respect to the parameters S_s, θ, α_L, α_T, and λ:

$$\frac{\partial I}{\partial S_s} = \int_0^T \int_{(\Omega)} \left(\frac{\partial f}{\partial S_s} + \frac{\partial h}{\partial t} \phi_1 \right) d\Omega \, dt,$$

$$\frac{\partial I}{\partial \theta} = \int_0^T \int_{(\Omega)} \left(\frac{\partial f}{\partial \theta} + \frac{\partial C}{\partial t} \phi_2 \right) d\Omega \, dt,$$

$$\frac{\partial I}{\partial \alpha_L} = \int_0^T \int_{(\Omega)} \left(\frac{\partial f}{\partial \alpha_L} + \frac{\partial D_{ij}}{\partial \alpha_L} \frac{\partial C}{\partial x_i} \frac{\partial \phi_2}{\partial x_i} \right) d\Omega \, dt,$$

$$\frac{\partial I}{\partial \alpha_T} = \int_0^T \int_{(\Omega)} \left(\frac{\partial f}{\partial \alpha_T} + \frac{\partial D_{ij}}{\partial \alpha_T} \frac{\partial C}{\partial x_i} \frac{\partial \phi_2}{\partial x_i} \right) d\Omega \, dt,$$

$$\frac{\partial I}{\partial \lambda} = \int_0^T \int_{(\Omega)} \left(\frac{\partial f}{\partial \lambda} + \phi_2 \right) d\Omega \, dt.$$

For the derivation of these equations and their numerical examples, refer to Sun and Yeh (1990a).

7.3.3 Identifiability

A major difficulty associated with the parameter identification of an aquifer is the *non-uniqueness* and *instability* of inverse solutions. This problem has already been pointed out by many authors (Neuman, 1973; Chavent, 1979; Yakowitz and Duckstein, 1980; Yeh and Sun, 1984; Carrera and Neuman, 1986a). If two or more parameters can lead to the same group of observations, we cannot judge which parameter is the real one from this group of observations alone. It is the problem of non-uniqueness. On the other hand, since observation error, model structure error, and calculation error always exist in practice, if a minor change in the observations causes a great change in the parameter to be determined, then we cannot find the true parameter either. This is the problem of instability.

Some simple examples of non-uniqueness and instability in the solution of the inverse problem of groundwater flow are given by Neuman (1973), Chavent (1979), Sun (1981) and Sun (1994). When solving the inverse problem of groundwater quality, the non-uniqueness resulting from the coexistence of advection and dispersion terms may become very significant.

An essential problem in parameter identification is the *identifiability* problem. The classical definition of identifiability (Kitamura and Nakagiri, 1977) requires that the unknown parameter be uniquely determined by the observation data obtained from an experimental design. According to this definition, the value of the unknown parameter and the set of observations must correspond to each other one by one. Of course, it is impossible to meet this demand in practice.

Chavent (1979) put forward another concept of identifiability, the *Least Squares Identifiability*. It only requires that the optimization problem in Eq. (7.2.5) have a unique solution, and this solution continuously depend on the observation data. This definition allows the existence of observation errors. However, it is still difficult to satisfy in practice. Yeh and Sun (1984) suggested a concept, *δ-identifiability*, which only requires that the predicted results not greatly deviate from the real results when the identified parameters is used for a prediction purpose. Various definitions of identifiability were reviewed by Chavent (1987). Sun and Yeh (1990b) presented the *Prediction Equivalence Identifiability* and *Management Equivalence Identifiability* for general coupled problems. Based on these ideas, hydrogeologists can design experiments and identify parameters in accordance with the accuracy demanded by the manager, so as to guarantee the reliability of model applications.

Now, let us use D to represent an experimental design and use vector $\mathbf{u}_D^*$ to denote the observations obtained from the design. The components of $\mathbf{u}_D^*$ are the observed values of hydraulic head and concentration at certain points and certain times. The output of the model corresponding to the observations are denoted by $\mathbf{u}_D(\mathbf{p})$, where $\mathbf{p}$ is the parameter used in the model. Moreover, we use a vector $\mathbf{g}$ to denote the model prediction objectives. Its components are the values of water head and concentration to be predicted at certain points and certain times. Obviously, $\mathbf{g}$ is a function of model parameters. The manager's demand for the accuracy of model predictions can be generally represented as:

$$\|\mathbf{g}(\mathbf{p}) - \mathbf{g}(\hat{\mathbf{p}})\|_G = \left\{ \sum_{k=1}^{K} w_{g,k} [\mathbf{g}_k(\mathbf{p}) - \mathbf{g}_k(\hat{\mathbf{p}})]^2 \right\}^{1/2} < \varepsilon, \qquad (7.3.50)$$

where $\hat{\mathbf{p}}$ is the real parameter, $\|\cdot\|_G$ the vector norm defined by the generalized least squares, $w_{g,k}$ $(k = 1, 2, \ldots, K)$ a set of weighting coefficients, K the dimension of vector $\mathbf{g}$, and $\varepsilon > 0$ the limit of error.

All parameters **p** satisfying the condition in Eq. (7.3.50) form a subset of the admissible set which is named the *predictive equivalent set* of $\hat{\mathbf{p}}$. Our question is: can we design an experiment, so that when we use the results of the experiment to solve the inverse problem, the parameters obtained will certainly belong to the predictive equivalent set of $\hat{\mathbf{p}}$? If the answer is positive, parameter $\hat{\mathbf{p}}$ is said to be prediction equivalence identifiable. By comparing it with the classical definition of identifiability mentioned above, we can see that this kind of extended identifiability does not require the uniqueness either for the inverse problem or for the optimization problem (7.3.12). The major advantage of prediction equivalence identifiability is that it may be satisfied in practice by limited quantity and quality of observed data.

We can prove that any parameter **p** satisfying

$$\|\mathbf{u}_D(\mathbf{p}) - \mathbf{u}_D(\hat{\mathbf{p}})\|_D = \left\{ \sum_{l=1}^{L} w_{D,l}[u_{D,l}(\mathbf{p}) - u_{D,l}(\hat{\mathbf{p}})]^2 \right\}^{1/2} < \delta \qquad (7.3.51)$$

must fall into the prediction equivalence set of $\hat{\mathbf{p}}$. If $\delta \geq 2\bar{\eta}$, then design D will provide sufficient data for the prediction equivalence identifiability. In Eq. (7.3.51), $\|\cdot\|_D$ is the vector norm defined by the generalized least squares; $w_{D,l}$ is the weighted coefficient, usually taken as $1/\sigma_l^2$; σ_l^2 is the variance of observed errors associated with the lth observation; L is the total number of observations, and $\bar{\eta}$ is the estimated upper bound of the norm of observation errors, it is the maximal value of

$$\|\boldsymbol{\eta}_D\|_D = \left(\sum_{l=1}^{L} w_{D,L}\eta_{D,l}^2 \right)^{1/2}, \qquad (7.3.52)$$

where $\boldsymbol{\eta}_D$ is the vector of the observation errors, i.e.,

$$\mathbf{u}_D^* = \mathbf{u}_D(\hat{\mathbf{p}}) + \boldsymbol{\eta}_D, \qquad (7.3.53)$$

its component $\eta_{D,l}$ is the observation errors associated with the lth observation.

In fact, when we solve the optimization problem in Eq. (7.2.5) to obtain parameter **p***, we must have

$$\|\mathbf{u}_D(\mathbf{p}^*) - \mathbf{u}_D^*\|_D < \bar{\eta}. \qquad (7.3.54)$$

As a result, we have

$$\|\mathbf{u}_D(\mathbf{p}^*) - \mathbf{u}_D(\hat{\mathbf{p}})\|_D \leq \|\mathbf{u}_D(\mathbf{p}^*) - \mathbf{u}_D^*\|_D + \|\boldsymbol{\eta}_D\|_D < 2\bar{\eta}, \qquad (7.3.55)$$

that is, **p*** satisfies condition (7.3.51). In other words, **p*** and $\hat{\mathbf{p}}$ are prediction equivalent. In the next section, we will demonstrate how to use condition (7.3.51) to design an experiment. Replacing the prediction vector in Eq. (7.3.50) by a management decision vector, we can similarly define management equivalence identifiability (Sun and Yeh, 1990b).

7.3.4 Experimental Design

Before we carry out pumping (or injection) and tracer injection experiments in the field, a design is needed to designate the locations and rates of pumping (or injection) wells, the tracer concentration of injected water, the number and locations of observation wells, the sampling frequency and experiment periods, and so on. A simple example has been given in Section 7.2.7.

In order to make a comparison between various designs and judge which is good and which is bad, we must have a criterion. Different criteria may lead to different optimal designs. A very common criterion is to minimize the uncertainty of the estimated parameters, i.e., to minimize the determinant of the covariance matrix of the estimated parameters. The design using this criterion is called the *D-Optimal Design*. Since the inverse matrix of the covariance matrix can be represented approximately by the *Fisher information matrix* (Bard, 1974; Silvey, 1980),

$$[\mathbf{M}]_D(\mathbf{p}) = [\mathbf{J}]_D^T [\mathbf{C}]_D^{-1} [\mathbf{J}]_D, \tag{7.3.56}$$

where $[\mathbf{J}]_D = \left[\dfrac{\partial \mathbf{u}_D}{\partial \mathbf{p}} \right]$ is the sensitivity matrix of observations with respect to the parameters and $[\mathbf{C}]_D$ is the covariance matrix of observation errors, the D-Optimal criterion is equivalent to maximize the determinant of the information matrix.

In recent years, the problem of optimizing experimental design for groundwater modeling has attracted much attention, because the reliability of a model closely depends on the quantity and quality of the observation data. The goal of optimal design is either to improve the reliability of the model as much as possible while keeping the expense in a reasonable range, or to reduce the cost of the experiment as much as possible while a certain reliability of the model is satisfied (Yeh and Sun, 1984; Carrera and Neuman, 1986c; Knopman and Voss, 1987; Hsu and Yeh, 1989; Sun and Yeh, 1990b). Among them, Yeh and Sun (1984), and Sun and Yeh (1990b) suggested a method of experimental design for satisfying the requirement of extended identifiabilities. It allows us to select the most economic design among all sufficient designs.

According to Eq. (7.3.51), to judge if an experimental design D can guarantee the prediction equivalence identifiability, we need to solve the following optimal problem:

$$\min \| \mathbf{u}_D(\mathbf{p}) - \mathbf{u}_D(\hat{\mathbf{p}}) \|_D, \tag{7.3.57}$$

subject to

$$\| \mathbf{g}(\mathbf{p}) - \mathbf{g}(\hat{\mathbf{p}}) \|_G \geq \varepsilon, \quad \mathbf{p} \in P_{ad}, \quad \hat{\mathbf{p}} \in P_{ad}. \tag{7.3.58}$$

If the minimum of Eq. (7.3.57) is greater than $2\bar{\eta}$, we can conclude that design D is *sufficient for the prediction equivalence identifiability* of parameter $\hat{\mathbf{p}}$. Since $\hat{\mathbf{p}}$ is unknown during the design phase, the above optimization problem should be solved for all possible p and $\hat{p}$. In Eq. (7.3.59), P_{ad} is the admissible set of parameters. Fortunately, in many cases, the minimum of Eq. (7.3.57) is not sensitive to $\hat{p}$. As a result, it may be sufficient to run trials for only a few estimated $\hat{p}$. An example of applying this method to design an experiment for controlling the contamination of an aquifer was given by Sun and Yeh (1990b). A complete discussion on experimental design for groundwater modeling can be found in Sun (1994).

7.4 Statistic Theory and Uncertainty Analysis

7.4.1 The Statistic Theory of Mass Transport in Porous Media

In Chapter 2, the advection-dispersion equation is derived from the microscopic level by spatial averaging over REV. In this procedure, the mechanical dispersion coefficient $\mathbf{D}$ is introduced to represent the mechanical dispersive flux:

$$\overline{C'V'} = -\rho \mathbf{D} \operatorname{grad}\left(\frac{C}{\rho}\right), \tag{7.4.1}$$

where $C' = \overline{C} - C$ and $V' = \overline{V} - V$ are deviations of microscopic concentration C and velocity V to their spatial averages $\overline{C}$ and $\overline{V}$, respectively. Equation (7.4.1) implies that the Fick's law is valid for mass transport in the macroscopic level.

The laboratory tests on rather homogeneous columns proved that the classical dispersion theory was very acceptable. The observed concentration distribution along a sand column can be well represented by the solution of the advection-dispersion equation with a constant dispersivity α_L in the range of a few centimeters.

In 1970s, hydrogeologists directly used the advection-dispersion equation, which was only verified on the laboratory scale, to simulate the mass transport on the local and regional scales. They were interested in finding the dispersivities of natural geological formations, such as aquifers. The "scale effect" problem, which we have mentioned in Sections 2.5 and 7.3, was soon discovered. As we have seen in Section 7.3, if we are limited in the classical dispersion theory, dispersivities would become arbitrary fitting parameters. To overcome this difficulty, a new theory, the *statistic theory* of porous media, was developed in early 1980s (Smith and Schwartz, 1980; Gelhar and Axness, 1983; Dagan, 1984; Yeh et al., 1985a, b; Sposito et al., 1986; Neuman et al., 1987). In the new theory, the heterogeneity of natural formations is described by *random functions*, and the effect of macro-dispersion is taken into account.

From the statistic theory, the Fick's law is considered as an asymptotic law, and the classical dispersion equation is valid only after the tracer travels a long distance or a long time. In recent years, some field tests are designed to verify the validity of the new theory. We have mentioned these tests in Section 7.2.3. In what follows, we will give a short introduction to the statistic theory. For a systematical discussion, the reader may refer to Dagan (1989). New developments in this area can be found in Neuman and Zhang (1990), Rubin (1990), Rubin (1991), Dagan and Neuman (1991), Rehfeldt and Gelhar (1992), Russo (1993), Naff (1994), and Kapoor and Gelhar (1994a, b).

7.4.2 The Heterogeneity of Natural Formations

Natural geological formations in local and regional scales are always heterogeneous. It is unrealistic to describe the detail of this kind of porous medium by any deterministic parameters. In the statistical framework, complex spatial structures are represented by random functions. For example, the hydraulic conductivity $K(\mathbf{x})$ of an aquifer can be viewed as a spatial random field. According to an analysis of field data, Freeze (1975) discovered that the probability distribution of $Y(\mathbf{x}) = \ln K(\mathbf{x})$ is close to a normal distribution. This argument was supported by the work of Hoeksema and Kitanidis (1985), in which data from the literature for about twenty aquifers in the United States were analyzed.

A normal random function $Y(\mathbf{x})$ may be completely characterized by its first two moments: the mathematical expectation (mean), $E[Y(\mathbf{x})]$, and the covariance function, $C_{YY}(\mathbf{x}_i, \mathbf{x}_j)$ between two locations $\mathbf{x}_i$ and $\mathbf{x}_j$. When $\bar{Y} = E[Y(\mathbf{x})]$ is a constant and $C_{YY}(\mathbf{x}_i, \mathbf{x}_j)$ only depends the separation vector $\mathbf{r} = \mathbf{x}_i - \mathbf{x}_j$, the random function $Y(\mathbf{x})$ is said to be *stationary* or *statistic homogeneous*. When the mean $\bar{Y}(\mathbf{x})$ is not a constant, we can express $Y(\mathbf{x})$ into two terms:

$$Y(\mathbf{x}) = \bar{Y}(\mathbf{x}) + f(\mathbf{x}), \qquad (7.4.2)$$

where $f(\mathbf{x})$ is called a *perturbation*. It represents the fluctuation of $Y(\mathbf{x})$ around its mean $\bar{Y}(\mathbf{x})$. Since $E[f(\mathbf{x})] \equiv 0$, $f(\mathbf{x})$ may be considered as a stationary random field.

In the literature, the following stationary three-dimensional covariance function is often adopted to represent the *spatial variability* of hydraulic conductivity:

$$C_{YY}(\mathbf{r}) = \sigma_Y^2 \exp[-(r_1^2/l_1^2 + r_2^2/l_2^2 + r_3^2/l_3^2)^{1/2}], \qquad (7.4.3)$$

where σ_Y^2 is the variance of $Y = \ln K$, (r_1, r_2, r_3) are the coordinates of $\mathbf{r}$, l_1, l_2, l_3 are called *correlation scales* in the three coordinate directions. When l_1, l_2, l_3 are not equal, the spatial variability is *anisotropic*. For the *isotropic* case, we have $l_1 = l_2 = l_3 = l_Y$, and Eq. (7.4.3) reduces to

$$C_{YY}(r) = \sigma_Y^2 \exp(-r/l_Y), \qquad (7.4.4)$$

where r is the distance of two locations. *Correlation length* l_Y is an important measure of the heterogeneous scale. When the distance of two locations is larger than l_Y, the correlation between them is very small. A realization of a stationary random field can be estimated by kriging when the mean, covariance function and some point measurements of the field are given. A complete discussion on *geostatistical methods* (kriging and co-kriging) for estimating stationary and non-stationary random fields can be found in de Marsily (1986).

In practice, most values of σ_Y^2 found in field tests are less than unity, however, cases of $\sigma_Y^2 > 10$ are also possible. Values of l_Y may change in a large range, from less than one meter to tens and even hundreds of meters (Gelhar, 1986).

7.4.3 Stochastic Advectian-Dispersion Equations and Macrodispersivities

In Chapter 2, we have derived the advection-dispersion equation

$$\frac{\partial C}{\partial t} = \frac{\partial}{\partial x_i}\left(D_{ij}\frac{\partial C}{\partial x_j}\right) - \frac{\partial}{\partial x_i}(V_i C). \quad (i,j = 1,2,3) \tag{7.4.5}$$

The dispersion term on the right-hand of Eq. (7.4.5) is obtained by considering only the velocity variability in the microscopic scale. When dealing with large-scale problems, we should consider the effect of the velocity variability in the *macroscopic scale*.

Assume that the concentration C and velocity V are both random fields, and let

$$C(\mathbf{x}, t) = \overline{C}(\mathbf{x}, t) + C'(\mathbf{x}, t), \tag{7.4.6}$$

$$V_i(\mathbf{x}, t) = \overline{V}_i(\mathbf{x}, t) + V_i'(\mathbf{x}, t), \quad (i = 1, 2, 3) \tag{7.4.7}$$

where the over-bar denotes mean values and the prime denotes perturbations. Substituting Eqs. (7.4.6) and (7.4.7) into Eq. (7.4.5), and taking the mean of the resulting equation, we can obtain a *mean equation* relating $\overline{C}$ and $\overline{V}_i$ as follows:

$$\frac{\partial \overline{C}}{\partial t} = \frac{\partial}{\partial x_i}\left[D_{ij}\frac{\partial \overline{C}}{\partial x_j}\right] - \frac{\partial}{\partial x_i}[\overline{V}_i\overline{C}] - \frac{\partial}{\partial x_i}[\overline{V_i'C'}]. \tag{7.4.8}$$

The last term on the right-hand of the above equation is generated by the variability of the macroscopic velocity. $\overline{C'V_i'}$ is called the *mean macrodispersive flux* or *effective dispersion flux*. We may assume that the macrodispersion is also Fickian in nature and can be expressed by (Gelhar and Axness, 1983):

$$\overline{C'V_i'} = -VA_{ij}\frac{\partial \overline{C}}{\partial x_j}, \tag{7.4.9}$$

where A_{ij} is the effective or *macroscopic dispersivity* tensor. Hereafter, D_{ij} will be called the *local dispersion coefficient*. The longitudinal and transverse dispersivities α_L and α_T will be called *local dispersivities* in order to differentiate them with the *macrodispersivities* A_{ij} defined in Eq. (7.4.9). Note that the mean equation Eq. (7.4.8) is still an advection-dispersion equation, and Eq. (7.4.9) is similar to Eq. (7.4.1), but the meaning of dispersion parameters are changed after upscaling.

Subtracting the mean equation (7.4.8) from Eq. (7.4.5), and only retaining first-order terms, we can obtain a *stochastic equation relating perturbations* C' and V_i' as follows

$$\frac{\partial C'}{\partial t} = \frac{\partial}{\partial x_i}\left[D_{ij}\frac{\partial C'}{\partial x_j}\right] - \frac{\partial}{\partial x_i}[V_i'\bar{C}] - \frac{\partial}{\partial x_i}[\bar{V}_iC']. \tag{7.4.10}$$

For unidirectional flow, we can assume that $\bar{V}_1 = V$ and $\bar{V}_2 = \bar{V}_3 = 0$. If we further assume that the fluid is incompressible, i.e., $\partial V_i/\partial x_i = 0$, we then have the mean equation

$$\frac{\partial\bar{C}}{\partial t} = V(\alpha_{ij} + A_{ij})\frac{\partial^2\bar{C}}{\partial x_i\partial x_j} - V\frac{\partial\bar{C}}{\partial x_1}, \tag{7.4.11}$$

where $\alpha_{11} = \alpha_L$, $\alpha_{22} = \alpha_{33} = \alpha_T$, and $\alpha_{ij} = 0$, when $i \neq j$. The perturbation equation (7.4.10) now becomes

$$\frac{\partial C'}{\partial t} = V\alpha_{ij}\frac{\partial^2 C'}{\partial x_i\partial x_j} - V\frac{\partial C'}{\partial x_1} - V_i'\frac{\partial\bar{C}}{\partial x_i}. \tag{7.4.12}$$

Before the asymptotic stage is achieved, the statistic theory shows that macrodispersivities A_{ij} are functions of time (Dagan, 1984). Gelhar and Axness (1983) considered the steady case. Using the spectrum method to solve the perturbation equation, and using Darcy's law, the relationship between C' and the variability of hydraulic conductivity can be found. For stochastic homogeneous and isotropic formations, Gelhar and Axness (1983) obtained the following expressions for the asymptotic macrodispersivities:

$$A_{11} = \sigma_Y^2 l_Y/\gamma^2, \tag{7.4.13a}$$

$$A_{22} = A_{33} = \frac{\sigma_Y^2\alpha_L}{15\gamma^2}\left(1 + \frac{4\alpha_T}{\alpha_L}\right), \tag{7.4.13b}$$

where $\gamma = 1 + \sigma_Y^2/6$. Along the same way, Rehfeldt and Gelhar (1992) extended the results of Eq. (7.4.13) to the unsteady case. Their paper showed that the macrodispersive flux in the unsteady case is larger than that in the steady case, because of the temporal variability in the hydraulic gradient.

7.4.4 Uncertainties of Groundwater Quality Models

There may be stochastic factors contained in the coefficients of governing equations, source/sink terms or boundary conditions of groundwater mathematical models. The following are a few examples:

1. The hydraulic conductivity K, porosity n, flow velocity V, dispersivity α_L, α_T and so forth in the governing equations are all obtained by measurements, data processing, or through solving inverse problems. Because of their uncertainties, they should be regarded as stochastic variables with certain statistical characteristics. Mathematical expectations, covariances, or confidence intervals are often used to describe these random fields.
2. The initial and boundary conditions may contain uncertainties caused by measurement errors, inaccurate geological inferences, and so forth. Sometimes, the boundary conditions themselves are stochastic processes. For example, the water level of a river, which is regarded as a boundary condition, may randomly vary.
3. The source/sink terms may also change randomly. For instance, rainfall infiltration, irregular pumping and recharge, and locations and times of contaminants entering into the aquifer may all vary randomly.

Since there are random components in the models, their solutions (head and concentration distributions) should be regarded as random functions with certain probability distributions. Therefore, even in the framework of the classical dispersion theory, we have to deal with *stochastic partial differential equations* (SPDE). To solve a SPDE means to find the probability distribution of its solution based on the probability distributions of input data. Generally, it is enough to obtain the first two moments, the mean and variance functions, of the solution.

To solve a SPDE, we must know the probability distributions of input parameters, at least their first two moments. If the input parameters are obtained by solving inverse problems based on the observed head and concentration, then we have to consider the uncertainties of the estimated parameters caused by the insufficient quantity and quality of the observation data. When the observation errors of concentration C and head h are normally distributed with zero means and independent of each other, the covariance matrix of the estimated parameters can be approximately expressed by the following equation (Bard, 1974; Wagner and Gorelick, 1987):

$$C_{\mathbf{pp}} \approx ([\mathbf{J}]_C^T [\mathbf{W}]_C [\mathbf{J}]_C + [\mathbf{J}]_h^T [\mathbf{W}]_h [\mathbf{J}]_h)^{-1}, \qquad (7.4.14)$$

where $[\mathbf{W}]_C$ and $[\mathbf{W}]_h$ have been defined in Eq. (7.3.15), and the sensitivity matrices $[\mathbf{J}]_C$ and $[\mathbf{J}]_h$ have been defined by Eq. (7.3.17).

Let us now discuss how to obtain the estimate for the uncertainties of solutions from the uncertainties of model parameters. If a linear relationship between the solution Φ (head or concentration) and the model parameters is assumed based on the *first-order approximation*, then it is easy to derive

$$E[\Phi(\mathbf{p})] \approx \Phi[E(\mathbf{p})], \qquad (7.4.15)$$

where $E[\Phi(\mathbf{p})]$ is the mathematical expectation of solution Φ of the equation and $\Phi[E(\mathbf{p})]$ is the solution obtained by substituting the expectation of the parameters into the equation. Using first-order approximation, we can also derive the following expression for the covariance matrix of solution Φ (Wagner and Gorelick, 1987):

$$C_{\Phi\Phi} \approx [\mathbf{J}]C_{\mathbf{pp}}[\mathbf{J}]^T, \qquad (7.4.16)$$

where $C_{\mathbf{pp}}$ is defined by Eq. (7.4.14) and $[\mathbf{J}] = [\partial\Phi/\partial\mathbf{p}]$ is the Jacobian. Equations (7.4.15) and (7.4.16) based on the first-order approximation are simple but too rough.

Tang and Pinder (1979) used the perturbation method to study the uncertainty of the solutions of advection-dispersion equations. A one-dimensional hydrodynamic dispersion equation can be written in the form of a linear operator:

$$LC = f, \qquad (7.4.17)$$

where operator

$$L \equiv \frac{\partial}{\partial t} + V\frac{\partial}{\partial x} - \frac{\partial}{\partial x}\left(D\frac{\partial}{\partial x}\right), \qquad (7.4.18)$$

and hydrodynamic dispersion coefficient $D = \alpha V$. Let D, V, and α be represented by

$$D = \bar{D} + D', \qquad (7.4.19a)$$

$$V = \bar{V} + V', \qquad (7.4.19b)$$

$$\alpha = \bar{\alpha} + \alpha', \qquad (7.4.19c)$$

where the overbar indicates the expected values and the prime indicates the perturbations from the expected values. By inserting these definitions into Eq. (7.4.18), operator L will be separated into two parts:

$$L = L_0 + L_1, \qquad (7.4.20)$$

where

$$L_0 = \frac{\partial}{\partial t} + \bar{V}\frac{\partial}{\partial x} - \frac{\partial}{\partial x}\left(\bar{D}\frac{\partial}{\partial x}\right), \qquad (7.4.21)$$

$$L_1 = V'\frac{\partial}{\partial x} - \frac{\partial}{\partial x}\left(D'\frac{\partial}{\partial x}\right). \qquad (7.4.22)$$

Equation (7.4.17) can then be expressed as

$$L_0 C + L_1 C = f. \qquad (7.4.23)$$

When using FDM or FEM to discretize this equation, we will use the same notation as in the continuous form for convenience. After discretization, we regard L_0 as the coefficient matrix which depends on $\bar{V}$, $\bar{\alpha}$, Δx, and Δt; L_1 as the coefficient matrix varying with V', α', Δx and Δt; f as a vector containing initial and boundary conditions; and C as the unknown solution vector. Then, Eq. (7.4.23) can also express a set of stochastic algebraic equations.

From (7.4.23), we arrive at

$$C = L_0^{-1} f - L_0^{-1} L_1 C, \tag{7.4.24}$$

where L_0^{-1} is the inverse matrix of L_0. Successive substitution of this equation yields

$$
\begin{aligned}
C &= L_0^{-1} f - L_0^{-1} L_1 (L_0^{-1} f - L_0^{-1} L_1 C) \\
&= L_0^{-1} f - L_0^{-1} L_1 L_0^{-1} f + L_0^{-1} L_1 L_0^{-1} L_1 C \\
&= L_0^{-1} f - L_0^{-1} L_1 L_0^{-1} f + L_0^{-1} L_1 L_0^{-1} L_1 (L_0^{-1} f - L_0^{-1} L_1 C) \\
&= L_0^{-1} f - L_0^{-1} L_1 L_0^{-1} f + L_0^{-1} L_1 L_0^{-1} L_1 L_0^{-1} f \\
&\quad - L_0^{-1} L_1 L_0^{-1} L_1 L_0^{-1} L_1 C.
\end{aligned}
\tag{7.4.25}
$$

Letting

$$M \equiv L_0^{-1} L_1, \quad F \equiv L_0^{-1} f, \tag{7.4.26}$$

then we can rewrite Eq. (7.4.25) as

$$C = F - MF + M^2 F - M^3 F + \cdots, \tag{7.4.27}$$

that is, concentration C is expressed in the form of a series. The condition of convergence of this series is that the norm of matrix M, $\|M\|$, should be less than 1. As M depends on Δx and Δt, this condition can always be satisfied when values of Δx and Δt are appropriately assigned.

Note that L_0 is deterministic, and $E(V') = E(\alpha') = 0$ is assumed. Thus, we have $E(L_0) = L_0$ and $E(L_1) = 0$. By ignoring the terms higher than the second order in Eq. (7.4.27), and taking mathematical expectations on both sides, we obtain

$$E(C) = L_0^{-1} f + L_0^{-1} E(L_1 L_0^{-1} L_1) E(C). \tag{7.4.28}$$

After rearrangement we can find that $E(C)$ satisfies

$$\{ L_0 - E(L_1 L_0^{-1} L_1) \} E(C) = f. \tag{7.4.29}$$

By calculating $E(C^2) - E^2(C)$, we may obtain

$$\sigma_C^2 = \mathrm{Var}(C) = L_0^{-1} E(L_0 L_0^{-1} L_1) E^2(C). \tag{7.4.30}$$

Because L_1 is a function of V' and α', these results show that the uncertainty of solutions may be inferred from the uncertainty of parameters V and α. In the numerical example given by Tang and Pinder (1979), the maximum of σ_C generally occurs near the front of concentration, i.e., the uncertainty at the concentration front is the greatest. In addition, they also discovered that the variance of the concentration is always less than the variance of the input parameter. In other words, the dispersion equation itself can decrease the uncertainty to some extent. From Eq. (7.4.29), it is worth to point out that $E(C)$ is not the solution of equation $L_0 E(C) = f$. This means that the solution obtained using the expected value of the parameters in a water quality model is not equal to the mathematical expectation of the solution. The same conclusion has also been reached by Sagar (1978) from his research on a finite element groundwater flow model. The inaccuracy of Eq. (7.4.15), which is based on first-order approximation, is obvious now.

The operator approach described above is a useful tool for moment analysis. Mclaughlin and Wood (1988) presented a general perturbation method for coupled groundwater flow and mass transport problems, in which mathematical expectation $E[\Phi]$, covariance matrix $C_{\Phi\Phi}$, and cross-covariance matrix $C_{\Phi p}$ are obtained by solving the corresponding moment equations.

In recent years, the *uncertainties of stochastic models* were considered by many researchers (Vomvoris and Gelhar, 1990; Dagan, 1990; Li and McLaughlin, 1991; Kapoor and Gelhar, 1994a). In study the uncertainties of stochastic models, the local dispersion is often neglected. With the assumption of unidirectional flow, we have the following stochastic equation for concentration C:

$$\frac{\partial C}{\partial t} + V\frac{\partial C}{\partial x_1} + V_i'\frac{\partial C}{\partial x_i} = 0. \tag{7.4.31}$$

The mean equation of Eq. (7.4.31) is

$$\frac{\partial \bar{C}}{\partial t} + V\frac{\partial \bar{C}}{\partial x_1} - VA_{ij}\frac{\partial^2 \bar{C}}{\partial x_i \partial x_j} = 0. \tag{7.4.32}$$

Following Kapoor and Gelhar (1994a), we define $S = C^2$. From Eq. (7.4.31), it is easy to find the following equation:

$$\frac{\partial S}{\partial t} + V\frac{\partial S}{\partial x_1} + V_i'\frac{\partial S}{\partial x_i} = 0. \tag{7.4.33}$$

Since

$$S = C^2 = (\bar{C} + C')^2 = \bar{C}^2 + C'^2 + 2\bar{C}C', \tag{7.4.34}$$

we have

$$\bar{S} = \bar{C}^2 + \sigma_C^2 \tag{7.4.35a}$$

and

$$S' = C'^2 + 2\bar{C}C' - \sigma_C^2. \tag{7.4.35b}$$

From Eqs. (7.4.9), (7.4.31) and (7.4.33), the following Fickian relationship can be obtained:

$$\overline{S'V_i'} = -VA_{ij}\frac{\partial \bar{S}}{\partial x_j}. \tag{7.4.36}$$

Substituting Eq. (7.4.35) into the above equation, we then have

$$\overline{C'^2 V_i'} = -VA_{ij}\frac{\partial \sigma_C^2}{\partial x_j}. \tag{7.4.37}$$

Using Eq. (7.4.37) and taking the mean of Eq. (7.4.33), we finally obtain

$$\frac{\partial \sigma_C^2}{\partial t} + V\frac{\partial \sigma_C^2}{\partial x_1} - VA_{ij}\frac{\partial^2 \sigma_C^2}{\partial x_i \partial x_j} = 2VA_{ij}\frac{\partial \overline{C}}{\partial x_i}\frac{\partial \overline{C}}{\partial x_j}. \tag{7.4.38}$$

This is a PDE satisfied by σ_C^2. It has the same form as Eq. (7.4.32) which is satisfied by $\overline{C}$, but with a source term on its right-hand side. After solving the mean equation (7.4.31), we will have $\overline{C}$, then the right-hand side of Eq. (7.4.38) can be calculated, and the variance σ_C^2 can be solved. If there are no uncertainties associated with the initial and boundary conditions of the original dispersion problem, we can set zero initial and boundary conditions for the variance equation (7.4.38).

Similar results were obtained in Kapoor and Gelhar (1994a) when the local dispersion term is also taken into account.

7.4.5 Conditional Simulations and Stochastic Inverse Problems

The *Monte-Carlo method*, or the so-called simulation method, is one of the most powerful methods for studying the solution uncertainties of stochastic differential equations. It can be easily realized on the computer. As long as we know the probability distribution of the uncertainty of parameters or subsidiary conditions, the expected value and variance of the solution can be obtained through a series of numerical simulation runs and the results are not constrained by the statistical assumptions of the solution.

The basic principle of the Monte-Carlo method is very simple: according to the expected values and variances of the given data, the computer generates a group of input data. Each member of the group is a "realization" of the parameters or subsidiary conditions when they are regarded as stochastic processes. Then, the common simulation program is used to obtain the solution of the equation for each realization, which is equivalent to a "simulation experiment" being completed by the computer. The computer is used to complete a large number of simulation experiments and obtain a group of solutions. Once the number of solutions is large enough to be statistically significant, the approximate values of mathematical expectation and variance of the solution can be easily obtained.

To be more exact, assume that only dispersivity α_L and α_T are random variables in a water quality model, and their mathematical expectations and variances are α_L^0, α_T^0 and σ_L^2, σ_T^2, respectively. The major steps of the Monte-Carlo method for this problem are as follows:

1. Quote or write a computer program for solving the water quality model, in which α_L, α_T are variable parameters;
2. Transfer a $(0,1)$ random number generated by the computer into a random sampling value from a normal distribution with center α_L^0 and variance σ_L^2. A realization of α_L is denoted by $\alpha_L(\xi)$. Similarly, we can define

$\alpha_T(\xi)$. Besides the assumption of a normal distribution, some authors suggest that α_L and α_T are lognormal distributed.

3. Substitute $\alpha_L(\xi)$ and $\alpha_T(\xi)$ (one "realization" of α_L and α_T) into the simulation model and solve it. We then obtain one realization of the concentration distribution. The concentration realization at node i and time t_k is denoted by $C_{i,k}(\xi)$;

4. Repeat Steps 2 and 3 for N times, we then have N concentration realizations at every node i and time t_k: $\{(C_{i,k}(\xi_r), r = 1, 2, \ldots, N\}$;

5. Adopt the following formulas to calculate the approximate mathematical expectation of concentration $C_{i,k}$:

$$E(C_{i,k}) \approx \frac{1}{N} \sum_{r=1}^{N} C_{i,k}(\xi_r), \qquad (7.4.39)$$

and its approximate variance:

$$\text{Var}(C_{i,k}) \approx \frac{1}{N} \sum_{r=1}^{N} [C_{i,k}(\xi_r) - E(C_{i,k})]^2. \qquad (7.4.40)$$

Thus we have obtained a measure of the uncertainty of the model solution.

From the steps mentioned above we know that the Monte-Carlo method is simple and convenient for computer use and is widely applicable. We can obtain the estimate of solution uncertainty caused not only by the uncertainties of dispersivities but also by that of flow velocity boundary conditions, as well as other parameters in the simulation model. Regardless of whether the problem is two-dimensional or three-dimensional, steady or unsteady, simple or complex, we only need to replace the simulation program according to the problem considered. The only shortcoming of the Monte-Carlo method is that it requires great computational effort, because hundreds, or thousands of simulation runs may be necessary to make Eqs. (7.4.39) and (7.4.40) have statistical sense.

Delhomme (1979) used the *kriging method* (de Marsily, 1986) to estimate the values of transmissivity at each node. At an observation node, the parameter value is determined directly by observed data. For each unobserved node, the expected value and variance of the parameter can be obtained by the unbiased estimation and minimized variance conditions. Thus, we can obtain an estimate of the uncertainty of model solution (hydraulic head) using the Monte-Carlo method with the parameter values estimated by kriging. This method is called the *condition simulation*. Nelson et al. (1987) gave a case study in which the method of condition simulation was used to obtain the uncertainty of a water quality model. Their study involves the following steps:

1. Use kriging to estimate the logarithmic distribution of transmissivity $(\ln T(\mathbf{x}))$ based on some local measurements (from local pumping tests) and use this distribution as the initial estimate;

2. Solve the inverse problem with the method suggested by Neuman (1980), and Clifton and Neuman (1982) to identify the logarithm transmissivity. That is, using the initial estimate of the parameter in combination with the generalized least squares method to obtain the logarithmic distribution of transmissivity (denoted as $\hat{Y} = \ln \hat{T}(x)$) and covariance matrix of the estimation errors (denoted as $[\mathbf{V}]$) simultaneously;
3. Carry out many simulations on the computer along the lines of the Monte-Carlo method. From the calculated $\hat{Y}$ and $[\mathbf{V}]$, a lot of possible values of transmissivity may be generated by calculating

$$Y = \hat{Y} + [\mathbf{M}]\xi, \tag{7.4.41}$$

where Y is a randomly generated "realization" of the logarithm transmissivity; $[\mathbf{M}]$ the upper triangular matrix of the LU decomposition of the covariance matrix $[\mathbf{V}]$, i.e., $[\mathbf{M}][\mathbf{M}]^T = [\mathbf{V}]$; and ξ is a sampling value of multidimensional normal distribution with zero mean and unit variance. Nelson et al. (1985) obtained 600 "realizations" in their case study;
4. With the 600 "realizations" of transmissivity, 600 "realizations" of head distribution can be obtained by solving the water flow model. Then, using Darcy's law, 600 "realizations" of velocity distribution can be calculated, that is,

$$V_{ij} = -\frac{T_{ij}}{bn}\mathrm{grad}(h_{ij}), \quad i = 1, 2, \ldots, 600; \quad j = 1, 2, \ldots, M \tag{7.4.42}$$

where V_{ij}, T_{ij}, h_{ij} are the ith "realizations" of velocity, transmissivity and hydraulic head, respectively, at node j; M the total number of nodes; b the thickness of the aquifer; and n the porosity;
5. Substituting the 600 "realizations" of velocity distribution successively in to a pure advection quality model to calculate the migration of the pollutant and its arrival time at the out-flow boundary, or at a fixed location, we can obtain 600 "realizations" of the solution.
6. With all the aforementioned data available, we can calculate the mean arrival time and statistical characteristics of the concentration distribution, including their variances and confidence intervals. As a result, a quantitative estimate of the uncertainties of the model has been obtained.

The *stochastic inverse problem* (SIP) involves the identification of mean and covariance functions of unknown parameters with the aid of head and concentration measurements. The SIP was first considered by Kitanidis and Vomvoris (1983). The mean value μ_Y of the $Y = \ln K$ field and covariance parameters σ_Y and l_Y in Eq. (7.4.4) were estimated by the maximum likelihood estimator (MLE). This SIP was further considered by Hoeksema and Kitanidis (1984, 1985), Dagan (1985), Rubin and Dagan (1987a, b) and Dagan and Rubin (1988). Some field applications were given in these papers. However, there are limitations associated with the previous research: the flow field was assumed to be stable and only ln K and head measurements were used for parameter identification. In Sun and Yeh (1992), the SIP for transient flow fields was solved, and an adjoint state method was developed for efficiently calculating the cross-covariance matrices between head observations at different times. The solution of a SIP includes the following steps:

1. Using kriging to estimate a set of initial values of μ_Y, σ_Y, and l_Y.
2. Calculating the covariance matrices $C_{\Phi\Phi}$ and $C_{\Phi Y}$ relevant to known observation locations and times, where Φ is the head.
3. Using MLE to improve the estimations of μ_Y, σ_Y, and l_Y iteratively with the aid of both Y and Φ measurements.
4. Using the same set of measurements and the co-kriging estimator to generate realization of the Y field. The so-obtained realization is regarded as the inverse solution.

For a complete discussion on SIP, readers may refer to Sun (1994). The solution of SIP can be easily extended to the case that both head and concentration observations are used for the identification of stochastic parameters.

Exercises

7.1. Assuming that there are two tracer components, α and β, in a confined aquifer, write the hydrodynamic system for this situation. Then, use a flow chart to describe the solution procedure.

7.2. Assuming that only the longitudinal dispersivity is unknown in a advection-dispersion model, formulate an inverse problem for this case and translate it into an optimization problem.

7.3. Average Eq. (7.2.32) along the y direction to find the expressions of "apparent velocity" and "apparent dispersion coefficient."

7.4. Derive the adjoint system for one-dimensional coupled flow and mass transport problems and find the expressions of $\partial h/\partial K$ and $\partial C/\partial K$.

7.5. Explain the differences between Fickian expressions Eqs. (7.4.1) and (7.4.9).

7.6. What terms are neglected in the derivation of the perturbation equation (7.4.10)?

7.7. Derive Eqs. (7.4.15) and (7.4.16) based on the first-order approximation.

7.8. Draw a flow chart to describe the procedure of Monte-Carlo simulation method for estimating the uncertainty of a water quality model.

7.9. Draw a flow chart to describe the procedure of conditional simulation method for estimating the uncertainty of a water quality model.

8
Applications of Groundwater Quality Models

8.1 Simulation and Prediction of Groundwater Pollution

8.1.1 The General Procedure of Studying Groundwater Pollution Problems

In the application of mathematical models to practical groundwater pollution problems, the following steps should be included:

1. Determination of Study Objectives and Tasks

The determination of study objectives and tasks should be based on practical problems under consideration. The problems often encountered are:

a. Analysis and prediction of the contamination tendency in water supply for factories, mines and municipalities;

b. Estimation of the impact on groundwater caused by waste water drainage and waste heaps;

c. Appraisal of the possible pollution of groundwater due to irrigation with waste water, and the residual pesticide and chemical fertilizer in the soil;

d. Study of the relationship between surface water pollution and groundwater pollution;

e. Research on the influence of artificial recharge on groundwater quality;

f. Estimation of the possibility and degree of salt water intrusion from the neighboring aquifers;

g. Study of the potential hazard created by radioactive waste stored underground, and so on.

Different problems will require different tasks, such as:

a. Defining the polluted region and preventing water sources from being polluted;

b. Designing the location of new wells;

c. Predicting the time required for self-purification of the polluted aquifers;

247

d. Determining the maximum permissible drawdown or the maximum permissible pumpage in order to prevent poor quality water from entering the supply aquifers;

e. Prescribing the quality standards for both artificial recharge water and waste water to be drained;

f. Providing the quantitative basis for water resources management and pollution remediation;

2. Field Survey

The objective of field survey is to collect field data concerning geology, hydrology and environment, which include the geometric shape and boundary conditions, lithology and heterogeneity, recharge and extraction, relationships between surface water and groundwater, historical and present states of pollution, range and intensity of pollution sources, and so forth. Field surveys can provide a basis for model selection.

3. Model Selection

The selection of a model depends on the results of field survey and accuracy requirements of the problem. A variety of water quality models and their selection criteria have been introduced in Section 7.1. If a distributed parameter water quality model is chosen, we have to construct a conceptual model and judge what parameters can be estimated by field surveys and what parameters should be estimated by field tests.

4. Field Tests

If sufficient historical data is not available, then pumping tests and tracer injection tests are needed at the site for determining the unknown parameters or calibrating the conceptual model. Based on the data of the field survey and the requirements of model calibration, an experiment design should be made. During the tests, the changes of water levels are observed and water samples are analyzed to obtain the concentrations of the chemical components that we are interested in. If multi-layer or three-dimensional water quality models are used, it is necessary to take water samples at different depths in order to obtain the changes of concentration with depth.

5. Programming and Data Compiling

Analyses of nearly all practical groundwater quality problems depend on the use of numerical methods. Therefore, it is necessary to prepare some suitable programs. According to the conditions of practical problems and the performance of the computers, hydrologists may select a ready-made program. Before using the program, data should be prepared according to the requirements of the program. For instance, the following input data are required for two-dimensional advection-dispersion models:

a. Geometric parameters of the aquifers (boundary shape, thickness, elevations of the roof and bottom);
b. Initial distributions of hydraulic head and concentration;
c. Locations and rates of the pumping wells (or injection wells), locations and intensities of the pollution sources;
d. Hydraulic relationships with adjacent aquifers and surface water;
e. Estimations of various hydrogeological parameters, including porosity, hydraulic conductivity, longitudinal and transverse dispersivities of the aquifer;
f. Historical observation data and the data obtained from field tests.

For some of the data mentioned above, we can use their estimated values first, and further determine more accurate values of them through model calibration.

6. Model Calibration and Reliability Analysis

Model calibration and reliability analysis have already been illustrated in Chapter 7. Hydraulic conductivity and porosity can be determined through the calibration of the flow model. However, if observed water quality data is available, better results may be obtained by simultaneously calibrating the flow model and the water quality model. The local dispersivities can be determined by fitting the observed results of tracer injection tests with the output curves of corresponding analytic solutions or numerical solutions. Based on these local values, distributed parameters in the whole region can be composed by kriging interpolation. Next, substitute these distributed parameters into the water quality model to simulate the existing pollution or tracer tests, and then check if the simulated results coincide with the observations. If not, the trial-and-error method or other optimal methods should be used to modify the parameters until a satisfactory fitting is achieved.

Reliability analysis of the final model should be conducted to estimate the statistical characteristics of model output (concentration distribution or arrival time) and examine if the requirements of accuracy are satisfied. It is better to verify the model by the data that has not been used for model calibration.

7. Prediction and Control

With a reliable mathematical model, predictions can be made according to the concrete tasks as mentioned above. The model can be used to predict the developing tendency of contamination, to compare the results between various remediation and control schemes, to draw the concentration contours at different time. It can also be taken as a component in a water resources management model.

8.1.2 Groundwater Pollution of Saturated Loose Aquifers

A significant amount of work has been done to predict groundwater pollution in saturated loose aquifers with advection-dispersion models and there have been some successful cases. In these cases, the study procedure mentioned above are generally adopted. The following is a brief introduction of two case studies. For more case studies, refer to Van der Heijde et al. (1985).

The Study of Groundwater Pollution in an Alluvial Aquifer in the Rocky Mountains of Colorado

This is a successful example of using the hydrodynamic dispersion model on a regional scale. The calculated results were provided by Konikow (1977) of the USGS. In this study, an advection-dispersion model was calibrated by observation data of over 30 years.

From 1943 to 1956, industrial waste water containing a high concentration of some chloride compound was disposed of in unlined ponds at a munitions factory. The waste water percolated from the ponds into the alluvial aquifers, and then spread out over several square miles, causing serious harm to the crops. Starting from 1956, the waste water has been disposed of in a reservoir lined with bitumen, and clean water has been injected into the original ponds to displace the polluted groundwater. As a result, the groundwater pollution has now been reduced in that region.

The following are some advantages of establishing an advection-dispersion model for the region:

1. The geological and hydrogeological conditions in the region are well known.
2. There is sufficient long-term observation data of water quality for model calibration.
3. There are historical records of waste water treatment in the munitions factory. Chloride can be regarded as a conservative tracer. Both adsorption and chemical reactions need not be considered.

In the first phase of the study, all available data was collected and analyzed. The data from about 200 wells, including observation wells, test wells, irrigation wells and domestic wells, was analyzed. Four maps were prepared from the data to show the elevations of the bedrock, the levels of water tables, the thickness of alluvial deposits, and the distribution of transmissivities. These maps illustrated that the alluvial deposit was an aquifer system which was inclined, discontinuous, and heterogeneous, with different groundwater flow directions.

In the second phase, a mathematical model of two-dimensional steady flow was developed to estimate the recharge of the aquifer and to compute the distribution of groundwater velocities. During the calibration of the flow model, it was interesting to find out that there were areas where some alluvial deposits were absent, and areas which were unsaturated most of the time. These areas formed internal barriers, which greatly affected the pattern of groundwater flows in the aquifers.

The study area was about 88 km^2, and the *iterative alternative direction implicit* (IADI) finite difference method was used to solve the groundwater flow equation. The whole area was divided into 25 × 38 square meshes, with the side of each square being 308 m long.

According to aerial survey photographs, the irrigation area occupied about 111 squares. It is interesting that the sink/source term, including the pure recharge rate of irrigation, rainfall infiltration, the leakage of ditches, and so forth, was determined by model calibration. Pumping rate from each well was 50 m^3/d on average. This procedure of calibration provided good reference for the establishment of a regional flow model. The flow directions in the whole domain, as calculated by the flow model, were not identical, and the velocities ranged from 0.3 to 6.1 m/d.

The third phase of the study was the building of a solute transport model. The two-dimensional advection-dispersion model was adopted and solved by the method of characteristics introduced in Section 4.2. The distribution of chloride concentration observed in 1956 was taken as the basis of model calibration. The end of 1943 was chosen as the initial time. The simulated results were compared with the observed data of 14 years. Through the model calibration, the porosity obtained was 0.3 and the longitudinal and transverse dispersivities were both 30 m. Then, a verification of the water quality model was made based on the data from 1956 to 1960, 1961 to 1968, and 1968 to 1972. The computed concentration distributions agreed fairly well with the observed distributions.

The fourth phase was using the calibrated model to predict the development of contamination and evaluating the remediation strategy. According to the prediction, chloride would be eventually drained from the aquifer of the simulated domain, but the recovery of the water quality to its original state would require at least several decades. In order to speed up the remediation, researchers suggested digging a drainage ditch at the northern boundary of the factory to extract the polluted water before it migrated from the factory to downstream. The effectiveness of the scheme may be pre-estimated by the model. The model results showed that it was almost impossible to remove all contaminants from most of the polluted aquifers at the Rocky Mountain munitions site. However, some portions of the aquifers may be affected by artificial plans, which may speed up the improvement of water quality in those aquifers. The feasibility of such plans largely depends on the hydraulic features of the aquifers, the types of contamination, and the pollution sources. To pre-estimate the effectiveness of a remediation plan, a reliable prediction of concentration changes must be made, which would rely on the model output. Thus, the role of the water quality models was fully affirmed.

A Study of the Mathematical Modeling for Groundwater
Contamination in Xi'an City

The purpose of this study was to build a groundwater quality model based
on the observed data of chlorine ions. The model was then used as a tool for
analyzing and predicting the tendency of groundwater contamination in that
city. Another purpose of the study was to explore the feasibility of building a
regional water flow model and water quality model using the long-term
observation data.

In the first stage, the data of hydrogeology and environmental hydrogeology were collected, analyzed and synthesized. This work was completed
by the First Hydrogeological Brigade of Shaanxi Province, China, which
provided necessary preparations for building the water quality model.
Through a great amount of work, both in the laboratory and in the field, the
following data were obtained:

1. The descriptions of regional geologic and hydrogeologic conditions, including the distribution of the phreatic aquifers and their thickness,
   hydraulic conductivities, storativities and hydraulic relationships with adjacent aquifers;
2. The distribution, quantity and utilization of the wells; the distributions of
   sewerages, waste-water reservoirs and polluted ponds; the historical and
   the present situation of sewage irrigation;
3. The original hydrochemical background in the groundwater and the present state of pollution; the types and sources of contaminants in the area;
4. The long-term data (from 1954 to 1959) about chlorine pollution in the
   groundwater shows that the historical pollution of the groundwater was
   caused by city sewage, while the current pollution is generated by sewer-
   age irrigation;
5. The historical records of water levels and water extraction. The maps of
   water level contours for two hydrological years (from 1978 to 1980) were
   compiled, and researches on the historical changes of the groundwater
   flow field in Xi'an City were also conducted.

In the second stage, a groundwater flow model of the Xi'an City was
developed. The simulated area was 570 km² and was divided into 195 triangular elements. The simulation model was solved by the multiple cell balance
method. It was impossible to make effective pumping tests to determine
the hydrogeologic parameters in such a large region, so the calibration of
the flow model could only rely on the long-term observation data. Through
simulating water level changes in two years (from April 1978 to April 1980),
hydraulic conductivity K, specific yield S_y, coefficient of rainfall infiltration λ
in 9 subdomains, and seepage Q from sewage reservoirs and ponds were
obtained by solving the inverse problem. Then, a steady flow field and the
distribution of average velocities were computed, in which the velocities were
obtained indirectly by Darcy's law.

In the third stage, a water quality model was built. The simulated area was 168 km², located in the northern suburb of Xi'an, and was divided into 307 triangular elements. The two-dimensional advection-dispersion model was selected as the simulation model. The upstream weighted multiple cell balance method was used to avoid the oscillation of numerical solutions. The parameters to be identified in the model including longitudinal dispersivity α_L, transverse dispersivity α_T, and the amount of irrigation return water in the sewage irrigation area. The model outputs were compared with the observed concentration data in 25 years (1959 to 1984) and unknown parameters were adjusted based on the comparison. Finally, a good fit was reached, where α_L and α_T were determined to be about 130 m and 48 m, respectively.

In the fourth stage, the calibrated model was used for prediction. Several possible projects of reducing groundwater pollution were simulated.

There are many things in common in the groundwater pollution research between the Xi'an City and the munitions factory in Colorado. In both cases, chloride ion was used as the tracer and long-term observation data was used to calibrate the flow model and water quality model. However, the aquifer of the Xi'an City consists of a larger area with more complex conditions.

8.1.3 Groundwater Pollution of Saturated-Unsaturated Aquifers

It is very important to study the solute transport in unsaturated zones. In practice, however, the use of the mathematical models is still very difficult owing to the non-linearity of the problem. In particular, it is difficult to obtain the input parameters. Advection-dispersion models of water quality in the saturated-unsaturated zone have been introduced in Section 7.1.2. For instance, a two-dimensional problem of a profile is governed by the following water flow equation

$$(\zeta + \beta S_s)\frac{\partial \psi}{\partial t} = \frac{\partial}{\partial x}\left[K(\psi)\frac{\partial \psi}{\partial x}\right] + \frac{\partial}{\partial z}\left[K(\psi)\frac{\partial \psi}{\partial z}\right] + \frac{\partial}{\partial z}K(\psi) + W, \quad (8.1.1)$$

and the water quality equation

$$\frac{\partial(\theta C)}{\partial t} = \frac{\partial}{\partial x}\left[\theta\left(D_{xx}\frac{\partial C}{\partial x} + D_{xz}\frac{\partial C}{\partial z} - CV_x\right)\right]$$
$$+ \frac{\partial}{\partial z}\left[\theta\left(D_{xz}\frac{\partial C}{\partial x} + D_{zz}\frac{\partial C}{\partial z} - CV_z\right)\right] + I, \quad (8.1.2)$$

subject to certain initial and boundary conditions. In Eq. (8.1.1), $\zeta = \partial\theta/\partial\psi$; θ is the soil moisture content; ψ is the pressure water head; β is equal to 1 in the saturated zone and 0 in the unsaturated zone, and $K(\psi)$ is the hydraulic conductivity in the saturated-unsaturated zone. In Eq. (8.1.2), the sink/source term, I, may contain the radioactive decay, adsorption, chemical reaction and other factors, and the factor of retardation may also be added when necessary.

The problem is how to couple the two equations. Let $\psi(x, z, t)$ be the solution of the nonlinear equation (8.1.1), then θ used in Eq. (8.1.2) must be calculated from ψ. The relationship between θ and ψ is nonlinear and should be determined from the observation data. There is an *empirical equation* (King, 1965):

$$\theta(\psi) = \theta_0 \left[\frac{\cosh(\psi/\psi_m)^\kappa - (\theta_0 - \theta_r)/(\theta_0 + \theta_r)}{\cosh(\psi/\psi_m)^\kappa + (\theta_0 - \theta_r)/(\theta_0 + \theta_r)} \right], \qquad (8.1.3)$$

where θ_0, ψ_m, κ are *curve fitting parameters* determined by the least squares method.

The following empirical equation is commonly used to express the relationship between hydraulic conductivity K and θ:

$$K(\theta) = \mu\theta^\eta, \quad (\eta > 0), \qquad (8.1.4)$$

where μ and η are also curve fitting parameters. When all of the curve fitting parameters are obtained, we can substitute Eq. (8.1.3) into Eq. (8.1.4) to get function $K(\psi)$. Other empirical formulas for determining the nonlinear functions $\theta(\psi)$ and $K(\psi)$ have been given by Van Genuchten (1980) and Kool et al. (1987).

Using Darcy's law:

$$V_x = -K(\psi)\frac{\partial\psi}{\partial x}, \quad V_z = -K(\psi)\left[\frac{\partial\psi}{\partial z} + 1\right], \qquad (8.1.5)$$

the distribution of velocity can be obtained from the known $K(\psi)$ and ψ.

The hydrodynamic dispersion coefficients, including molecular diffusion, can be calculated by the following formulas:

$$D_{xx} = \alpha_L \frac{V_x^2}{V} + \alpha_T \frac{V_z^2}{V} + D_0 a e^{b\theta}$$

$$D_{zz} = \alpha_T \frac{V_x^2}{V} + \alpha_L \frac{V_z^2}{V} + D_0 a e^{b\theta} \qquad (8.1.6)$$

$$D_{xz} = (\alpha_L - \alpha_T)\frac{V_x V_z}{V},$$

where D_0 is the molecular diffusion coefficient of solute in water, and a and b are empirical constants. Table 8.1 lists some typical values of parameters appearing in the above formulas (quoted from Pickens and Gillham, 1980).

The empirical formula of Eq. (8.1.3) together with Eqs. (8.1.5) and (8.1.6) have provided necessary input information for the water quality equation (8.1.2). Thus, we have an iterative calculation system. The finite element discretization equations for Eqs. (8.1.1) and (8.1.2) may be represented as:

$$[\mathbf{A}]\psi + [\mathbf{B}]\frac{d\psi}{dt} + \mathbf{E} = 0 \qquad (8.1.7)$$

TABLE 8.1. Values of parameters.

Parameter	Symbol	Value
Saturated hydraulic conductivity	K^S	0.35 cm/min
Saturated water content	θ^S	0.3
Specific storage coefficient	S_s	~ 0
Parameter of $\theta(\psi)$	θ_0	0.30
	ψ_m	-38.71 cm
	θ_r	0.09
	κ	-2.85
Parameter of $K(\theta)$	μ	282.2 cm/min
	η	5.561
Diffusion coefficient of solute molecules in free water	D_0	1.2×10^{-3} cm^2/min
Constants of diffusivity	a	0.003
	b	10
Longitudinal dispersivity	α_L	2.0 cm
Transverse-dispersivity	a_T	0.4 cm

and

$$[F]C + [G]\frac{dC}{dt} + H = 0. \qquad (8.1.8)$$

The coefficient matrix $[F]$ contains $K(\psi)$, which depends on the unknown pressure head ψ, so Eq. (8.1.7) is nonlinear; while Eq. (8.1.8) is linear with respect to the unknown function C, although the coefficient matrix contains θ, V_x, V_z, D_{xx}, D_{xz}, D_{zz}, which are dependent on ψ.

The term of time derivative in Eq. (8.1.7) may be approximated by the implicit finite difference, so the equation is transformed into a set of nonlinear algebraic equations as follows:

$$\left([A] + \frac{1}{\Delta t}[B]\right)\{\psi\}_{k+1} = \frac{1}{\Delta t}[B]\{\psi\}_k - \{E\}. \qquad (8.1.9)$$

As solution $\{\psi\}_k$ at time t_k is known, the solution $\{\psi\}_{k+1}$ at time $k+1$ can be obtained by solving this system of equations. Actually, we can first use an extrapolation method to estimate the solution $\{\psi\}^{(0)}_{k+1/2}$ at $t_{t+1/2}$, thus the coefficient matrix in Eq. (8.1.9) can be determined. Then $\{\psi\}^{(1)}_{k+1}$ can be solved and the estimation of $\{\psi\}_{k+1/2}$ can be improved. We may take

$$\{\psi\}^{(1)}_{k+1/2} = \frac{\{\psi\}_k + \{\psi\}^{(1)}_{k+1}}{2} \qquad (8.1.10)$$

to recalculate the coefficient matrix of Eq. (8.1.9) and solve Eq. (8.1.9) again to obtain an improved solution $\{\psi\}^{(2)}_{k+1}$. This iterative procedure is repeated until it converges. Finally, the solution at time t_{k+1}, $\{\psi\}_{k+1}$, will be obtained.

After ψ_{k+1} is obtained, we can solve for the concentration distribution. Using the backward finite difference approximation to replace the time derivative term in Eq. (8.1.8), the equation can be turned into a system of linear algebraic equations:

$$\left([\mathbf{F}] + \frac{1}{\Delta t}[\mathbf{G}]\right)\{\mathbf{C}\}_{k+1} = \frac{1}{\Delta t}[\mathbf{G}]\{\mathbf{C}\}_k - \{\mathbf{H}\}. \qquad (8.1.11)$$

All coefficients depending on ψ in the above equation are expressed by $\{\psi\}_{k+1/2}$, which is a mean of the values $\{\psi\}_k$ and $\{\psi\}_{k+1}$. Based on the known $\{\mathbf{C}\}_k$, we can solve Eq. (8.1.11) to obtain the concentration distribution at the next time step $\{\mathbf{C}\}_{k+1}$, thereby complete the calculation of one time step.

Numerical simulation is a powerful tool for understanding and analyzing solute transport in the unsaturated zone. In practical applications, it is very difficult to obtain reliable input parameters because it requires experimental studies and theoretical analyses both in the field and in the laboratory. Xiao (1984) presented some theoretical results for solute transport in the unsaturated zone. In a saturated-unsaturated model presented by Yang (1988), the immobile water phase was considered. The inverse problem of groundwater flow and mass transport in the unsaturated zone was considered by Kool et al. (1987), Kool and Parker (1988), and Mishra and Parker (1989).

Groundwater flow in the unsaturated zone can be seen as a special case of *two-phase flow* (Bear, 1979). The governing equations consist of mass balance equations for both air and water phases. In recent years, more complicated cases, including various multicomponent transport in multiphase flow, were considered. For example, Falta et al. (1992) presented a multidimensional integral finite difference method for modeling the steam displacement of non-aqueous phase liquid (NAPL) contaminants in shallow subsurface systems, where three flowing phases (gas, aqueous, and NAPL) and three mass components (air, water and an organic chemical) are considered. In this model, the effects of adsorption and heat transport are also included.

8.1.4 Groundwater Pollution of Fractured Aquifers

The problem of *solute transport in fractured aquifers* is currently a hot subject. It is interesting in theoretical study and important in practical application.

Because of the existence of fractures, the solute distribution depends on both the microstructure and the macrostructure of the fractures in a large degree. Generally speaking, along fractures the solute moves faster and propagates farther than it does in a loose rock formation. Figure 8.1 shows the process of solute transport in a fractured aquifer. In the fractures, the solute transport is controlled by both advection and dispersion, and is often dominated by advection. It should not be ignored that porous blocks among the fractures can store the solute. In the porous blocks, the flow velocity is very small, so the solute transport is usually dominated by dispersion and molecular diffusion.

FIGURE 8.1. Solute transport in fractured aquifers.

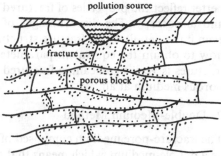

Three methods, the equivalent porous medium method, the double-medium method, and the discrete fracture method, can be used for simulating fractured aquifers. Now let us consider the use of these methods in the study of solute transport.

1. The Equivalent Porous Medium Method

This method regards the fractured medium as a porous medium, i.e., the phenomena which occur in the fractured medium and in the porous medium are described by the same physical models (including physical laws and parameters) and mathematical models.

In many practical cases, large and small fractures crisscross the aquifer and the majority of them are filled by fine-grain media. The fine grain media causes the groundwater to flow very slowly. Thus, the fractured medium can be regarded as a porous medium on the macroscopic scale and good results can be obtained by using porous medium models. The parameters, such as the hydraulic conductivity, porosity and water head are all defined in the macroscopic level. The representative elementary volume (REV) which is used for averaging these parameters is much bigger than individual fractures.

Knowledge of the fracture structure is not necessary when using this method to build mathematical models for fractured aquifers. The effect of the fracture system is manifested through heterogeneity and anisotropy and will be reflected in the macroscopic parameters.

When building the water quality model for a fractured aquifer, the equivalent pore medium method can still be used under certain conditions, as shown by some practical examples. However, physical processes in fractured media and porous media are more different for solute transport than those for groundwater flow. We should remember that the applicability of this type of models is limited.

The numerical solutions previously given are all for porous media. Thus, it is unnecessary to say more when we use the equivalent model to solve the problems of fractured media. It is important to note that the flow model should be well calibrated so that the distribution of average velocities can better reflect the properties of fractured porous media, such as heterogeneity and anisotropy. The disadvantages of the equivalent method are evident: even if the details of the fracture structures are known, we still do not know how to obtain the equivalent parameters, and moreover, when the effect of fractures is very significant, a fractured medium may not be equivalent to a porous medium at all.

2. Double-Medium Method

The fracture-pore medium composed of fractures and pores may be regarded as a double-medium, which means that two kinds of continuum, the fractured medium and the porous medium, exist simultaneously in the same spatial region. Both have their own physical parameters. The fractured medium has significant hydraulic conductivity and weak storage, and hence, large velocities. On the contrary, the porous medium has weak conductivity and significant storage, and hence, small velocities. Therefore, there are two groups of quite different parameters related to the fractured and the porous media at each spatial point. Meanwhile, there are exchanges of water and solute between the two media.

It is easy to derive a mathematical expression for such an idealized physical model. The governing equation for solute transport in the fractured medium is still the advection-dispersion equation:

$$\frac{\partial C}{\partial t} = \frac{\partial}{\partial x_i}\left(D_{ij}\frac{\partial C}{\partial x_j}\right) - V_i\frac{\partial C}{\partial x_i} + I + \left(\frac{1-\varepsilon}{\varepsilon}\right)\Gamma, \quad (i = j = 1, 2, 3) \quad (8.1.12)$$

where D_{ij} and V_i are the components of the hydrodynamic dispersion coefficient and the average velocity, respectively, and C is the concentration distribution, all of which are macroscopic average parameters; I is the source and sink term, which may contain the pumping and injection of water, radioactive decay, chemical reactions, and adsorption. The last term on the right-hand side of Eq. (8.1.12) denotes the exchange of solute between fractures and pores, where ε is the fracture porosity, i.e., the volume occupied by fractures in per unit volume of aquifer; and Γ represents the rate of solute transport from the fractured medium to the porous medium per unit volume of aquifer. Thus, the dimensions of Γ are M/L^3T.

In order to solve for Γ, Bibby (1981), Huyakorn et al. (1983) made an idealized treatment to the geometric structures of fractured media. They assumed a structure of porous blocks and fractures like the one in Figure 8.2, in which the flows in the porous blocks are one-dimensional, and the solute enters into the fractures through the interfaces of the porous blocks. The concentration distributions in the porous blocks are expressed as C'. From the middle plane of each porous block, a local one-dimensional coordinate z' can be established, as shown in Figure 8.3. Then, from the definition of Γ, we have

$$\Gamma = \left(V'C' - D'\frac{\partial C'}{\partial z'}\right)\bigg|_{z'=a}\frac{2\cdot l\cdot d}{2a\cdot l\cdot d} = \frac{1}{a}\left(V'C' - D'\frac{\partial C'}{\partial z'}\right)\bigg|_{z'=a}, \quad (8.1.13)$$

FIGURE 8.2. An idealized geometric structure of fractures-pores.

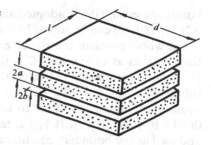

FIGURE 8.3. One-dimensional flow and solute transport in a porous block.

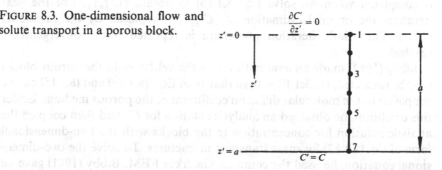

where V' is the one-dimensional velocity in the porous blocks, D' is the dispersion coefficient and the definitions of a, l and d are shown in Figure 8.2. The rate of the volume occupied by fractures can be seen from the figure as

$$\varepsilon = b/(a + b). \tag{8.1.14}$$

In the porous blocks, C' must satisfy the following one-dimensional advection-dispersion equation:

$$\frac{\partial C'}{\partial t} = \frac{\partial}{\partial z'}\left(D'\frac{\partial C'}{\partial z'}\right) - V'\frac{\partial C'}{\partial z'} \tag{8.1.15}$$

subject to the following subsidiary conditions:

$$\begin{cases} C' = C_0', & t = 0; \\ C' = C, & z' = a; \\ \dfrac{\partial C'}{\partial z'} = 0, & z' = 0. \end{cases} \tag{8.1.16}$$

We first solve the one-dimensional problem to obtain C', then substitute C' into Eq. (8.1.13) to determine Γ and substitute it into Eq. (8.1.12). With appropriate initial and boundary conditions, the concentration distribution in the continuous fractured medium can be solved by the usual way. Because the average velocity in the fractured media is large, the numerical methods for advection-dominated problems are often used to avoid oscillations of the numerical solution. For instance, Noorishad and Mehran (1982), and

Huyakorn et al. (1983) adopted the upstream weighted FEM (see Section 6.2). Eq. (8.1.15) can be solved by either FEM or FDM.

The whole problem can be solved by an iteration procedure. Assume that the solutions at time n are $\{C'\}_n$ (the concentration in porous blocks) and $\{C\}_n$ (the concentration in fractures), and the solutions at time $n + 1$ are to be solved. For this purpose, we first extrapolate an estimation $\{C'\}_{n+1}$, and substitute it into Eq. (8.1.13) to obtain Γ. Then, Γ is substituted into Eq. (8.1.12) to yield $\{C\}_{n+1}$. $\{C\}_{n+1}$ is taken as the concentration at the interface and used as the boundary condition (8.1.16) of Eq. (8.1.15). The first iteration is completed when we solve Eq. (8.1.15) to obtain $\{C'\}_{n+1}$. For the next iteration, the original estimation of $\{C'\}_{n+1}$ is replaced by the calculated values. The whole iteration procedure is repeated until convergence is reached.

Bibby (1981) made an assumption that the velocities in the porous blocks may be neglected, i.e., let $V' = 0$, so that D' in Eqs. (8.1.13) and (8.1.15) can be simplified to the molecular diffusion coefficient of the porous medium. Under this condition, he obtained an analytic solution for C', and then coupled the analytic solution for concentration in the blocks with the two-dimensional form of Eq. (8.1.12) for mass transport in fractures. To solve the two-dimensional equation, he used the common Galerkin FEM. Bibby (1981) gave an example of using the method in a limestone aquifer located in the eastern Kent district of England.

The disadvantage of the double-medium method is its undetermination in physics. When we measure the concentration at a point, we can only obtain one value and we do not know whether this value represents the concentration in the fractured medium, or in the porous medium. Thus, it is difficult to solve the inverse problem with the double-medium model. The advantage of the double-medium method is that the knowledge of the fracture structure is not required.

3. Discrete Fracture Method

If there are some large fractures and solution grooves in the aquifer, the assumption of continuum will not be valid. In that case, we should use the discrete fracture method to consider the effects caused by these large fractures. The discrete fracture method is a numerical method in which mass conservation equations are built up for fractured elements and porous elements, respectively. For instance, in Figure 8.4, the elements marked with oblique lines are fractured elements, and the others are porous elements. The former have large transmissivities and small storage coefficients, and the latter have the opposite properties. Transport through fractured elements is advection-dominated, while in porous elements it is dispersion-dominated. The exchanges of water and solute between the two obey Darcy's law and Fick's law. Therefore, the general governing equation is still the advection-dispersion equation, differing only in that different parameters are taken for different elements. As long as the fractures are divided into elements

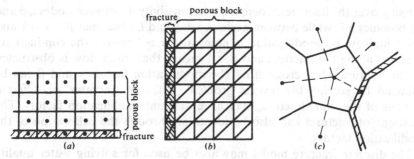

FIGURE 8.4. Some partition methods for discrete fractured elements.

FIGURE 8.5. The nodes through which a fracture passes and the exclusive subdomain of node i.

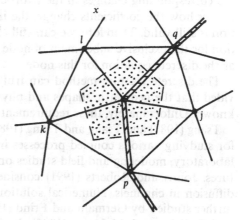

according to their actual positions and geometric sizes, the various numerical methods mentioned before can still be used. For instance, Rasmuson et al. (1982) adopted a spatial discretization shown in Figure 8.4(a). Narasimhan (1982) used a spatial discretization shown in Figure 8.4(c). They adopted the irregular grid FDM with upstream weights added. Grisak and Pickens (1980) used the discretization in Figure 8.4(b) and the Galerkin FEM. Sun and Yeh (1983) adopted the same discretization as Figure 8.4(b), but with the upstream weighted multiple cell balance method.

Sun et al. (1989) proposed a method for simulating the groundwater flow in fractured aquifers. It has been applied to water resource evaluation and mine water prediction in limestone aquifers. This method is similar to the irregular grid FDM given by Narasimhan (1982), but there are some differences between the two methods.

In Sun et al. (1989), the domain of the aquifer is divided into triangular elements as usual. Figure 8.5 shows node i and its neighboring nodes. Assume that a fracture passes through nodes p, l and q.

In building the water balance equation for node i, we introduce two modification factors, one is α, which is along the fractures; the other is β, which is crossing over the fractures. Then, the transmissivity T between nodes i, p and i, q becomes αT, while between nodes i, j, i, k and i, l becomes βT. $\alpha = 1$ and $\beta = 1$ indicate no modification is necessary; $\alpha > 1$ means the conductivity increases along the fracture; and $\beta < 1$ means that water flow is obstructed by the fracture when crossing it. The modification factors α and β can be obtained by solving the inverse problem. Different fractures, or different sections of the same fracture, may have different modification factors. The locations of fractures can also be identified through the calibration of the modification factors.

The discrete fracture model may also be used for solving water quality problems. Because the transmissivities are modified, the flow velocities in the exclusive subdomain of node i will also change, which, in turn, causes the corresponding changes in the hydrodynamic dispersion coefficients. No matter how the coefficients change, the integrated mass conservation equation is still valid. Therefore, we can still establish a mass conservation equation for the exclusive subdomain of node i, just as we did before, and take it as the discrete equation for this node.

The discrete fracture method can truly reflect the actual condition, provided that the geometric shapes and physical properties of the fractures are known. Unfortunately, these requirements are rarely satisfied in practice.

Tsang (1987) and Tsang and Tsang (1987) proposed several channel models for studying various coupled processes in fractured media. They conducted laboratory, modeling and field studies on solute and heat transport in fractures. Johns and Roberts (1991) considered the effects of adsorption and diffusion in channels. Numerical solutions of discrete fracture models were further studied by Germain and Frind (1989), Rasmussen and Evans (1989), and Sudicky and McLaren (1992). In Sudicky and McLaren (1992), the Laplace transform Galerkin (LTG) method (Sudicky, 1989) was adopted because it can avoid time stepping and permit the use of a relatively coarse grid. Thus, it is well-suited for solving large scale and long-term contaminant transport problems. Bear et al. (1993) presented a comprehensive review of the state of the art of flow and concentration transport in fractured rocks, from theories, numerical methods to field tests.

8.2 Seawater Intrusion

8.2.1 The Problem of Seawater Intrusion

In coastal areas, groundwater usually flows into the sea. As the specific gravity of seawater is larger than that of fresh water, the seawater rests under the fresh water like a wedge, as shown in Figure 8.6.

If the fresh water in a confined or unconfined aquifer is overpumped, the hydraulic head of the fresh water may be dropped drastically. As a result, a

FIGURE 8.6. A sketch of the bottom, interface, and groundwater table.

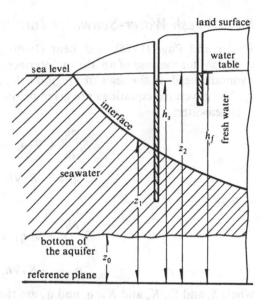

series of problems will be generated, e.g., the seawater may intrude the land, the fresh water in the wells may be salinated, and the aquifers and cultivated horizons may be destroyed. Thus, seawater intrusion is a special problem of groundwater pollution. The development and utilization of groundwater in coastal areas must be managed in a rational way in order to prevent seawater intrusion. If seawater intrusion has happened, measures should be taken to remedy it, such as the control of water supply, artificial recharge, and the building of hydraulic barriers. For this purpose, it is necessary to use mathematical models to simulate the movements of both seawater and fresh water and make quantitative estimations of the effects of various management decisions and engineering installations.

Seawater and fresh water may dissolve and mix with each other, so there must be a *transition zone* between them due to the effect of hydrodynamic dispersion. Through this zone, the water changes from saline to fresh. The widths of the transition zones in the coastal areas measured by some researchers show that they vary greatly. In some areas, the width of the zone is only on the order of 10 m, which can be neglected when compared to the total thickness of the aquifer. In this case it can be considered that there exists an *abrupt interface* between seawater and fresh water. In other areas, however, the width of the transition zone may exceed 100 m with no clear interface. Therefore, there are two different ways for describing the problem of seawater intrusion: when it is assumed that there is an abrupt interface, it is only necessary to calculate the location of the interface; otherwise, both the location of the transition zone and the salt concentration distribution in the zone should be calculated.

8.2.2 Fresh Water-Seawater Interfaces

Pinder and Page (1976), and Bear (1979) gave a complete mathematical statement for the case of an abrupt interface. Assume that the distribution of hydraulic heads in the seawater region is h_s, and in the fresh water region, is h_f. The governing equations are given below:

In seawater:

$$S_s \frac{\partial h_s}{\partial t} + \nabla \cdot \mathbf{q}_s = 0, \tag{8.2.1a}$$

$$\mathbf{q}_s = -K_s \nabla h_s, \tag{8.2.1b}$$

in fresh water:

$$S_f \frac{\partial h_f}{\partial t} + \nabla \cdot \mathbf{q}_f = 0, \tag{8.2.2a}$$

$$\mathbf{q}_f = -K_f \nabla h_f, \tag{8.2.2b}$$

where S_s and S_f, K_s and K_f, $\mathbf{q}_s$ and $\mathbf{q}_f$ are the specific storage coefficients (or specific yields if in unconfined aquifer), hydraulic conductivities and Darcy's velocities in seawater region and fresh water region, respectively.

The gradient operator ∇ is defined as:

$$\nabla = \frac{\partial}{\partial x}\mathbf{i} + \frac{\partial}{\partial y}\mathbf{j} + \frac{\partial}{\partial z}\mathbf{k},$$

where $\mathbf{i}$, $\mathbf{j}$, and $\mathbf{k}$ are unit vectors along the coordinate axes.

Assume that the equation of the sharp interface is

$$F(x, y, z, t) = 0. \tag{8.2.3}$$

Pressure balance must be satisfied on the interface, that is

$$\rho_f(h_f - z_1) = \rho_s(h_s - z_1), \tag{8.2.4}$$

where ρ_s and ρ_f are the densities of seawater and fresh water, respectively; and z_1 is the elevation of the interface. From Eq. (8.2.4), we have

$$z_1 = \frac{\rho_s h_s - \rho_f h_f}{\rho_s - \rho_f},$$

which can be written as

$$z_1 = \rho_s^* h_s - \rho_f^* h_f; \quad \rho_s^* = \frac{\rho_s}{\rho_s - \rho_f}; \quad \rho_f^* = \frac{\rho_f}{\rho_s - \rho_f}. \tag{8.2.5}$$

If h_s and h_f can be solved from Eqs. (8.2.1) and (8.2.2), then the elevation of the interface z_1 may be obtained from Eq. (8.2.5). Thus, the equation of the interface can be expressed by

$$F(x, y, z, t) \equiv z - z_1(x, y, t) = 0. \tag{8.2.6}$$

Because the seawater region and the fresh water region have a common free boundary, which is an unknown interface, governing equations (8.2.1) and (8.2.2) cannot be solved individually. This case is very similar to the case of an unknown water table that is between the unsaturated zone and the saturated zone. To solve Eqs. (8.2.1) and (8.2.2), the interface boundary condition must be defined.

The interface is always formed by identical fluid molecules. Thus, we have $DF/Dt \equiv 0$. From the relationship between the total derivative and partial derivatives, we have the following equations:

$$\frac{\partial F}{\partial t} + \mathbf{V}_s \cdot \nabla F = 0 \qquad (8.2.7a)$$

and

$$\frac{\partial F}{\partial t} + \mathbf{V}_f \cdot \nabla F = 0, \qquad (8.2.7b)$$

where $\mathbf{V}_s$ and $\mathbf{V}_f$ are the porous velocities in seawater and fresh water, respectively. According to Eqs. (8.2.1b) and (8.2.2b), they satisfy

$$n\mathbf{V}_s = -K_s \nabla h_s \qquad (8.2.8a)$$

and

$$n\mathbf{V}_f = -K_f \nabla h_f, \qquad (8.2.8b)$$

where n is the porosity. Substituting Eq. (8.2.5) into Eq. (8.2.6), we have:

$$F(x, y, z, t) \equiv z - \rho_s^* h_s + \rho_f^* h_f = 0, \qquad (8.2.9)$$

then, substituting Eqs. (8.2.9) and (8.2.8) into Eq. (8.2.7), we obtain

$$n\rho_f^* \frac{\partial h_f}{\partial t} - n\rho_s^* \frac{\partial h_s}{\partial t} - K_s [\nabla z - \rho_s^* \nabla h_s + \rho_f^* \nabla h_f] \cdot \nabla h_s = 0 \quad (8.2.10a)$$

and

$$n\rho_f^* \frac{\partial h_f}{\partial t} - n\rho_s^* \frac{\partial h_s}{\partial t} - K_f [\nabla z - \rho_s^* \nabla h_s + \rho_f^* \nabla h_f] \cdot \nabla h_f = 0. \quad (8.2.10b)$$

These two equations are the boundary conditions that the interface between seawater and fresh water should satisfy. They are nonlinear partial differential equations with respect to h_s and h_f, and therefore, are very difficult to solve even with numerical methods.

Let us recall that when dealing with unconfined aquifers, we adopted *Dupuit's assumptions* in order to avoid the free surface boundary condition of the water table. Under these assumptions, the three-dimensional problem of a unconfined flow can be simplified to a two-dimensional one through *averaging in the vertical direction*, so that the governing equation no longer contains the boundary condition of water table (Bear, 1979). A similar method can be used to deal with the interface between seawater and fresh water.

Let the elevation of aquifer bottom be z_0, the elevation of the interface be z_1, the elevation of the fresh water table be z_2 (if the aquifer is confined, z_2 will be the elevation of the aquifer roof). All of which are functions of x, y and t, as shown in Figure 8.6.

Integrating (8.2.1a) along the z direction in the seawater region, we obtain

$$\int_{z_0}^{z_1} \left(\nabla \cdot \mathbf{q}_s + S_s \frac{\partial h_s}{\partial t} \right) dz = 0. \qquad (8.2.11)$$

Using the Leibniz formula, this equation can be transformed into

$$\nabla' \cdot \int_{z_0}^{z_1} \mathbf{q}_s' \, dz - \mathbf{q}_s'|_{z_1} \cdot \nabla' z_1 + \mathbf{q}_s'|_{z_0} \cdot \nabla' z_0 + q_{sz}\Big|_{z_0}^{z_1}$$

$$+ S_s \left[\frac{\partial}{\partial t} \int_{z_0}^{z_1} h_s \, dz - h_s|_{z_1} \frac{\partial z_1}{\partial t} \right] = 0, \qquad (8.2.12)$$

where ∇' is a gradient operator on the xy plane and $\mathbf{q}_s'$ is the projection of $\mathbf{q}_s$ on plane xy, i.e.,

$$\nabla' \equiv \frac{\partial}{\partial x}\mathbf{i} + \frac{\partial}{\partial y}\mathbf{j}, \qquad (8.2.13a)$$

$$\mathbf{q}_s' = q_{sx}\mathbf{i} + q_{sy}\mathbf{j}, \qquad (8.2.13b)$$

where q_{sx}, q_{sy}, and q_{sz} are the components of $\mathbf{q}_s$ on the three coordinate axes. The means of h_s and $\mathbf{q}_s'$ along the z direction are defined as

$$\bar{h}_s = \frac{1}{b_s} \int_{z_0}^{z_1} h_s \, dz, \qquad (8.2.14a)$$

$$\bar{\mathbf{q}}_s' = \frac{1}{b_s} \int_{z_0}^{z_1} \mathbf{q}_s' \, dz, \qquad (8.2.14b)$$

where $b_s = (z_1 - z_0)$ is the thickness of the seawater layer. Suppose the change of h_s along the vertical direction is very small, then we have

$$h_s|_{z_1} \approx \bar{h}_s. \qquad (8.2.15)$$

Consequently, the term in the square brackets on the left-hand side of Eq. (8.2.12) can be translated into

$$\frac{\partial}{\partial t} \int_{z_0}^{z_1} h_s \, dz - h_s|_{z_1} \frac{\partial z_1}{\partial t} = \frac{\partial}{\partial t}(b_s \bar{h}_s) - \bar{h}_s \frac{\partial b_s}{\partial t} = b_s \frac{\partial \bar{h}_s}{\partial t}. \qquad (8.2.16)$$

Substituting this equation into Eq. (8.2.12), we obtain

$$\nabla' \cdot (b_s \bar{\mathbf{q}}_s') + q_{sz}|_{z_1} - q_{sz}|_{z_0} - \mathbf{q}_s'|_{z_1} \cdot \nabla' z_1 + \mathbf{q}_s'|_{z_0} \cdot \nabla' z_0 + S_s b_s \frac{\partial \bar{h}_s}{\partial t} = 0. \quad (8.2.17a)$$

Using the same method, Eq. (8.2.1b) may be averaged along the vertical direction to yield

$$\mathbf{q}_s' b_s + K_s b_s \cdot \nabla' \bar{h}_s = 0, \qquad (8.2.17b)$$

where the hydraulic conductivity K_s is assumed to be independent of the depth. Equations (8.2.17a) and (8.2.17b) are the two-dimensional equations resulting from the vertical averaging of the seawater equations (8.2.1a) and (8.2.1b), with the assumption of Eq. (8.2.15).

Using exactly the same derivation and assumption, the fresh water equations (8.2.2a) and (8.2.2b) can be transformed into the following two-dimensional equations:

$$\nabla' \cdot (b_f \bar{\mathbf{q}}_f') + q_{fz}|_{z_2} - q_{fz}|_{z_1} - \mathbf{q}_f'|_{z_2} \cdot \nabla' z_2 + \mathbf{q}_f'|_{z_1} \cdot \nabla' z_1 + S_f b_f \frac{\partial \bar{h}_f}{\partial t} = 0,$$

(8.2.18a)

and

$$\mathbf{q}_f' b_f + K_f b_f \cdot \nabla' \bar{h}_f = 0,$$ (8.2.18b)

where the subscript f means fresh water and the meanings of the notations are the same as those used in Eq. (8.2.17).

Using the averaged $\bar{h}_s$ and $\bar{h}_f$, the elevation of the interface z_1 can be expressed as

$$z_1 = \rho_s^* \bar{h}_s - \rho_f^* \bar{h}_f,$$ (8.2.19)

so the equation of the interface is

$$F(x, y, z, t) \equiv z - z_1(x, y, t) = z - \rho_s^* \bar{h}_s + \rho_f^* \bar{h}_f = 0.$$ (8.2.20)

Because $DF/Dt \equiv 0$, the velocity of particles moving in the z direction on the interface is

$$V_z|_{z_1} = \left[\frac{\partial z_1}{\partial x} V_x + \frac{\partial z_1}{\partial y} V_y + \frac{\partial z_1}{\partial t} \right]_{z_1};$$

or

$$q_{sz}|_{z_1} = n \left[\mathbf{V}' \cdot \nabla' z_1 + \frac{\partial z_1}{\partial t} \right]_{z_1},$$ (8.2.21)

where $\mathbf{V}' = V_x \mathbf{i} + V_y \mathbf{j}$. Substituting Eq. (8.2.21) into Eq. (8.2.17a), we obtain

$$\nabla' \cdot (\bar{\mathbf{q}}_s' b_s) + (q_{sz} - n V_{sz})|_{z_1} - q_{sz}|_{z_0} - (\mathbf{q}_s' - n \mathbf{V}_s')|_{z_1} \cdot \nabla' z_1$$
$$+ n \frac{\partial z_1}{\partial t} + S_s b_s \frac{\partial \bar{h}_s}{\partial t} = 0.$$ (8.2.22)

If the bottom of the aquifer is impermeable, $q_{sz}|_{z_0} = 0$. The values of $(q_{sz} - n V_{sz})|_{z_1}$ and $(\mathbf{q}_s' - n \mathbf{V}_s') \cdot \nabla' z_1$ represent the flux across the boundary, i.e., the source and sink term, written as Q_s. Thus, Eq. (8.2.22) can be simplified as

$$\nabla' \cdot (\bar{\mathbf{q}}_s' b_s) + S_s b_s \frac{\partial \bar{h}_s}{\partial t} + n \left(\rho_s^* \frac{\partial \bar{h}_s}{\partial t} - \rho_f^* \frac{\partial \bar{h}_f}{\partial t} \right) + Q_s = 0.$$

Substituting Eq. (8.2.17b) into this equation, we obtain the following equation for the average hydraulic head $\bar{h}_s$ of seawater

$$-\nabla' \cdot (K_s b_s \cdot \nabla' \bar{h}_s) + S_s b_s \frac{\partial \bar{h}_s}{\partial t} + n\left(\rho_s^* \frac{\partial \bar{h}_s}{\partial t} - \rho_f^* \frac{\partial \bar{h}_f}{\partial t}\right) + Q_s = 0. \quad (8.2.23)$$

Using a parallel derivation, we can also obtain the following equation for the average hydraulic head $\bar{h}_f$ of fresh water

$$-\nabla' \cdot (K_f b_f \cdot \nabla' \bar{h}_f) + S_f b_f \frac{\partial \bar{h}_f}{\partial t} - n\left(\rho_s^* \frac{\partial \bar{h}_s}{\partial t} - \rho_f^* \frac{\partial \bar{h}_f}{\partial t}\right)$$

$$+ \alpha n \frac{\partial \bar{h}_f}{\partial t} + Q_f = 0, \quad (8.2.24)$$

where the source and sink term covers pumping, recharge and infiltration of rainfall; and α is a coefficient, $\alpha = 0$ for confined water while $\alpha = 1$ for phreatic water. For the sake of simplifying the notation, $\bar{h}_s$ and $\bar{h}_f$ are still written as h_s and h_f, but it should be remembered that they are only functions of x, y, and t. Equations (8.2.23) and (8.2.24) can be expressed in the scalar form as

$$S_s b_s \frac{\partial h_s}{\partial t} + n\left(\rho_s^* \frac{\partial h_s}{\partial t} - \rho_f^* \frac{\partial h_f}{\partial t}\right) - \frac{\partial}{\partial x}\left(K_{sx} b_s \frac{\partial h_s}{\partial x}\right)$$

$$- \frac{\partial}{\partial y}\left(K_{sy} b_s \frac{\partial h_s}{\partial y}\right) + Q_s = 0 \quad (8.2.25)$$

and

$$S_f b_f \frac{\partial h_f}{\partial t} - n\left(\rho_s^* \frac{\partial h_s}{\partial t} - \rho_f^* \frac{\partial h_f}{\partial t}\right) + \alpha n \frac{\partial h_f}{\partial t} - \frac{\partial}{\partial x}\left(K_{fx} b_f \frac{\partial h_f}{\partial x}\right)$$

$$- \frac{\partial}{\partial y}\left(K_{fy} b_f \frac{\partial h_f}{\partial y}\right) + Q_f = 0. \quad (8.2.26)$$

To solve the two equations, appropriate initial and boundary conditions must be given. Boundary conditions along the coast are not easy to determine. It is incorrect to let h_f be equal to the level of seawater along the coast, because the fresh water flows towards the sea. As a result, the boundary line should be the seepage surface of the fresh water. If the overflow of the fresh water can be determined, the coast line may be regarded as a flux boundary. Of course, the overflow is, in fact, also unknown, but it might be determined as a part of the parameter identification by solving the inverse problem.

Bear (1979) suggested the adoption of the third-type of boundary condition. Along the coast line, for the fresh water equation, it is defined as

$$\frac{h_f}{\alpha} + [\rho_s^* h_s - \rho_f^* h_f] \frac{\partial h_f}{\partial n} = 0, \quad (8.2.27)$$

and for the seawater equation as

$$\frac{h_s}{\beta} + [D - z_0 + \rho_s^* h_s - \rho_f^* h_f] \frac{\partial h_s}{\partial n} = 0, \qquad (8.2.28)$$

where D is the elevation of sea level, z_0 is the elevation of the aquifer bottom, and α and β are parameters to be identified.

8.2.3 Numerical Methods for Determining the Location of Interfaces

The discrete form of Eqs. (8.2.25) and (8.2.26) can be obtained using the standard procedure of the Galerkin FEM. First, the two equations are re-written into the following operator form

$$L_s(h_s, h_f) \equiv 0 \quad \text{and} \quad L_f(h_s, h_f) \equiv 0. \qquad (8.2.29)$$

Define the trial functions

$$h_s = \sum_{j=1}^{N} H_{sj}(t)\phi_j(x, y) \qquad (8.2.30a)$$

and

$$h_f = \sum_{j=1}^{N} H_{fj}(t)\phi_j(x, y) \qquad (8.2.30b)$$

as the approximations of h_s and h_f, where $\{\phi_j(x, y), j = 1, 2, \ldots, N\}$ is a set of basis functions. The $\phi_j(x, y)$ are also used as the weighting functions based on the principle of Galerkin FEM, so we will have the following weighted residual equations:

$$\int_{(\Omega)} L_s(h_s, h_f)\phi_j \, d\Omega = 0, \quad (i = 1, 2, \ldots, N); \qquad (8.2.31a)$$

$$\int_{(\Omega)} L_f(h_s, h_f)\phi_j \, d\Omega = 0, \quad (i = 1, 2, \ldots, N). \qquad (8.2.31b)$$

Substituting Eq. (8.2.29) into the above two equations and using Green's formulas to eliminate the terms with second-order derivatives, we then have

$$\int_{(\Omega)} \left[\left(K_{sx} b_s \frac{\partial \hat{h}_s}{\partial x} \frac{\partial \phi_i}{\partial x} + K_{sy} b_s \frac{\partial \hat{h}_s}{\partial y} \frac{\partial \phi_i}{\partial y} \right) + (n\rho_s^* + S_s b_s) \frac{\partial \hat{h}_s}{\partial t} \phi_i \right.$$
$$\left. - n\rho_f^* \frac{\partial \hat{h}_f}{\partial t} \phi_i + Q_s \phi_i \right] d\Omega - \int_{(\Gamma)} g_s \phi_i \, dS = 0, \quad (i = 1, 2, \ldots, N) \qquad (8.2.32a)$$

and

$$\int_{(\Omega)} \left[\left(K_{fx} b_s \frac{\partial \hat{h}_f}{\partial x} \frac{\partial \phi_i}{\partial x} + K_{fy} b_s \frac{\partial \hat{h}_f}{\partial y} \frac{\partial \phi_i}{\partial y} \right) + (\alpha n + n\rho_f^* + S_f b_s) \frac{\partial \hat{h}_f}{\partial t} \phi_i \right.$$
$$\left. - n\rho_s^* \frac{\partial \hat{h}_s}{\partial t} \phi_i + Q_f \phi_i \right] d\Omega - \int_{(\Gamma)} g_f \phi_i \, dS = 0, \quad (i = 1, 2, \ldots, N). \qquad (8.2.32b)$$

In the above equations, g_s and g_f are given by the flux boundary conditions in the seawater and fresh water equations, respectively. After substituting Eq. (8.2.30) into Eq. (8.2.32), the obtained equations can be arranged into the following matrix form:

$$[A]H + [B]\frac{dH}{dt} + F = 0, \qquad (8.2.33)$$

where the elements of matrices $[A]$ and $[B]$ are

$$A_{ij}$$
$$= \begin{bmatrix} \int_{(\Omega)} \left(K_{sx} b_s \frac{\partial \phi_i}{\partial x} \frac{\partial \phi_j}{\partial x} + K_{sy} b_s \frac{\partial \phi_i}{\partial y} \frac{\partial \phi_j}{\partial y} \right) d\Omega & 0 \\ 0 & \int_{(\Omega)} \left(K_{fx} b_f \frac{\partial \phi_i}{\partial x} \frac{\partial \phi_j}{\partial x} + K_{fy} b_f \frac{\partial \phi_i}{\partial y} \frac{\partial \phi_j}{\partial y} \right) d\Omega \end{bmatrix},$$
$$(8.2.34)$$

$$B_{ij} = \begin{bmatrix} \int_{(\Omega)} (n\rho_s^* + S_s b_s) \phi_i \phi_j \, d\Omega & -\int_{(\Omega)} n\rho_f^* \phi_i \phi_j \, d\Omega \\ -\int_{(\Omega)} n\rho_s^* \phi_i \phi_j \, d\Omega & \int_{(\Omega)} (\alpha n + n\rho_f^* + S_f b_f) \phi_i \phi_j \, d\Omega \end{bmatrix}. \qquad (8.2.35)$$

The components of vector F, H and dH/dt are, respectively:

$$F_i = \left\{ \begin{array}{c} \int_{(\Omega)} Q_s \phi_i \, d\Omega - \int_{(\Gamma)} g_s \phi_i \, dS \\ \int_{(\Omega)} Q_f \phi_i \, d\Omega - \int_{(\Gamma)} g_f \phi_i \, dS \end{array} \right\} \qquad (8.2.36)$$

and

$$h_i = \left\{ \begin{array}{c} h_{si} \\ h_{fi} \end{array} \right\}, \qquad \frac{dh_i}{dt} = \left\{ \begin{array}{c} \dfrac{dh_{si}}{dt} \\ \dfrac{dh_{fi}}{dt} \end{array} \right\}. \qquad (8.2.37)$$

Divide the domain into elements and choose the basis functions, then all coefficients in Eq. (8.2.33) can be computed. For example, we can use triangular elements and linear basis functions, or quadrilateral elements and bilinear basis functions.

From Eqs. (8.2.34) to (8.2.36), we can see that the coefficients in Eq. (8.2.33) depend on the thickness of the seawater b_s and that of the fresh water b_f, while b_s and b_f depend on the hydraulic heads h_s and h_f, respectively. As a result, Eq. (8.2.33) is a system of nonlinear equations and can be solved only by an iteration method. The backward difference may be used to discrete the time derivative to turn Eq. (8.2.33) into a system of algebraic equations:

$$[A]_{n+1/2}^m H_{n+1}^{m+1} + \frac{1}{\Delta t}[B]_{n+1}^m (H_{n+1}^{m+1} - H_n) + F = 0, \qquad (8.2.38)$$

where subscript n denotes the nth time step, superscript m denotes the mth iteration, Δt the length of the time step, H_{n+1} the hydraulic heads to be solved at time $n + 1$, and H_n the known vector of hydraulic heads at time n. Several iterations may be required for each time step. During each iteration, the values of the coefficients are replaced with new values obtained from the last iteration until convergence is achieved.

Pinder and Page (1976) used this method to consider the problem of seawater intrusion in an aquifer near Long Island, New York. The domain considered was divided into 144 triangular elements, with a total of 85 nodes. The observation data from 1973 to 1975 in seven wells were used to calibrate the model. The inputs of the model include the known rainfall infiltration and artificial recharge. The flow boundary conditions along the coast were determined during the model calibration. The calibrated model was then used to predict the changes of the seawater-fresh water interface in the case of increased extraction.

Mercer et al. (1980) adopted the FDM to solve Eqs. (8.2.25) and (8.2.26). They also tried various iteration methods for solving the resulting nonlinear difference equations, and concluded that the method of blocked linear successive over relax iteration was the most effective. In their paper, an example is given which considers a problem of sewage irrigation in a coastal aquifer in Hawaii. The sewage was injected into the seawater zone at the lower part of the aquifer. The objective of the study was to estimate the seawater-fresh water interface caused by the sewage. The simulated area covered 13 km². The parameters of the model were calibrated by the observed steady-state heads.

8.2.4 Determination of the Transition Zones

When the transition zone is relatively wide, *advection-dispersion models* must be used to describe the problem of seawater intrusion. In this case, the unknown variable is the distribution of salt concentration and not the location of the interface. The salinity decreases gradually from the seawater region to the fresh water region, and thus a natural transition zone is formed.

If the impact of salt concentration on groundwater flow is neglected, then the problem of seawater intrusion will be the same as that of groundwater contamination mentioned in Section 8.1. For instance, Gupta and Yapa (1982) adopted the advection-dispersion model in their study of seawater intrusion near Bangkok, Thailand, without considering the impact of concentration changes on flow velocity. Hydraulic conductivity, porosity and dispersivity are determined through fitting the observed chloride ion concentration curves. Using this model, they found that a major cause of salination of the fresh water aquifer was the intrusion of marine water from adjacent aquifers, in addition to the direct seawater intrusion.

Owing to the remarkable difference between the density of the seawater and that of the fresh water, significant calculation errors would occur if the seawater was considered as a tracer. Strictly speaking, the problem of seawater intrusion should refer to the general case mentioned in Section 7.1, i.e., the fluid flow may be affected by the change of fluid density. Therefore, the water flow equation and water quality equation cannot be solved separately. Instead, they must be solved iteratively through the state equations. The Galerkin FEM for solving this kind of problem will be introduced below.

Let us consider a two-dimensional seawater intrusion problem on the xz plane, where the x axis is horizontal and the z axis is vertically upward. For the sake of simplicity, the media of aquifers are assumed to be isotropic. This problem is governed by:

1. Dispersion equation

$$\frac{\partial C}{\partial t} = \frac{\partial}{\partial x}\left(D_{xx}\frac{\partial C}{\partial x} + D_{xz}\frac{\partial C}{\partial z}\right) + \frac{\partial}{\partial z}\left(D_{zx}\frac{\partial C}{\partial x} + D_{zz}\frac{\partial C}{\partial z}\right)$$

$$- \frac{\partial}{\partial x}(V_x C) - \frac{\partial}{\partial z}(V_z C), \tag{8.2.39}$$

2. Continuity equation of an incompressible fluid

$$\left[\frac{\partial(\rho V_x)}{\partial x} + \frac{\partial(\rho V_z)}{\partial z}\right] = 0, \tag{8.2.40}$$

3. Two movement equations

$$V_x = -\frac{k}{\mu n}\frac{\partial p}{\partial x}, \tag{8.2.41}$$

$$V_z = -\frac{k}{\mu n}\left(\frac{\partial p}{\partial z} + \rho g\right). \tag{8.2.42}$$

4. State equation

$$\rho = \rho_0 + EC, \tag{8.2.43}$$

where ρ_0 is the density of fresh water; E can be taken as 0.7; C is the salt concentration determined by Eq. (8.2.39); and viscosity μ is assumed to be a constant.

The general procedure for solving these equations has been presented in Section 7.1. Assume that the solution at time t is known. First, estimate the distribution of density ρ at time $t + \Delta t$, then simultaneously solve for the distributions of ρ, V_x, V_z and C at time $t + \Delta t$. Finally, use Eq. (8.2.43) to update the distribution of ρ. This procedure is repeated until convergence is achieved. Thus, the major problem is how to simultaneously solve Eqs. (8.2.39) to (8.2.42).

In order to use the Galerkin method, the four equations are expressed as:

$$L_1(C, p, V_x, V_z) \equiv \frac{\partial C}{\partial t} + \frac{\partial}{\partial x}(V_x C) + \frac{\partial}{\partial z}(V_z C) - \frac{\partial}{\partial x}\left(D_{xx}\frac{\partial C}{\partial x} + D_{xz}\frac{\partial C}{\partial z}\right)$$

$$- \frac{\partial}{\partial z}\left(D_{zx}\frac{\partial C}{\partial x} + D_{zz}\frac{\partial C}{\partial z}\right) = 0, \tag{8.2.44a}$$

$$L_2(C, p, V_x, V_z) \equiv \frac{\partial}{\partial x}(\rho V_x) + \frac{\partial}{\partial z}(\rho V_z) = 0, \tag{8.2.44b}$$

$$L_3(C, p, V_x, V_z) \equiv V_x + \frac{k}{\mu n}\frac{\partial p}{\partial x} = 0, \tag{8.2.44c}$$

$$L_4(C, p, V_x, V_z) \equiv V_z + \frac{k}{\mu n}\left(\frac{\partial p}{\partial z} + \rho g\right) = 0. \tag{8.2.44d}$$

We may adopt the following approximate solutions:

$$\hat{C}(x, z, t) = \sum_{j=1}^{N} C_j(t)\phi_j(x, z), \tag{8.2.45a}$$

$$\hat{p}(x, z, t) = \sum_{j=1}^{N} p_j(t)\phi_j(x, z), \tag{8.2.45b}$$

$$\hat{V}_x(x, z, t) = \sum_{j=1}^{N} V_{x,j}(t)\phi_j(x, z), \tag{8.2.45c}$$

$$\hat{V}_z(x, z, t) = \sum_{j=1}^{N} V_{z,j}(t)\phi_j(x, z), \tag{8.2.45d}$$

where $\{\phi_i(x, z), i = 1, 2, \ldots, N\}$ are basis functions. When the basis functions are chosen, the problem of solving approximate solutions will become a problem of determining $4 \times N$ coefficients: $\{C_i(t), p_i(t), V_{x,i}(t), V_{z,i}(t)\}$, $(i = 1, 2, \ldots, N)$. According to the Galerkin method, these coefficients may be determined by solving the $4 \times N$ equations as follows

$$\iint_{(\Omega)} L_1(\hat{C}, \hat{p}, \hat{V}_x, \hat{V}_z)\phi_i \, dx \, dz = 0, \tag{8.2.46a}$$

$$\iint_{(\Omega)} L_2(\hat{C}, \hat{p}, \hat{V}_x, \hat{V}_z)\phi_i \, dx \, dz = 0, \tag{8.2.46b}$$

$$\iint_{(\Omega)} L_3(\hat{C}, \hat{p}, \hat{V}_x, \hat{V}_z)\phi_i \, dx \, dz = 0, \tag{8.2.46c}$$

$$\iint_{(\Omega)} L_4(\hat{C}, \hat{p}, \hat{V}_x, \hat{V}_z)\phi_i \, dx \, dz = 0. \tag{8.2.46d}$$

$$(i = 1, 2, \ldots, N)$$

We should note that the latter $3 \times N$ equations do not explicitly contain $\{C_i\}$, therefore $3 \times N$ coefficients $\{p_i\}$, $\{V_{x,i}\}$ and $\{V_{z,i}\}$ may be obtained first from these equations, and then substituted into the first N equations to obtain $\{C_i\}$. Substituting the last three equations of Eqs. (8.2.44), and (8.2.45), into the last $3 \times N$ equations of Eq. (8.4.46) correspondingly, we will have

$$\iint_{(\Omega)} \sum_{j=1}^{N} \left[\frac{\partial}{\partial x}(\rho\phi_j)\phi_i V_{x,j} + \frac{\partial}{\partial z}(\rho\phi_j)\phi_i V_{z,i} \right] dx\, dz = 0, \quad (8.2.47a)$$

$$\iint_{(\Omega)} \sum_{j=1}^{N} \left[\phi_i\phi_j V_{x,j} + \frac{k}{\mu n}\frac{\partial\phi_j}{\partial x}\phi_i p_j \right] dx\, dz = 0, \quad (8.2.47b)$$

$$\iint_{(\Omega)} \sum_{j=1}^{N} \left[\phi_i\phi_j V_{z,j} + \frac{k}{\mu n}\frac{\partial\phi_j}{\partial z}\phi_i p_j + \frac{k}{\mu n}\rho g\phi_i \right] dx\, dz = 0. \quad (8.2.47c)$$

$$(i = 1, 2, \ldots, N)$$

After rearrangement, these equations can be reduced to

$$[H]\pi + R = 0, \quad (8.2.48)$$

where

$$\pi_i = \{P_i, V_{x,i}, V_{z,i}\}^T, \quad (8.2.49a)$$

$$R_i = \left\{ 0, 0, \iint_{(\Omega)} \frac{k}{\mu n}\rho g\phi_i\, dx\, dz \right\}^T, \quad (8.2.49b)$$

$$[H_{ij}] = \iint_{(\Omega)} \begin{bmatrix} 0 & \frac{\partial}{\partial x}(\rho\phi_j)\phi_i & \frac{\partial}{\partial z}(\rho\phi_j)\phi_i \\ \frac{k}{\mu n}\frac{\partial\phi_j}{\partial x}\phi_i & \phi_j\phi_i & 0 \\ \frac{k}{\mu n}\frac{\partial\phi_j}{\partial z}\phi_i & 0 & \phi_j\phi_i \end{bmatrix} dx\, dz. \quad (8.2.49c)$$

It should be noted that i and j in these equations all vary from 1 to N, so π and R are $3N$-order column vectors, and $[H]$ is a $3N \times 3N$ square matrix.

Substitute Eqs. (8.2.44a) and (8.2.45a) into Eq. (8.2.46a) to obtain

$$\iint_{(\Omega)} \sum_{j=1}^{N} \left[\phi_i\phi_j\frac{\partial C_j}{\partial t} + \frac{\partial}{\partial x}(\hat{V}_x\phi_j)\phi_i C_j + \frac{\partial}{\partial z}(\hat{V}_z\phi_j)\phi_i C_j \right.$$

$$\left. - \frac{\partial}{\partial x}\left(D_{xx}\frac{\partial\phi_j}{\partial x} + D_{xz}\frac{\partial\phi_j}{\partial z} \right)\phi_i C_j - \frac{\partial}{\partial z}\left(D_{zx}\frac{\partial\phi_j}{\partial x} + D_{zz}\frac{\partial\phi_j}{\partial z} \right)\phi_i C_j \right] dx\, dz$$

$$= 0, \quad (i = 1, 2, \ldots, N)$$

then use Green's formula to eliminate the terms of second-order derivatives to obtain a set of ordinary differential equations,

$$[A]C + [B]\frac{dC}{dt} + F = 0, \tag{8.2.50}$$

where

$$A_{ij} = \iint_{(\Omega)} \left[D_{xx}\frac{\partial\phi_i}{\partial x}\frac{\partial\phi_j}{\partial x} + D_{xz}\frac{\partial\phi_i}{\partial x}\frac{\partial\phi_j}{\partial z} + D_{zx}\frac{\partial\phi_i}{\partial z}\frac{\partial\phi_j}{\partial x} + D_{zz}\frac{\partial\phi_i}{\partial z}\frac{\partial\phi_j}{\partial z} \right.$$

$$\left. + \hat{V}_x\phi_i\frac{\partial\phi_j}{\partial x} + \hat{V}_z\phi_i\frac{\partial\phi_j}{\partial z} + \phi_i\phi_j\frac{\partial\hat{V}_x}{\partial x} + \phi_i\phi_j\frac{\partial\hat{V}_z}{\partial z} \right] dx\,dz, \tag{8.2.51a}$$

$$B_{ij} = \iint_{(\Omega)} \phi_i\phi_j \, dx\,dz, \tag{8.2.51b}$$

$$F_i = \int_{(\Gamma_2)} g_2\phi_i \, dS. \tag{8.2.51c}$$

This derivation procedure is the same as that discussed in Section 5.1.

To obtain a system of algebraic equations, the time derivative in Eq. (8.2.50) is approximated by the backward difference, which can be combined with Eq. (8.2.48) to yield:

$$\begin{cases} [H]\pi_{t+\Delta t} + R = 0, & (8.2.52a) \\[2mm] ([A] + [B]/\Delta t)C_{t+\Delta t} = \dfrac{[B]}{\Delta t}C_t - F. & (8.2.52b) \end{cases}$$

When the domain (Ω) is divided into elements and the basis functions are chosen, all of the coefficients in the above equations can be computed. Segol et al. (1975) gave some examples for using rectangular elements, bilinear basis functions, and quadratic isoparametric FEM.

Desai and Contractor (1977) gave a numerical example using the above procedure to predict the position of the seawater front in a coastal aquifer. Huyakorn et al. (1987) presented a three-dimensional FEM for seawater intrusion problems. Galeati et al. (1992) used a modified method of characteristics (see Section 6.3) to solve the seawater intrusion problems. Both the increase of solution accuracy and the decrease of computational effort were considered.

8.3 Groundwater Quality Management Models

8.3.1 Groundwater Hydraulic Management Models

The basic concepts of groundwater management have been introduced in Section 1.2, which include the concepts of decision variables and state variables, as well as the relationships between prediction problems and manage-

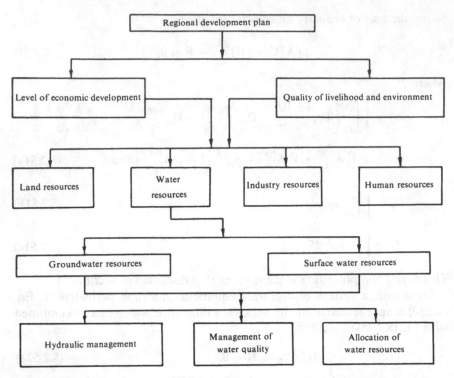

FIGURE 8.7. The role and position of groundwater management in regional development.

ment problems. A mathematical model of water flow and water quality, in which the decision variables are considered as model parameters, can be used to solve the prediction problem and answer whether or not a decision is feasible. The management problem is selection of the optimal decision that can best meet the presented objectives from all feasible decisions.

Groundwater management models can be divided into groundwater hydraulic management models, groundwater quality management models and the models in combination with resources allocation, economic development and policy evaluation. Figure 8.7 shows the role and position of groundwater management in regional economic development.

The *hydraulic management model* is the simplest kind of management model. The decision variables mainly include the locations and rates of extraction and injection wells, hydraulic heads and hydraulic gradients in some parts of the domain, and so forth. The management objectives usually involve minimizing the drawdown in the extraction wells, maximizing water supply, forming an optimal hydraulic barrier, and so forth. In this kind of model, the economy consideration is realized through controlling the hydraulic variables. For instance, more water supply will promote economic

FIGURE 8.8. Water is pumped out from the aquifer and transported to the user.

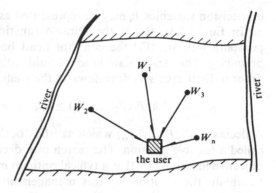

development, and less drawdowns in the wells can reduce the pumping cost.

A very typical example is shown in Figure 8.8. There are M pumping wells W_i ($i = 1, 2, \ldots, m$) in the aquifer, and the purpose is to find a scheme which minimizes the cost of water supply under the constraints that the total pumping yield must satisfy the users' demand (D) and the steady hydraulic heads in the wells must be no less than the designated values h_i^* ($i = 1, 2, \ldots, m$).

Let us assume that for each well, the cost of the water supply is proportional to the amount of water pumped, then the total cost for water supply will be

$$Z = \sum_{i=1}^{m} c_i P_i, \qquad (8.3.1)$$

where P_i is the pumping rate of well W_i, c_i is a constant depending on the water transportation distance. The decision variables are pumping rates P_1, $P_2, \ldots, P_m$, which must satisfy the constraint of water demand:

$$\sum_{i=1}^{m} P_i \geq D, \qquad (8.3.2)$$

where D is the proposed water demand. Next, for each set of $(P_1, P_2, \ldots, P_m)$, we can use the following governing equation:

$$\frac{\partial}{\partial x}\left(T\frac{\partial h}{\partial x}\right) + \frac{\partial}{\partial y}\left(T\frac{\partial h}{\partial y}\right) + \sum_{i=1}^{m} P_i \delta(x - x_i)\delta(y - y_i) = 0 \qquad (8.3.3a)$$

and additional boundary conditions

$$h|_{\Gamma_1} = H_0, \qquad (8.3.3b)$$

$$\frac{\partial h}{\partial n}\bigg|_{\Gamma_2} = 0 \qquad (8.3.3c)$$

to solve for the steady hydraulic heads h_i ($l = 1, 2, \ldots, m$) in the m wells, which are the state variables of the problem. To emphasize the dependence of heads on decision variables, h_i may be represented as $h_i(P_1, P_2, \ldots, P_m)$, $i = 1, 2, \ldots, m$. In Eq. (8.3.3), $\delta(\cdot)$ is the Dirac-δ function, (x_i, y_i) the coordinates of pumping well W_i, (Γ_1) the constant head boundary and (Γ_2) the no-flow boundary. The steady state heads should satisfy the following constraint in order to limit excessive drawdown in the wells:

$$h_i(P_1, P_2, \ldots, P_m) \geq h_i^*, \quad (i = 1, 2, \ldots, m). \tag{8.3.4}$$

A decision $(P_1, P_2, \ldots, P_m)$ which satisfies both Eq. (8.3.2) and Eq. (8.3.4) is called a feasible decision. The search of a decision, which can minimize the objective function (8.3.1), is a typical optimal management problem. In order to obtain the solution for this management problem, the flow model of Eq. (8.3.3) must be imbedded into a nonlinear optimization program.

Since linear optimization problems are easier to solve than nonlinear problems, we would like to approximately linearize the constraint, Eq. (8.3.4). Let h_i^0 represent the hydraulic head of well W_i before pumping, and s_i the steady drawdown of the well, we then have

$$h_i(P_1, P_2, \ldots, P_m) = h_i^0 - s_i(P_1, P_2, \ldots, P_m). \tag{8.3.5}$$

Substitute this equation into the inequality (8.3.4), then it becomes a constraint with respect to drawdown:

$$s_i(P_1, P_2, \ldots, P_m) \leq h_i^0 - h_i^*. \tag{8.3.6}$$

Because $s_i(0, 0, \ldots, 0) = 0$, which means that there is no drawdown as long as there is no pumping, the inequality (8.3.6) can then be expressed as

$$s_i(P_1, P_2, \ldots, P_m) - s_i(0, 0, \ldots, 0) \leq b_i, \tag{8.3.7}$$

where $b_i = h_i^0 - h_i^*$. The linear terms of the Taylor expansion can be used to replace the left-hand side of this equation approximately to obtain

$$\frac{\partial s_i}{\partial P_1} P_1 + \frac{\partial s_i}{\partial P_2} P_2 + \cdots + \frac{\partial s_i}{\partial P_m} P_m \leq b_i, \tag{8.3.8}$$

where all the partial derivatives $\partial s_i / \partial P_j$ ($j = 1, 2, \ldots, m$) are taken values at point $(0, 0, \ldots, 0)$. The partial derivative $\partial s_i / \partial P_j$ is the sensitivity coefficient of drawdown of the ith well, which is under the influence of pumping from the jth well. Thus, it is also called the *drawdown influence coefficient*, which represents the drawdown of the ith well caused by a unit of pumping from the jth well only. When considering all the wells, the constraints of Eq. (8.3.8) can be rewritten in the following vector and matrix form:

$$[A]\{P\} \leq \{b\}, \tag{8.3.9}$$

where $\{P\}$ is a vector of decision variables, $\{P\} = (P_1, P_2, \ldots, P_m)^T$; $\{b\} =$

$(b_1, b_2, \ldots, b_m)^T$; and

$$[\mathbf{A}] = \begin{bmatrix} \dfrac{\partial s_1}{\partial P_1} & \dfrac{\partial s_1}{\partial P_2} & \cdots & \dfrac{\partial s_1}{\partial P_m} \\[2mm] \dfrac{\partial s_2}{\partial P_1} & \dfrac{\partial s_2}{\partial P_2} & \cdots & \dfrac{\partial s_2}{\partial P_m} \\[2mm] \cdots & \cdots & \cdots & \cdots \\[2mm] \dfrac{\partial s_m}{\partial P_1} & \dfrac{\partial s_m}{\partial P_2} & \cdots & \dfrac{\partial s_m}{\partial P_m} \end{bmatrix} \qquad (8.3.10)$$

is called the *drawdown influence matrix*, all the elements of which can be obtained by running the simulation model Eq. (8.3.3) m times. During the jth simulation ($j = 1, 2, \ldots, m$), let the pumping yield from the jth well be ΔP_j and with no pumping at the other wells, then the steady drawdowns of all wells can be determined as $s_i(0, 0, \ldots, \Delta P_i, \ldots, 0)$. Thus, we have

$$\frac{\partial S_i}{\partial P_j} \approx \frac{\Delta S_i}{\Delta P_j} = \frac{S_i(0, 0, \ldots, \Delta P_j, \ldots, 0)}{\Delta P_j}. \qquad (8.3.11)$$

Equation (8.3.3) can be solved by a numerical method.

The management problem mentioned above is now transformed into a problem of linear programming as

$$\min Z = \sum_{i=1}^{m} c_i P_i \qquad (8.3.12a)$$

subject to constraints:

$$P_1 + P_2 + \cdots + P_m \geq D, \qquad (8.3.12b)$$

$$[\mathbf{A}]\{\mathbf{P}\} \leq \{\mathbf{b}\}, \qquad (8.3.12c)$$

$$P_1 \geq 0, \quad P_2 \geq 0, \ldots, \quad P_m \geq 0. \qquad (8.3.12d)$$

Let us now consider the problem in a deeper and more practical way. For example, let the unit cost of water supply in each well i be a linear function of its drawdown:

$$c_i = c_{i1} + c_{i2} s_i. \qquad (8.3.13)$$

Since the drawdown at well i can be expressed by

$$s_i = a_{i1} P_1 + a_{i2} P_2 + \cdots + a_{im} P_m,$$

where $a_{ij} = \partial s_i / \partial P_j$ are elements of the influence matrix $[\mathbf{A}]$, Eq. (8.3.13) can be rewritten as

$$c_i = c_{i1} + c_{i2} \sum_{j=1}^{m} a_{ij} P_j, \qquad (8.3.14)$$

and the objective function of Eq. (8.3.1) is then translated into

$$Z = \sum_{i=1}^{m} c_{i1} P_i + \sum_{i=1}^{m} \sum_{j=1}^{m} c_{i2} a_{ij} P_i P_j, \tag{8.3.15}$$

which is a quadratic function of decision variables $(P_1, P_2, \ldots, P_m)$. The management problem, which requires finding the minimum of Eq. (8.3.15) with constraints Eqs. (8.3.12b) to (8.3.12d), is a problem of quadratic programming.

A simple example of hydraulic management can be found in Bear (1979). Remson and Gorelick (1980) studied the optimal control problem of hydraulic gradients. Haidari (1982) and Willis (1983) demonstrated some practical applications of the influence matrix method. Willis and Finney (1985) considered the hydraulic management of phreatic aquifers and proposed a quasi-linearization solution.

8.3.2 Management of Groundwater Pollution Sources

The problem of groundwater quality management has been raised in recent years. For instance, in many cases, aquifers are both the sewage disposal sites and the water supply sources. How do we determine the time and quantity of waste water injection to keep the sources up to the water quality criteria? Also, in order to prevent the pollution of the sources, how do we assign the water quality of recharged waste water and, at the same time, minimize the cost of waste water treatment? To solve these problems we should combine the management model with a prediction model of water quality, and consider them as a problem of mathematical programming.

Gorelick and Remson (1982) extended the influence matrix method to study the pollution source management problems. They considered the simple one-dimensional problem shown in Figure 8.9. There were three ditches of water sources (A, B, C) and three ditches of sewage injection (I, II, III) in the region. The water from the left river recharges the aquifer. The solute

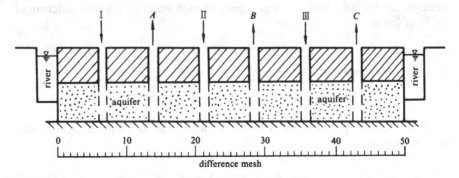

FIGURE 8.9. A sketch for the management of a one-dimensional pollution source and the finite difference meshes for simulation.

concentration in the river and the initial solute concentration in the aquifer were all assumed to be 100 mg/L. For simplicity, the groundwater velocity and the dispersion coefficient were assumed to be constants and unaffected by pumping and injection. When sewage is injected, the solute concentration in the aquifer will increase. The requirement of the pollution source management was to provide an optimal schedule for the sewage injection (the time and quantity of each injection ditch) so that the total injection reaches the maximum, and also, that the solute concentration in the source ditches does not exceed 250 mg/L at any time. In this problem, the management period was divided into three time intervals, each of which was 200 days, and it was required that the management scheme should be obtained for each time interval.

The governing equation of water quality for this problem is

$$\frac{\partial C}{\partial t} = D\frac{\partial^2 C}{\partial x^2} - V\frac{\partial C}{\partial x} + \frac{C'W}{bn}, \tag{8.3.16}$$

where C' is the concentration of the injected water, W the injection rate per unit area of aquifer, b the thickness of the aquifer, and n the effective porosity. The additional initial and boundary conditions are

$$C(x, 0) = C_0, \quad 0 < x < L; \tag{8.3.17a}$$

$$C(0, t) = C_0, \quad t > 0; \tag{8.3.17b}$$

$$\frac{\partial C}{\partial x} = 0, \quad x = L, \quad t > 0. \tag{8.3.17c}$$

Values of parameters used in the example are

$$L = 5000 \text{ m}, \quad V = 2 \text{ m/d};$$

$$D = 12 \text{ m}^2/\text{d}, \quad b = 10 \text{ m};$$

$$n = 0.2, \quad C_0 = 100 \text{ mg/L}.$$

This one-dimensional model can be solved by the FDM. The finite difference mesh of which is shown in Figure 8.9.

The injection rate, $C'W$, in the model is a decision variable to be determined. Assume that the rates in the three injection ditches at the three time intervals are $q_{1,I}, q_{2,I}, \ldots, q_{3,III}$, respectively, and are renumbered as $q_1, q_2, \ldots, q_9$ for convenience. The objective of management is to maximize the total injection, i.e.,

$$\max Z = \sum_{i=1}^{I} q_i, \tag{8.3.18}$$

where $I = 9$, which is equal to the number of injection ditches multiplied by the number of the management intervals. The constraints are that the concentrations in the source ditches do not exceed the specified value at any time. During the simulation, time step $\Delta t = 10$ day is taken, so there are 60 time steps in the total management period of 600 days. The concentrations in the three source ditches at the 60 time steps are numbered as $C_1, C_2, \ldots, C_J$, $J = 180$, which is equal to the number of the source ditches times the number of the time steps. These state variables depend on decision variables $q_1, q_2, \ldots, q_I$. The constraints stated above can be expressed as

$$C_j(q_1, q_2, \ldots, q_I) \leq C^*, \quad (j = 1, 2, \ldots, J), \tag{8.3.19}$$

where $C^* = 250$ mg/L. Subtracting the concentration for the case of no injection $C_j(0, 0, \ldots, 0) = C_0$ from both sides of the inequality (8.3.19), we then have

$$C_j(q_1, q_2, \ldots, q_I) - C_j(0, 0, \ldots, 0) \leq \bar{C}, \tag{8.3.20}$$

where $\bar{C} = C^* - C_0 = 150$ mg/L. The left-hand side of the inequality (8.3.20) may be approximated by the linear terms of Taylor expansion to yield

$$\frac{\partial C_j}{\partial q_1} q_1 + \frac{\partial C_j}{\partial q_2} q_2 + \cdots + \frac{\partial C_j}{\partial q_I} q_I \leq \bar{C}, \tag{8.3.21}$$

where $\partial C_j / \partial q_i$ is the sensitivity coefficient of concentration C_j with respect to injection rate q_i, and is also called the influence coefficient. By combining Eq. (8.3.18) with Eq. (8.3.21) and the nonnegative constraints on the decision variables, the water quality management model is finally formulated into a linear programming problem with objective function (8.3.18) and constraints:

$$[\mathbf{R}] \{\mathbf{q}\} \leq \{\bar{\mathbf{C}}\}, \tag{8.3.22a}$$

$$q_1 \geq 0, \quad q_2 \geq 0, \ldots, \quad q_I \geq 0, \tag{8.3.22b}$$

where $\{\mathbf{q}\} = (q_1, q_2, \ldots, q_I)^T$ is a vector of decision variables, and matrix

$$[\mathbf{R}] = \begin{bmatrix} \dfrac{\partial C_1}{\partial q_1} & \dfrac{\partial C_1}{\partial q_2} & \cdots & \dfrac{\partial C_1}{\partial q_I} \\[2mm] \dfrac{\partial C_2}{\partial q_1} & \dfrac{\partial C_2}{\partial q_2} & \cdots & \dfrac{\partial C_2}{\partial q_I} \\[2mm] \cdots & \cdots & \cdots & \cdots \\[2mm] \dfrac{\partial C_J}{\partial q_1} & \dfrac{\partial C_J}{\partial q_2} & \cdots & \dfrac{\partial C_J}{\partial q_I} \end{bmatrix} \tag{8.3.23}$$

is called the concentration influence matrix. Its elements are $r_{ji} = \partial C_j / \partial q_i$, which can be expressed by the difference approximation as

$$\frac{\partial C_j}{\partial q_i} \approx \frac{C_j(0, \ldots, 0, \Delta q_i, 0, \ldots, 0) - C_j(0, 0, \ldots, 0)}{\Delta q_i}, \tag{8.3.24}$$

TABLE 8.2. The optimal management decision.

Ditch No.	Optimal injection rates for each ditch (kg/d)		
	First interval (0–200 d)	Second interval (200–400 d)	Third interval (400–600 d)
I	$q_1 = 13.3$	$q_2 = 10.7$	$q_3 = 13.0$
II	$q_4 = 12.1$	$q_5 = 11.2$	$q_6 = 5.2$
III	$q_7 = 13.2$	$q_8 = 10.5$	$q_9 = 5.9$

FIGURE 8.10. The curves of concentration changes in water source ditches under the optimal decision.

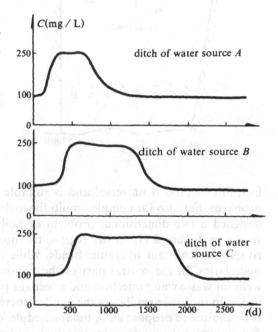

where $C_j(0,\ldots,0,\Delta q_i,0,\ldots,0)$ can be obtained by solving the simulation problem in Eq. (8.3.16) subject to the conditions of Eq. (8.3.17) and all the corresponding source and sink terms are assigned to be zero, except q_i. Thus, all the elements of matrix $[R]$ can be determined by only solving the simulation problem I times. The last step is to solve the linear program (8.3.22). The solution for the problem is listed in Table 8.2.

Figure 8.10 shows the obtained curves of solute concentrations versus time in the three source ditches by making an inference from this decision, i.e., taking these q_i as the source and sink terms to solve the prediction problem defined by Eqs. (8.3.16) and (8.3.17). We can see from the figure that the solute concentrations in the ditches do not exceed $C^* = 250$ mg/L at any time.

We have introduced the application of the influence matrix method to groundwater quality management through an example. As a matter of

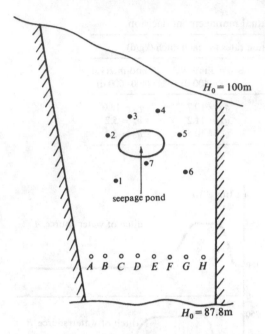

FIGURE 8.11. A sketch of two-dimensional pollution source management.
• = wells for injecting waste water;
○ = wells of water supply.

fact, this method is universal and is suitable not only for one-dimensional problems, but also for complex multidimensional problems. Gorelick (1982) designed a two-dimensional problem of pollution source management, as shown in Figure 8.11. In the region, the north and south boundaries are rivers with constant hydraulic heads, while the east and west are no-flow boundaries. In the central part of the region, there are seven controllable wells for waste water injection and a seepage pond which can dilute the waste water. In the lower reaches of the aquifer, there is a row of water supply wells. It is required to propose an optimal schedule for waste water injection, which can maximize the waste water injection, and meanwhile keep the concentrations of the contaminants in the water supply wells located at the lower reaches of the aquifer within the standard ($C^* \leq 250$ mg/L). We can still use the same procedure to derive the problem of linear programming in Eq. (8.3.22). Only the concentration influence matrix needs to be calculated with a program for two-dimensional water quality simulation. In Gorelick (1982), the method of characteristics was used to avoid numerical dispersion. In the paper, Gorelick also pointed out that when there are many time intervals of management, the number of decision variables will become very large. In that case, using the dual linear programming can save the computational effort and increase the stability of the solution.

8.3.3 Project Models of Groundwater Quality Management

When we design a plan for the development of a groundwater unit, many aspects should be considered. For instance, we may wish to select the locations of pumping wells and determine the pumping schedule that can minimize the drawdowns of the wells, while satisfying the water demand. Also, in order to reduce the cost of recharge and sewage treatment, we may wish to choose the locations of artificial recharge wells and the recharge schedules to minimize the pressure in the recharge wells. Obviously, designing a good plan before regional development will bring more benefit in the future. Management models for these purposes are more complicated than pollution management models mentioned before.

Assume that the whole management procedure is divided into several time intervals and it is required to determine the values of decision variables for each interval. For instance, in the kth interval, the decision variables of a project model may include: (1) extraction rate Q_i^k for node i; (2) injection rates $-Q_j^k$ for node j; (3) solute concentrations of injection wells $\hat{C}_j^k$. The state variables may consist of: (1) hydraulic heads h_i^k; and (2) concentrations C_i^k. During the simulation, one management interval needs to be divided into several time steps. The concentrations corresponding to different time steps in the kth interval are denoted by C_i^{kl}, where l indicates the lth time step.

The relationship between state variables and decision variables can be determined by the water flow and water quality models. Using FEM or FDM, the models may be expressed in the following discrete forms:

$$[E]h + [G]\frac{dh}{dt} + F = 0, \qquad (8.3.25)$$

$$[A]C + [B]\frac{dC}{dt} + H = 0, \qquad (8.3.26)$$

where h and C are vectors of state variables. Vectors of decision variables, Q and $\hat{C}$, are included in the source and sink vectors, F and H. In order to clearly express the dependence relationships between state and decision variables, Eqs. (8.3.25) and (8.3.26) in the kth management interval may be written as

$$P_k(h^k, Q^k) = 0, \qquad (8.3.27)$$

$$R_k(C^k, Q^k, \hat{C}^k) = 0. \qquad (8.3.28)$$

After the time derivatives in Eqs. (8.3.25) and (8.3.26) are approximated by finite difference, both P_k and R_k are algebraic operators.

The objectives of groundwater hydraulic and water quality management usually contain:

$$(1) \qquad Z_1 = \max_k \min_i h_i^k, \quad i \in \pi, \qquad (8.3.29)$$

where π is a set of pumping wells. This objective is to reduce the cost of pumping as much as possible, and meanwhile to prevent the harmful consequences, such as poor quality water intrusion and land subsidence, from occurring.

$$(2) \qquad Z_2 = \min_k \max_j h_j^k, \quad j \in \psi, \qquad (8.3.30)$$

where ψ is a set of injection wells. This objective is to minimize the cost of water injection.

$$(3) \qquad Z_3 = \max_k \min_j \hat{C}_j^k, \quad j \in \psi, \qquad (8.3.31)$$

This objective aims at saving the expense for the treatment of recharge water. The total objective may be a certain weighted combination of these objectives.

The constraints of this problem include:

1. The total extraction must meet the demand, i.e.,

$$\sum_{i \in \pi} Q_i^k = D^k, \qquad (8.3.32)$$

where D^k represents the water demand in the kth interval.

2. The total recharge water must reach the amount of recharge, i.e.,

$$\sum_{j \in \psi} Q_j^k = R^k, \qquad (8.3.33)$$

where R^k is the amount of waste water to be injected into the aquifer during the kth interval.

3. Extraction and injection should not surpass the capability of the equipment, i.e.,

$$Q_i^k \le Q_i^*, \quad Q_j^k \le Q_j^*, \qquad (8.3.34)$$

$$(i \in \pi) \qquad (j \in \psi)$$

where Q_i^* and Q_j^* are the maximum capabilities of pumping and injection, respectively.

4. Water quality standard in the extraction wells should be satisfied, i.e.,

$$C_i^{k_1} \le C^*, \quad i \in \pi, \qquad (8.3.35)$$

where C^* is the given standard of water quality.

Eqs. (8.3.29) to (8.3.35) compose a multi-objective programming problem. It can be divided into two correlated subproblems. If hydraulic management is the major objective, we can first solve the subproblem of water quantity defined by Eqs. (8.3.29) and (8.3.30) in conjunction with the constraints (8.3.32) to (8.3.34). Its solution provides the scheme of water quantity allocation for each of the management intervals. Then, take the scheme as the source and sink term to obtain the flow velocity distribution. Finally, solve the management subproblem of water quality as defined in Eq. (8.3.31) with the additional constraints (8.3.35). Willis (1979) gave a hypothetical example to illustrate the use of the above method in building models for groundwater hydraulic management and quality management.

8.3.4 Conjunctive Use and Basin Water Resources Planning

For a developing arid and semi-arid basin, water resources planning problem is a major concern. Since local groundwater cannot satisfy the growing water demand, conjunctive use of groundwater, surface water, imported water and reclaimed water must be considered. To find the optimal water resources management decision thus becomes a very challenging topic. For this kind of management problem, decision variables may include:

- Extraction locations and rates in different years and seasons;
- Recharge locations and rates in different years and seasons;
- Amounts of imported water from different sources in different years and seasons;
- Amounts of using reclaimed water in different years and seasons;
- Amounts of using and recharging surface water bodies in different years and seasons.

Constraints may include:

- Groundwater level and quality constraints;
- Imported water quantity and quality constraints;
- Reclaimed water quantity and quality constraints;
- Surface water quantity and quality constraints;
- Physical constraints (capacities of wells and pipe lines).

Management objectives may include:

- Minimizing the total operation cost;
- Maximizing the quantity of supply water for basin development;
- Maintaining or improving the quality of supply water;
- Protecting the basin groundwater.

Obviously, there are conflicts among these objectives. It is impossible to achieve them simultaneously. We can only find a compromise solution based on the multiobjective decision making theory (Chankong and Haimes, 1983). Yeh et al. (1992) accomplished a multiobjective water resources management model for the upper Santa Ana basins, Southern California. In this project, minimizing the total operation cost, minimizing the total mass of TDS and minimizing the total mass of nitrogen were considered as management objectives. An integrated program was developed which included a two-dimensional groundwater flow and mass transport model for the saturated zone, a flow and mass transport model for the unsaturated zone, a river flow and mass transport model, a planning model and a non-linear programming model. The planning model was used to form objective functions, set constraints, and generate trade-off relations for any pairs of objectives.

In Yeh et al. (1995), a similar basin water resources planning problem was considered for the Hemet basin, Riverside, California. The purpose of this project was to answer if existing data in the basin was sufficient for constructing a reliable model for solving the proposed management problem. In the planning model, two recharge locations were designed for storing imported water and reclaimed water in wet seasons, and blocking the high TDS water flowing into the basin through its boundary. The optimal solution of the management model can provide a recharge schedule for the planning period and determine the amounts of recharge water coming from different sources.

8.3.5 *Remediation of Polluted Aquifers*

The self-purification of polluted aquifers usually requires a very long time. To speed up purification, we can pump out the polluted water and inject clean water, or use some chemical or biological methods to remediate the polluted aquifers. Optimal remediation design includes two contradictory objectives: one is to achieve the optimal remediation results, and the other is to minimize the cost. Therefore, we may have the following management problems:

(1) Under the condition of a fixed remediation budget (remediation capability), look for a plan (decision) that can produce the best remediation results.

(2) Under the condition that certain remediation results should be achieved, seek the minimum cost alternative.

The mathematical statements of the two problems are given below. Assume that the domain of the aquifer (Ω) is divided into M elements. Let

$$Z_1 = \sum_{m=1}^{M} W_m \theta_m V_m C_m, \qquad (8.3.36)$$

where W_m is the weighting coefficient, θ_m the porosity, V_m the volume of element m, and C_m the contaminant concentration of element m after the remediation is finished. If all $W_m = 1$, Z_1 will be the total content of contaminants in the domain (Ω) when the remediation is ended. The cost of operating this remedial plan can be expressed as

$$Z_2 = \sum_{i=1}^{I} \alpha_i q_i^+ + \sum_{j=1}^{J} \beta_j q_j^-, \qquad (8.3.37)$$

where I and J are the numbers of the potential injection and extraction wells, respectively, q_i^+ the injection rate of injection well i, α_i the cost of injecting a unit volume of water, q_j^- the pumping rate of extraction well j, and β_j is the cost of pumping a unit volume of water. If the cost of well construction is neglected, Z_2 may be used to represent the total cost of the remediation. When the decision variables of the remediation problem are expressed by vectors:

$$\mathbf{q}^+ = (q_1^+, q_2^+, \dots, q_I^+) \quad \text{and} \quad \mathbf{q}^- = (q_1^-, q_2^-, \dots, q_J^-), \qquad (8.3.38)$$

the problem of minimizing the remaining contaminant subject to a cost constraint may be stated as

$$\min Z_1(\mathbf{q}^+, \mathbf{q}^-) \qquad (8.3.39)$$

subject to constraints:

$$Z_2(\mathbf{q}^+, \mathbf{q}^-) \leq \bar{B}, \tag{8.3.40}$$

$$0 \leq \mathbf{q}^+ \leq \mathbf{Q}^+, \quad 0 \leq \mathbf{q}^- \leq \mathbf{Q}^-, \tag{8.3.41}$$

where $\bar{B}$ is the limited remediation budget, $\mathbf{Q}^+$ and $\mathbf{Q}^-$ are the upper limits of $\mathbf{q}^+$ and $\mathbf{q}^-$, respectively, which are determined by the capabilities of the equipment and limitations of hydraulic conditions.

The alternative formulation of minimizing the cost subject to residual contaminant standards may be stated as

$$\min Z_2(\mathbf{q}^+, \mathbf{q}^-) \tag{8.3.42}$$

subject to constraints:

$$Z_1(\mathbf{q}^+, \mathbf{q}^-) \leq \bar{C}, \tag{8.3.43}$$

$$0 \leq \mathbf{q}^+ \leq \mathbf{Q}^+, \quad 0 \leq \mathbf{q}^- \leq \mathbf{Q}^-, \tag{8.3.44}$$

where $\bar{C}$ is the maximum allowable contaminant concentration in the whole aquifer when the remediation is completed. Equation (8.3.43) may also be transformed into

$$C_k(\mathbf{q}^+, \mathbf{q}^-) \leq \bar{C}_k, \quad (k = 1, 2, \ldots, K), \tag{8.3.45}$$

where C_k is the concentration of the kth subdomain and $\bar{C}_k$ is the given standard of water quality in the subdomain.

Both formulations for the optimal remediation of aquifers are classified as nonlinear programming problems. $Z_1(\mathbf{q}^+, \mathbf{q}^-)$ and C_k in Eq. (8.3.45) can be computed by an established water quality model. During the process of solving the nonlinear programming, $\dfrac{\partial Z_1}{\partial \mathbf{q}^+}$ and $\dfrac{\partial Z_1}{\partial \mathbf{q}^-}$ $\left(\text{or } \dfrac{\partial C_k}{\partial \mathbf{q}^+}, \dfrac{\partial C_k}{\partial \mathbf{q}^-} \right)$ also need to be calculated, which can be obtained with the aid of the water quality model and using finite difference or variational methods (see Section 7.3.2).

Gorelick et al. (1984) combined a contaminant transport model with a non-linear programming technique to solve the aquifer reclamation problem. Ahlfeld et al. (1988a, b) gave a detailed introduction to the numerical solutions for the above two types of problems, as well as some examples.

8.3.6 The Reliability of Management Models

The solution of a management model depends on the established water flow and water quality models. In the simulation models, however, there are always uncertainties associated with model structure and model parameters. We should estimate the effects of these uncertainties on the optimal management decision so as to determine the reliability of the management model.

The Monte-Carlo method provides the most general way for solving this kind of problem. According to the known statistic characteristics of input parameters, the computer can generate many realizations of the input parameters. By solving the management problem for each realization, a great number of realizations of management decisions are obtained, then the expected value and the variance of the optimal decision can be calculated, and consequently, its confidence interval can be estimated (see Section 7.4.4).

Another approach is to formulate the management problem into a stochastic, rather than deterministic problem. Wagner and Gorelick (1987) modified deterministic constraint (8.3.43) into the following stochastic constraint:

$$\text{Prob}\{Z_1 \leq \bar{C}\} \geq \pi, \tag{8.3.46}$$

that is, when the model parameters are uncertain, the probability of the variable Z_1 satisfying constraints $Z_1 \leq \bar{C}$ must not be smaller than π, where $0 < \pi < 1$ is the preset reliability level. If Z_1 is normally distributed, Eq. (8.3.46) can be rewritten as

$$\frac{\bar{C} - m[Z_1]}{\sigma[Z_1]} \geq F^{-1}(\pi), \tag{8.3.47}$$

where $m[Z_1]$ and $\sigma[Z_1]$ are the mathematical expectation and standard deviation of Z_1, respectively, and $F^{-1}(\pi)$ is the value of the standard normal cumulative distribution corresponding to the reliability level π. After Eq. (8.3.47) is used to replace Eq. (8.3.43), the previous management problem can be expressed as

$$\min Z_2(\mathbf{q}^+, \mathbf{q}^-) \tag{8.3.48}$$

subject to constraints:

$$m[Z_1] + F^{-1}(\pi)\sigma[Z_1] \leq \bar{C}, \tag{8.3.49}$$

and

$$0 \leq \mathbf{q}^+ \leq \mathbf{Q}^+, \quad 0 \leq \mathbf{q}^- \leq \mathbf{Q}^-. \tag{8.3.50}$$

The first term on the left-hand side of Eq. (8.3.49) is the deterministic part of the constraint, while the second term is an additional demand created by the uncertainty of the simulation model. When there is no uncertainty on Z_1, the latter is equal to zero. Otherwise, when the standard deviation of Z_1 becomes larger, or a higher reliability level is required (π becomes larger), the value of the term will be larger and the constraints will be more difficult to satisfy. $m[Z_1]$ and $\sigma[Z_1]$ in Eq. (8.3.49) can be obtained by solving the water quality equation with stochastic parameters (see Section 7.4). Once $m[Z_1]$ and $\sigma[Z_1]$ are obtained, the method for solving the stochastic management problem will be the same as that of the deterministic one.

Kaunas and Haimes (1985), and Massmann and Freeze (1987a, b) considered the effects of inaccurate dispersion coefficients on the management decisions from the point of view of a multi-objective management, and estimated the risk of the decisions. Wagner and Gorelick (1989) further considered the solution of water quality management and the requirement for observation data when the transmissivity is regarded as a random field. Andricevic and Kitanidis (1990) used the stochastic dual control method to improve the estimation of parameters during the remediation of aquifer pollution, and thereby, raising the reliability of management.

Using the concept of Management Equivalence Identifiability presented by Sun and Yeh (1990b), we can find another way to estimate the uncertainty of the optimal management decision. In the Hemet project (Yeh and Sun, 1994) mentioned in Section 8.3.4, we found that the uncertainties in hydraulic and dispersion parameters only cause a small uncertainty in the management solution (the relative error of the total operation cost is less then 1%), while the uncertainties in boundary inflow, boundary outflow and precipitation infiltration may cause significant uncertainty in the management solution (the relative error of the total operation cost may be as large as 28%). An efficient data collection strategy was then suggested for improving the reliability of the management model.

8.3.7 Economic and Political Models in Groundwater Management

Models of this kind generally relate to a very large scale and a good many variables. The considerations with regard to economy, policies, and even the law, are directly represented by the models. For instance, in order to develop the agricultural economy in a region, what policies shall we adopt to manage the water resources (including the water rate, water law, water allocation, and so on) so as to achieve the best agricultural economic development, and at the same time, maintain the environment? According to the review by Gorelick (1983), such models can be grouped into three types: the hydraulic economic response models, the models of groundwater simulation linked with economy management and the multilevel management models. Detailed discussions of these models are beyond the scope of this book, so only a very brief introduction is given below in this section.

Hydraulic Economic Response Models

Hydraulic economic response models are the direct extension of groundwater hydraulic management models, which include the agricultural economic management and surface water allocation. For instance, the following objective function may be presented:

$$Z = \sum_{i=1}^{n} (c_S^i Q_S^i + c_P^i Q_P^i), \qquad (8.3.51)$$

where i is the number of the management intervals, c_S^i and c_P^i the costs of surface water and groundwater per ton at the ith interval, respectively, and Q_S^i and Q_P^i are the consumptions of surface water and groundwater during the ith interval, respectively. The management objective is to minimize the total cost of Z. The constraints include: total water demand must be satisfied, the drawdowns of groundwater should not exceed a certain level, and the water extracted from the reservoir must be limited within a certain amount.

In the agricultural economic hydraulic management model considered by Morel–Seytoux et al. (1980), the irrigation water comes from both the pumping of groundwater and the diversion of surface water. The goal of the model is to minimize the cost of the total water consumption, and meanwhile, guarantee the water demands of various crops and soil conditions. In the model, if we assume that the yields of the crops depend on the irrigation level and the pumping cost is dependent on the drawdown, then the objective function becomes a quadratic function with respect to the decision variables, because the drawdown can be expressed as a linear function of pumping rates using the influence matrix method mentioned before. Thus, the management problem becomes a quadratic programming problem.

The water demands of agriculture are stochastic variables because of the stochastic nature of precipitation. Consequently, various stochastic management models were proposed. Maddock (1974), and Emigdio and Flores (1976) studied stochastic management models for river-aquifer systems, and the latter also gave an example of groundwater basin management in Mexico.

Maddock and Haimes (1975) studied a regional groundwater allocation-charges system by using the influence matrix method associated with agricultural economy. The quota of water supply for the farm is determined by the model, overuse of water will be punished, and saving on water will be rewarded. The management problem is reduced to a quadratic programming problem with the goal of maximizing the net income of the farm. The decision variables are the pumping rates, while the relations between drawdown and pumping rate are determined by the influence matrix.

Study of groundwater management in combination with surface water has started in China. Han (1985) proposed a management model for a subsurface water reservoir in Shimenzhai, with the purpose of supplying as much water as possible from both the surface and the subsurface reservoirs. The model is reduced to a linear programming problem and its calculation results provide an optimal operation scheme for both the subsurface and the surface reservoirs. The General Observation Station of the Geological Bureau of Shaanxi Province and Shandong University (1990) accomplished a research on groundwater management for Xi'an City, Shaanxi Province. The research contained two contradictory goals: maximize the water supply and minimize the land subsidence.

Groundwater Simulation-Economic Management Models

The management goals of combined groundwater simulation-economic management models directly reflect economic aspects. Models of this kind can be used to maximize the net income of agricultural economy in a farm, a county or a groundwater basin. In these models, the kinds of crops planted, or the areas sown are taken as the decision variables, from which the allocations of both surface water and groundwater are determined. The simulation model is not used to form constraints for the economic model, but the two models are connected. The procedure is to estimate the pumping cost for each subregion based on the current hydraulic conditions, and solve the economic management model to get a pumping scheme. The pumping scheme is then inputted into the simulation model to re-estimate the pumping cost of the next management period. With this method, we can use the linear programming to solve complex economic models, or even to include the social and legal factors in the management model in order to determine the optimal allocation of water resources. In addition, because the simulation model is not included in the constraints, the programming problem will not become nonlinear.

This type of model has been used in practice. Young and Bredehoeft (1972) considered the problem of using groundwater resources to gain maximum agricultural net income. Bredehoeft and Young (1983) made further considerations that the management goals should not only be to maximize the net annual income, but also to minimize the variance of the income in order to guarantee reliable income. The development of an optimal groundwater policy can provide an stable and timely supply of water to agriculture and reduce the variance of income.

Multilevel Management Models

Currently, the development and applications of water resources management models are very wide in scope, covering not only hydraulic and environmental aspects, but also economic, political and legal concerns. As a result, water resources management is really a large system problem. It usually has numerous objectives. The optimal management scheme is the result of trade-offs among the objectives according to certain principles that reflect the preferences of policy-makers. For example, a compromise must be made between economic development and resources conservation, or between the short-term benefits and the long-term development. On the other hand, policy-makers may be organized in multilevels, from the junior supervisors of individual units to the middle and senior policy-makers holding responsibilities. As a result, the decision and execution of the optimal scheme is inevitably hierarchical, and may not be carried out in the most mathematically efficient way. In this case, the problem may need to be solved several times by a trial and error method until nearly optimal plan is found.

Haimes and his co-partners have done a lot of work and made great contributions to the development of this field in the past two decades. They not only developed the mathematical theories and methods of multi-objective decisions and hierarchical decisions, but also applied them to the management of water resources. These contents will not be discussed in this book and those interested may refer to the books written by Haimes et al. (1975), Haimes (1977), Haimes and Tarvainen (1981), Chankong and Haimes (1983). Friedman et al. (1984) also gave an overall review on models of this kind.

Exercises

8.1. Use a flow chart to describe the procedure of building a groundwater quality model.

8.2. Formulate a one-dimensional model to simulate mass transport in a saturated-unsaturated zone along the vertical direction. Then, design a flow chart to describe its solution procedure when the FDM is used.

8.3. From the point of view of two-phase flow, derive the governing equations for mass transport in the unsaturated zone.

8.4. Use a numerical method to solve the problem of mass transport in a single fracture presented in Section 3.2.4.

8.5. Design a water quality management problem which may be met in practice.

8.6. Assuming that the remediation cost is proportional to the pumping and the recharging rates q^+ and q^- in Eq. (8.3.42), use the influence matrix method to transform the remediation problem defined by Eqs. (8.3.42) to (8.3.44) into a mathematical programming problem that can be solved iteratively by Linear Programming.

Conclusions

This book gives a systematic discussion on the theories and methods of contaminant transport in porous media. Emphases are placed on how to construct mathematical models and use numerical solutions. Model calibration and model applications are also discussed in certain depth.

To date, however, there are still many open problems in this area. The author would like to propose the following research subjects to readers who wish to further study the mathematical modeling of groundwater quality.

1. Probe into the mechanism of hydrodynamic dispersion through laboratory and field experiments, and look for deep understanding of dispersion phenomena in different scales. Thereby, provide theoretical and experimental bases for improving the existing mathematical models.
2. Study adsorption, ion exchange, and biochemical reactions of different components in different soils and determine relative parameters.
3. Accurately determine groundwater velocities. One problem is how to measure them directly in the field. Another problem is how to accurately determine the hydraulic conductivity K and effective porosity n, then use Darcy's law to indirectly calculate them. Since indirect calculation is indispensable in the prediction of water quality changes in a transient flow field, it is also important to raise the accuracy of calculating gradient field from the head distribution.
4. Develop efficient numerical solutions for three-dimensional models, because only three-dimensional models can properly describe real cases. Develop numerical methods for solving the problems of multicomponent transport in multiphase flow. Establish convenient and applicable software that can solve very large scale problems with the aid of parallel algorithms.
5. Develop applicable numerical methods that can decrease both numerical dispersion and overshoot for complicated three-dimensional dispersion problems.
6. Present new concepts and methods for the parameter identification of advection-dispersion equations. Quantitatively estimate the model uncertainty, and study the methods of model validation.

7. Problems of mass transport in fractured aquifers have not been well solved yet. This kind of problem is of practical significance, although it is quite difficult.
8. Find the explanation and determine the range of the "scale effect" phenomenon.
9. Develop the theory and applications of stochastic models, including the solution of both forward and inverse problems.
10. Study the computer aided experimental design methods for parameter identification, for model prediction, and for management, because the reliability of model applications is mainly determined by the quantity and quality of observation data.

Appendix A
The Related Parameters in Modeling Mass Transport in Porous Media

Name	Notation	Dimension	Definition
Fluid density	ρ	ML^{-3}	Fluid mass per unit volume, which depends on pressure and temperature
Fluid gravity rate (specific weight)	γ	$ML^{-2}T^{-2}$	Fluid weight per unit volume, $\gamma = \rho g$, where g is acceleration of gravity
Fluid specific gravity	δ		Ratio of fluid density to pure water density at $4°C$
Hydrodynamic viscosity	μ	$ML^{-1}T^{-1}$	Ratio of shear stress per unit area to velocity gradient, mainly depending on temperature
Compression coefficient of fluid	β	$M^{-1}LT^2$	$\beta = -\dfrac{1}{\rho}\dfrac{d\rho}{dp}$, where p is the pressure
Molecular diffusion coefficient in fluid	D_d	L^2T^{-1}	Ratio between diffusion flux and concentration gradient
Grain size	d	L	In the method of hydrometer analysis, the grain size is defined as the diameter of a spheroidal grain which has the same decline velocity of water with the grain
Porosity	n		$n = \dfrac{U_U}{U}$, where U_U is the pore volume of REV, and U the total volume of REV
Specific surface	M	L^{-1}	$M = \dfrac{A_s}{U}$, A_s is the total surface area of REV, U the total volume of REV
Volumetric compression coefficient	α	$M^{-1}LT^2$	$\alpha = -\dfrac{1}{U}\dfrac{dU}{dp}$, where U is the volume, p is the pressure

Name	Notation	Dimension	Definition
Permeability (tensor)	$\mathbf{k}$	L^2	$\mathbf{k} = \mathbf{K}\mu/\rho g$, where $\mathbf{K}$ is the hydraulic conductivity
Hydraulic conductivity (tensor)	$\mathbf{K}$	LT^{-1}	The ratio between Darcy's velocity and hydraulic gradient. It is a tensor in an anisotropic medium.
Tortuosity (tensor)	$\mathbf{T}$		Proportionality constant between external force component F_i (on abscissa x_i) acted on the fluid volume of a point in porous space and F_j^δ which is the projected component of this force along the flow line of the point, $F_i T_{ij} = F_j^\delta$, T_{ij} are components of $\mathbf{T}$
Transmissivity of confined aquifer	T	$L^2 T^{-1}$	$T = Kb$, where b is the thickness of aquifer
Storage coefficient	S		The volume of water released from aquifer storage in unit horizontal area when the mean piezometric head on the perpendicular line of aquifer drops by one unit
Specific storage coefficient	S_s	L^{-1}	The volume of water released from aquifer storage of unit volume when the water head drops by a unit, therefore $S = S_s b$, $S_s = \rho g(\alpha + n\beta)$
Longitudinal dispersivity of isotropic porous medium	α_L	L	Proportionality constant of longitudinal dispersion coefficient (aligned with the flow direction) to average velocity
Transverse dispersivity of isotropic porous medium	α_T	L	Proportionality constant of transverse dispersion coefficient (perpendicular to the flow direction) to average velocity
Mechanical dispersion coefficient (tensor)	$\mathbf{D}'$	$L^2 T^{-1}$	$D_{ij}' = \alpha_T V \delta_{ij} + (\alpha_L - \alpha_T) V_i V_j / V$, where D_{ij}' is a component of $\mathbf{D}'$; $\delta_{ij} = 1$, when $i = j$; $\delta_{ij} = 0$, when $i \neq j$; V_i and V_j are velocity components, V the average velocity
Molecular diffusion coefficient (tensor) in porous medium	$\mathbf{D}''$	$L^2 T^{-1}$	$\mathbf{D}'' = D_d \mathbf{T}$
Hydrodynamic diffusion coefficient (tensor)	$\mathbf{D}$	$L^2 T^{-1}$	$\mathbf{D} = \mathbf{D}' + \mathbf{D}''$
Mean porous velocity	$\mathbf{V}$	LT^{-1}	Macroscopic mean value of microvelocities in REV of porous medium
Darcy's velocity	$\mathbf{q}$	LT^{-1}	$\mathbf{q} = n\mathbf{V}$

Name	Notation	Dimension	Definition
Fluid pressure in porous medium	p	$ML^{-1}T^2$	Macroscopic mean value of microscopic fluid pressure at REV of porous medium
Total hydraulic head	h	L	$h = Z + \dfrac{p}{\gamma}$, where Z is potential head, $\dfrac{p}{\gamma}$ is the piezometric head
Concentration of solute	C	ML^{-3}	Solute mass contained in per unit volumetric solution

Appendix B
A FORTRAN Program for Solving Coupled Groundwater Flow and Contaminant Transport Problems

The attached FORTRAN program, **MCB2D.FOR**, can be used to simulate two-dimensional groundwater flow and mass transport in confined or unconfined aquifers. It is useful not only for learning this book, but also for solving practical problems.

B.1 Features and Assumptions

The program has the following features:

1. Triangular elements are used to discrete the flow region (see Section 5.1.2).
2. Discrete equations are derived by the multiple cell mass balance method (see Section 5.2). Both local and global mass conservations can be maintained.
3. When solving the advection-dispersion equation, appropriate upstream weights are added adaptively according to the value of local Peclet number (see Section 6.2.3).
4. Only non-zero elements in the coefficient matrices of the discrete equations are stored (see Section 5.4.2). The required storage space is minimized.
5. Boundary conditions, infiltrations, pumping and recharge rates and water quality of recharged water may vary with time.
6. A modified iteration method, the selected-node iteration method (see Section 5.4.2), is used to solve the discrete equations. The iteration procedure is usually convergent faster than other iteration methods.
7. The groundwater flow equation and groundwater quality equation are solved by the same subroutines. The construction of the program is compact.
8. The program is of good universality. To solve different problems, generally, we only need to change data files without any change in the program.

9. It is easy to be modified and extended. For instance, it can become the Galerkin FEM, mass concentrated FEM or integrated FDM programs by only changing a few constants (see Section 5.2.3).

10. The program can combine with an inverse solution program to solve parameter identification problems (Sun, 1994), or combine with an optimization program to solve groundwater management problems.

The program is based on the following assumptions:

1. When solving a groundwater flow problem, the aquifer may be either isotropic or anisotropic. When solving a groundwater quality problem, however, the aquifer is assumed to be isotropic. The longitudinal and transverse dispersivities are used in the program (see Section 2.5.2.)

2. Both Darcy's law and Fick's law are valid.

3. Dispersion coefficients are proportional to velocity.

4. The flow region is saturated (confined or unconfined) and the flow direction is basically horizontal.

5. The fluid density ρ is approximately a constant and the changes of solute concentrations will not affect the flow field, i.e., the transport belongs to the tracer case (see Section 7.1.1).

6. The temperature of the groundwater is nearly constant.

7. The adsorption can be described by the linear equilibrium isotherm relationship. The retardation coefficient is used (see Section 2.6.2).

B.2 Program Structure

The program consists of a main program and seventeen subroutines:

1. Main program—allocates the memory space and calls subroutines **DATAG** and **MCB2D**.

2. Subroutine **DATAI**—reads initial conditions.

3. Subroutine **DATAB**—reads boundary conditions.

4. Subroutine **DATAG**—reads geometric information.

5. Subroutine **DATAW**—reads pumping and recharge locations and rates.

6. Subroutine **DATAS**—reads rates and quality of infiltration water.

7. Subroutine **DATAP**—reads parameters, such as hydraulic conductivity, storage coefficient, porosity, longitudinal and transverse dispersivities, retardation factor, and so on.

8. Subroutine **DATAO**—reads observation locations and times.

9. Subroutine **DATAC**—reads all computation control numbers, such as relaxation factors, convergence criteria, weighting coefficients, initial size of time step and so on.

10. Subroutine **FDB**—forms boundary conditions for each time step.

11. Subroutine **FDW**—forms pumping and recharge data for each time step.

12. Subroutine **FDS**—forms sink/source data for each time step.

13. Subroutine **COEFF**—forms coefficient matrices of discretized water flow and mass transport equations.
14. Subroutine **FVD**—calculates velocity distributions and hydrodynamic dispersion coefficients from hydraulic head distributions.
15. Subroutine **WEIT**—generates upstream weighted coefficients for all nodes.
16. Subroutine **SNIM**—solves the discretization equations of water flow and mass transport equations to obtain the distributions of hydraulic head and solute concentration.
17. Subroutine **MCB2D**—calls above subroutines to form a complete solution procedure.
18. Subroutine **FOB**—calculates predicted head and concentration values for designated observation locations and times.

B.3 Input Data Files

The user is required to prepare the following input data files:

1. **DATAN**—all control integers.
2. **DATAG**—geometric data.
3. **DATAI**—initial conditions.
4. **DATAB**—boundary conditions.
5. **DATAW**—pumping and recharge data.
6. **DATAS**—quantity and quality of infiltration and return water.
7. **DATAP**—model parameters.
8. **DATAO**—observation locations and times.
9. **DATAC**—computation control numbers.

Control integers in the file **DATAN** include:

NP	the total number of nodes.
NE	the total number of elements.
NB	the maximum number of neighboring nodes around a node.
NZON	the total number of zones for distributed parameters.
IHC	setting IHC = 1 when only solving groundwater flow problems; setting IHC = 2 when solving coupled flow and mass transport problems.
IST	setting IST = 0 for a transient flow filed; setting IST = 1 for a steady flow field.
IPR	the output control number.
NW	the total number of extraction and injection wells.
NTW	the total number of time periods during which the extraction and injection rates are constant.
NHB	the total number of given head boundary nodes.

NCB   the total number of given concentration boundary nodes.
NTB   the total number of times for which the boundary values are changed.
NS   the total number of zones for a distributed sink/source.
NTS   the total number of time periods during which the distributed sinks/sources are constant.
NOBW   the total number of observation wells.
NOBT   the total number of observation times.
ISO   setting ISO = 0 for isotropic aquifers, setting ISO = 1 for anisotropic aquifers when only solving groundwater flow problems.

Input data in the file **DATAG** include:

X   the x coordinate of each node.
Y   the y coordinate of each node.
TR   the top-rock elevation of each node. For an unconfined aquifer, it is the elevation of ground surface. When the head is higher than TR at a node, the aquifer is considered to be confined at that location. Otherwise, be unconfined.
BR   the bed-rock elevation of each node.
IJM   the node numbers of each element.
NPZON   the zone number of each element for distributed parameters.
NSZON   the zone number of each element for distributed sinks/sources.

Input data in the file **DATAI** include:

H00   the initial head of each node.
C00   the initial concentration of each node.

Input data in the file **DATAB** include:

TIME   the time at which the values of given boundary head and given boundary concentration are measured.
NODE   the node number of a boundary node.
HEAD   the measured value of head at a boundary node.
CONCEN   the measured value of concentration at a boundary node.

Note that the boundary values of an arbitrary time used in the program are calculated by linear interpolation. Thus, the first value of TIME must be 0 (initial time), and the boundary values must coincide with their initial values (as given in the file DATAI). The final value of TIME must be larger than or equal to the total simulation period.

Input data in the file **DATAW** include:

TIME   the time at which the extraction and injection rates or the concentration of injected water change their values.
ELEMENT   the element number where a well is located.

X   the x coordinate of a well.
Y   the y coordinate of a well.
RATE   the extraction or injection rate of a well during a time period.
                The value is positive for injection and negative for extraction.
CONC   the concentration of injected water of a well during a time
                period. It is unnecessary to give such a value for an extraction
                well.

Note that the RATE and CONC used in the program for an arbitrary time
t are equal to their values at TIME(k), when TIME(k − 1) < t ≤ TIME(k).
Thus, the final value of TIME must be larger than or equal to the total
simulation period.

Input data in the file **DATAS** include:

TIME   the time at which the rate and/or concentration of a distributed
        sink/source change their values.
RATE   the volume of water entering to groundwater through per unit area
        of aquifer. The value is positive for a source and negative for a sink.
CONC   the concentration of source water. It is unnecessary to give such a
        number for a sink.

Note that the RATE and CONC used in the program for an arbitrary time
t are equal to their values at TIME(k), when TIME(k − 1) < t ≤ TIME(k).
Thus, the final value of TIME must be larger than or equal to the total
simulation period.

Input data in the file **DATAP** include:

ZTK   the hydraulic conductivity of each zone.
ZSE   the storage coefficient of each zone.
ZPR   the porosity of each zone.
ZAL   the longitudinal dispersivity of each zone. When solving a ground-
        water flow problem in an anisotropic aquifer, it is used to represent
        the hydraulic conductivity in the x direction, K_{xx}.
ZAT   the transverse dispersivity of each zone. When solving a groundwater
        flow problem in an anisotropic aquifer, it is used to represent the
        hydraulic conductivity in the y direction, K_{yy}.
ZAD   the linear equilibrium isotherm adsorption factor of each zone de-
        fined by CM = ZAD*CF, where CM is the solute concentration
        in solid matrix and CF the concentration in fluid. When solving a
        groundwater flow problem in an anisotropic aquifer, it is used to
        represent the cross hydraulic conductivity, K_{xy}.
RDC   the radioactive decay coefficient for the whole region.
DFU   the molecular diffusion coefficient in porous media.

Input data in the file **DATAO** include:

ELEMENT   the element number where an observation well is located.
X   the x coordinate of an observation well.
Y   the y coordinate of an observation well.
TIME   the observation times.

Note that the values of head and concentration at observation wells are calculated from their nodal values by the finite element interpolation method. The final value of observation TIME should be equal to the total simulation period.

Input data in the file **DATAC** include:

DTI   the size of the initial time step.
DTF   the factor of increasing the size of time step.
DTL   the maximal size of time step.
ALF   the relaxation coefficient in the iteration solution.
EPS   the criterion of convergency.
PT   the upstream weighting coefficient.

Note that DTF, DTL, ALF, and EPS may have different values when solving the flow subproblem and the mass transport subproblem. Usually, it is unnecessary to change the values in this file for solving different problems. When the user finds that there are oscillations around a concentration front, he or she may try to increase the value of PT. Decrease the value of DTL may cause an increase of computation time.

In order to help the user to form these input data files, an auxiliary program **FINPUT.FOR** is attached. After all control integers are designated in the file **DATAN**, formats of all other input data files can be generated automatically by running this program. The user then can follow the prompts given in these fills to fill in all problem related data. The decimal location is denoted for each data. The user may incorporate a mesh generation subroutine into the program to form node coordinates and node numbers of each element.

B.4 Output Data Files

After running the program **MCB2D.FOR**, three output files will be generated:

1. **RESUL**—the computational results of the program.
2. **DRAWC**—data for drawing contours.
3. **DRAWT**—data for drawing time-head (time-concentration) curves.

The output of the program is controlled by integer IPR in the data file **DATAN**. First, we can set IPR = 0 to check all input data. Then set IPR = 1 to check the discretization pattern. After that, we can set IPR = 2 to print the distributed solutions, or set IPR = 3 for only printing the solutions corresponding to the designated observation locations and observation times.

In any case, files **DRAWC** and **DRAWT** will be automatically generated after running the program. The former can be used to draw head and/or concentration contours for all designated observation times, the latter can be used to draw time-head and/or time-concentration curves for all designated observation wells.

B.5 Floppy Disk and Demo Problem

A floppy disk which contains the source programs **MCB2D.FOR** and **FINPUT.FOR**, a demo problem with all input data files, and instructions of use, can be obtained from the author. Readers of this book can use the programs for teaching and learning purposes. It is not allowed, however, to use the programs for business purposes or for developing new software for business uses.

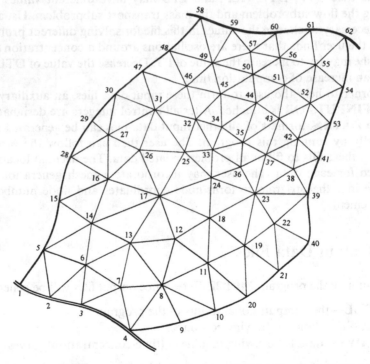

FIGURE B.1. Aquifer configuration and finite element discretization of the demo problem.

The demo problem is designed to simulate a groundwater remediation procedure. A confined aquifer between two rivers (Figure B.1) was polluted by the upstream river. In order to protect the downstream river, we must remove the pollution from the upstream river and clean up the aquifer. After the water in the upstream river is clean, we can pump out the polluted water from the aquifer. In this situation, an efficient remediation project may be found with the help of a model.

Before using **MCB2D** to solve a problem designed by the user, it is better to do some exercises to familiar with the input and output of the program. In each exercise, the user is required to change a few input data, run the program and observe the simulation results. The following exercises are suggested:

1. Use the data given in file **DATAG** to draw the geometry of the aquifer and the zonation pattern. Use the data given in file **DATAB** to find the boundary nodes and boundary conditions. Use the data given in file **DATAO** to find the locations of observation wells and observation times. Use the data given in file **DATAW** to find the locations and rates of pumping wells.
2. Switch IHC, IST and IPR in file **DATAN** to different values (IHC = 1, 2; IST = 0, 1; IPR = 0, 1, 2, 3) and observe corresponding results.
3. Answer what is the minimum rate that the maximum concentration in all observation wells will be less than 200 (mg/L) after 1000 days? (the dimension system is assumed to be meter, gram, and day.)
4. Move the pumping well to different locations and observe corresponding results.
5. Assuming that the water level in the upstream river decreases from 105m to 100m in the first 10 days and then keeps the same level, change file **DATAB** to simulate this situation.
6. Change the values of parameters in file **DATAP** and observe corresponding results.
7. Assuming that the extracted water will be treated on the ground and then recharged into the aquifer by a well, the concentration of injection water will be 100 (mg/L) and the recharge procedure will begin at $t = 50$ days, modify the file **DATAW** to simulate this procedure and observe corresponding results.

All input data files and simulation results of the demo problem are given below.

File: DATAN

```
CONTROL INTEGERS
      NP          NE          NB        NZON        IHC        IST        IPR
      62          97           9           3           2          0          3

CONTROL INTEGERS
      NW         NTW         NHB         NCB        NTB         NS        NTS
       2           2           9           5          2          2          2

CONTROL INTEGERS
    NOBW        NOBT         ISO
      11           5           0
```

File: DATAG

NODE COORDINATES, ELEVATIONS OF TOP AND BED ROCKS

NODE	X COORD.00	Y COORD.00	TR.00	BR.00
1	0.	390.	50.	0.
2	150.	330.	50.	0.
3	270.	240.	50.	0.
4	490.	100.	50.	0.
5	100.	540.	50.	0.
6	310.	440.	50.	0.
7	470.	310.	50.	0.
8	680.	290.	50.	0.
9	740.	100.	50.	0.
10	930.	150.	50.	0.
11	930.	380.	50.	0.
12	790.	540.	50.	0.
13	550.	520.	50.	0.
14	360.	640.	50.	0.
15	220.	770.	50.	0.
16	470.	860.	50.	0.
17	720.	750.	50.	0.
18	970.	610.	50.	0.
19	1120.	440.	50.	0.
20	1090.	200.	50.	0.
21	1270.	300.	50.	0.
22	1280.	540.	50.	0.
23	1210.	700.	50.	0.
24	1000.	810.	50.	0.
25	870.	960.	50.	0.
26	660.	990.	50.	0.
27	500.	1090.	50.	0.
28	310.	1000.	50.	0.
29	440.	1160.	50.	0.
30	570.	1280.	50.	0.
31	760.	1180.	50.	0.
32	950.	1110.	50.	0.
33	1030.	930.	50.	0.
34	1090.	1050.	50.	0.
35	1180.	960.	50.	0.
36	1150.	820.	50.	0.
37	1290.	890.	50.	0.
38	1410.	800.	50.	0.
39	1480.	660.	50.	0.
40	1450.	440.	50.	0.
41	1550.	840.	50.	0.
42	1450.	960.	50.	0.
43	1380.	1070.	50.	0.
44	1280.	1160.	50.	0.
45	1100.	1230.	50.	0.
46	930.	1290.	50.	0.
47	730.	1370.	50.	0.
48	890.	1470.	50.	0.
49	1070.	1380.	50.	0.
50	1270.	1310.	50.	0.
51	1440.	1230.	50.	0.
52	1540.	1130.	50.	0.
53	1640.	1010.	50.	0.
54	1710.	1180.	50.	0.
55	1580.	1320.	50.	0.

56	1430.	1360.	50.	0.
57	1270.	1410.	50.	0.
58	1070.	1600.	50.	0.
59	1200.	1550.	50.	0.
60	1400.	1510.	50.	0.
61	1590.	1500.	50.	0.
62	1780.	1420.	50.	0.

NODE NUMEBERS OF EACH LEEMENT AND ZONATION PATTERNS

ELEMENT	I	J	K	NPZON	NSZON
1	1	2	5	1	1
2	2	6	5	1	1
3	2	3	6	1	1
4	3	7	6	1	1
5	3	4	7	1	1
6	4	8	7	1	1
7	4	9	8	1	1
8	9	10	8	1	1
9	8	10	11	1	1
10	8	11	12	1	1
11	8	12	13	1	1
12	7	8	13	1	1
13	7	13	6	1	1
14	6	13	14	1	1
15	5	6	14	1	1
16	5	14	15	1	1
17	15	14	16	1	1
18	16	14	17	1	1
19	14	13	17	1	1
20	13	12	17	1	1
21	17	12	18	1	1
22	12	11	18	1	1
23	11	19	18	1	1
24	11	20	19	1	1
25	10	20	11	1	1
26	20	21	19	1	1
27	19	21	22	2	2
28	21	40	22	2	2
29	19	22	23	2	2
30	18	19	23	2	2
31	18	23	24	2	2
32	17	18	24	2	2
33	17	24	25	2	2
34	17	25	26	2	2
35	16	17	26	2	2
36	16	26	27	2	2
37	15	16	28	1	1
38	28	16	27	2	2
39	28	27	29	2	2
40	29	27	30	2	2
41	27	30	31	2	2
42	27	26	31	2	2
43	26	25	31	2	2
44	25	32	31	2	2
45	25	33	32	2	2
46	25	24	33	2	2
47	24	36	33	2	2
48	24	23	36	2	2
49	23	37	36	2	2
50	33	36	35	2	2
51	36	37	35	2	2

52	23	38	37	2	2
53	23	39	38	2	2
54	23	22	39	2	2
55	22	40	39	2	2
56	38	39	41	2	2
57	38	41	42	2	2
58	37	38	42	2	2
59	37	42	43	2	2
60	35	37	43	2	2
61	35	43	44	2	2
62	34	35	44	2	2
63	34	44	45	2	2
64	33	35	34	2	2
65	33	34	32	2	2
66	32	34	45	2	2
67	32	45	46	2	2
68	32	31	46	2	2
69	31	46	47	2	2
70	30	31	47	2	2
71	47	46	48	3	2
72	46	49	48	3	2
73	46	45	49	3	2
74	45	50	49	3	2
75	45	44	50	3	2
76	50	44	51	3	2
77	44	43	51	3	2
78	43	52	51	3	2
79	43	42	52	3	2
80	42	53	52	3	2
81	42	41	53	3	2
82	52	53	54	3	2
83	52	54	55	3	2
84	51	52	55	3	2
85	56	51	55	3	2
86	50	51	56	3	2
87	50	56	57	3	2
88	49	50	57	3	2
89	49	57	59	3	2
90	49	59	58	3	2
91	48	49	58	3	2
92	57	60	59	3	2
93	57	56	60	3	2
94	56	61	60	3	2
95	56	55	61	3	2
96	55	62	61	3	2
97	55	54	62	3	2

//

File: DATAI

INTIAL CONDITIONS OF HEAD AND CONCENTRATION

NODE	H00.00	C00.00
1	100.00	0.00
2	100.00	0.00
3	100.00	0.01
4	100.00	0.00
5	100.94	0.00
6	101.06	0.01
7	100.97	0.06
8	101.74	0.00
9	101.69	0.00
10	102.41	0.00
11	102.74	0.02
12	102.70	0.06
13	102.04	0.44
14	102.17	0.18
15	102.38	0.18
16	102.35	0.12
17	103.03	2.71
18	103.30	0.38
19	103.30	0.23
20	102.89	0.01
21	103.36	0.08
22	103.55	1.57
23	103.67	3.98
24	103.57	3.50
25	103.60	31.26
26	103.35	16.82
27	103.33	1.35
28	103.03	0.71
29	103.35	47.06
30	103.65	166.42
31	103.79	122.14
32	103.96	207.43
33	103.78	33.68
34	104.04	190.09
35	104.00	95.36
36	103.76	18.61
37	104.01	56.66
38	103.98	44.98
39	103.82	14.59
40	103.57	1.06
41	104.18	175.43
42	104.28	200.97
43	104.38	290.30
44	104.44	477.43
45	104.38	672.01
46	104.25	794.40
47	104.03	726.68
48	104.40	1542.65
49	104.55	1385.72
50	104.63	1258.84
51	104.61	840.76
52	104.54	613.75
53	104.44	455.10
54	104.67	1147.75
55	104.79	1359.42

```
        56          104.78                1520.98
        57          104.78                1859.78
        58          105.00                2000.00
        59          105.00                2000.00
        60          105.00                2000.00
        61          105.00                2000.00
        62          105.00                2000.00
//
```

File: DATAB

**** GIVEN HEAD BOUNDARY CONDITION ****

THE NUMBER OF TIME STEP IS: 1
              MEASURE TIME.00
                      0.00
        NUMBER   NODE   HEAD.00
           1   1   100.
           2   2   100.
           3   3   100.
           4   4   100.
           5   58   105.
           6   59   105.
           7   60   105.
           8   61   105.
           9   62   105.

THE NUMBER OF TIME STEP IS: 2
              MEASURE TIME.00
                   2000.00
        NUMBER   NODE   HEAD.00
           1   1   100.
           2   2   100.
           3   3   100.
           4   4   100.
           5   58   105.
           6   59   105.
           7   60   105.
           8   61   105.
           9   62   105.

**** GIVEN CONCENTRATION BOUNDARY CONDITION ****

THE NUMBER OF TIME STEP IS: 1
              MEASURE TIME.00
                      0.00
        NUMBER   NODE   CONCEN.00
           1   58   0.
           2   59   0.
           3   60   0.
           4   61   0.
           5   62   0.

THE NUMBER OF TIME STEP IS: 2
              MEASURE TIME.00
                   2000.00
        NUMBER   NODE   CONCEN.00
           1   58   0.
           2   59   0.
           3   60   0.
           4   61   0.
           5   62   0.

//

File: DATAP

```
**** INITIAL VALUES OF PARAMETERS ****
ZONE     ZTK.       ZSE.       ZPR.     ZAL.      ZAT.      ZAD.
  1      10.        0.0001     0.2      30.       10.       0.2
  2      20.        0.0002     0.25     40.       15.       0.15
  3      30.        0.0003     0.15     60.       25.       0.1

THE RADIOACTIVE COEFFICIENT
     RDC.00000      DFU.00000
       0.            0.
//
```

File: DATAW

TIME STEPS OF EXTRACTION AND INJECTION
        STEP   TIME.00
          1   1000.
          2   2000.

LOCATIONS OF PUMPING AND RECHARGE WELLS
      NUMBER   IN ELEMENT   X COORD.0   Y COORD.0
          1   50   1150.   820.
          2   38   500.   1090.

PUMPING AND RECHARGE RATES, CONC. OF RECHARGE WATER

WELL NUMBER IS   1
        STEP   RATE.00   CONC.00
          1   -8000.   0.
          2   -6000.   0.

WELL NUMBER IS   2
        STEP   RATE.00   CONC.00
          1   -5000.   0.
          2   -2000.   0.
    //

File: DATAS

```
TIME STEPS FOR INPUTTING SINK/SOURCE RECORDS
     STEP          TIME.00
      1            1000.
      2            2000.

RATES AND CONC. OF SINK/SOURCE TERMS

THE NUMBER OF SINK/SOURCE ZONE          1
     STEP          RATE.00          CONC.00
      1            0.008              0.
      2            0.006              0.

THE NUMBER OF SINK/SOURCE ZONE          2
     STEP          RATE.00          CONC.00
      1            0.005              0.
      2            0.004              0.
//
```

File: DATAO

LOCATIONS OF OBSERVATION WELLS

NUMBER	IN ELEMENT	X COORD.0	Y COORD.0
1	6	490.	100.
2	7	680.	290.
3	9	930.	380.
4	22	970.	610.
5	31	1000.	810.
6	48	1150.	820.
7	49	1290.	890.
8	59	1380.	1070.
9	78	1440.	1230.
10	85	1430.	1360.
11	94	1400.	1510.

OBSERVATION TIMES

NUMBER	TIME.00
1	50.
2	100.
3	200.
4	500.
5	1000.

//

File: DATAC

```
COMPUTATIONAL PARAMETERS FOR FLOW PROBLEM
  DTI.000     DTF.00     DTL.00     ALF.00   EPS.0000
   0.001       1.25       10.0        1.1      0.0001

COMPUTATIONAL PARAMETERS FOR TRANSPORT PROBLEM
  PT.0000     DTF.00     DTL.00     ALF.00   EPS.0000
   0.001       1.25        1.0        1.1      0.0001
//
```

Output File: RESUL

***** Principal control integers *******

IHC= 2   IST= 0   IPR= 3

THE OBSERVATION TIME IS: 50.00

NUMBER	HEAD	CONCENTRATION
1	100.00	0.00
2	101.65	0.00
3	101.86	0.02
4	101.18	0.32
5	100.65	4.48
6	98.95	33.13
7	101.44	81.05
8	103.12	370.79
9	103.92	921.93
10	104.43	1257.03
11	105.00	0.00

THE OBSERVATION TIME IS: 100.00

NUMBER	HEAD	CONCENTRATION
1	100.00	0.00
2	101.65	0.00
3	101.86	0.02
4	101.18	0.24
5	100.65	5.85
6	98.95	48.74
7	101.44	110.35
8	103.12	448.94
9	103.92	947.60
10	104.43	1037.72
11	105.00	0.00

THE OBSERVATION TIME IS: 200.00

NUMBER	HEAD	CONCENTRATION
1	100.00	0.01
2	101.64	0.00
3	101.86	0.01
4	101.18	0.02
5	100.64	10.04
6	98.95	85.17
7	101.43	179.93
8	103.12	576.73
9	103.92	895.92
10	104.43	707.66
11	105.00	0.00

THE OBSERVATION TIME IS: 500.00

NUMBER	HEAD	CONCENTRATION
1	100.00	0.01
2	101.65	0.00
3	101.86	0.68
4	101.18	2.09
5	100.65	35.09
6	98.95	200.71
7	101.44	390.34
8	103.12	651.75
9	103.92	507.01
10	104.43	232.25
11	105.00	0.00

THE OBSERVATION TIME IS: 1000.00

NUMBER	HEAD	CONCENTRATION
1	100.00	0.02
2	102.08	0.11
3	102.65	0.18
4	102.38	10.64
5	102.08	90.14
6	100.75	254.65
7	102.57	453.83
8	103.77	300.89
9	104.31	123.67
10	104.64	41.98
11	105.00	0.00

B.6 Source Programs

Program: **MCB2D.FOR**

```
C
C
C*********************************************************************
C
C                          Main program
C
C*********************************************************************
C
      DIMENSION A(80000), N(10000), CHA(70)
      IMPLICIT REAL*8 (A-H, O-Z)
      MA=80000
      MN=10000
C
      OPEN(1,FILE='DATAN')
      OPEN(2,FILE='DATAG')
      OPEN(3,FILE='DATAI')
      OPEN(4,FILE='DATAB')
      OPEN(5,FILE='DATAP')
      OPEN(6,FILE='RESUL')
      OPEN(7,FILE='DATAW')
      OPEN(8,FILE='DATAS')
      OPEN(9,FILE='DATAO')
      OPEN(10,FILE='DATAC')
      OPEN(11,FILE='DRAWC')
      OPEN(12,FILE='DRAWT')
C
      WRITE (*, 101)
      WRITE (*, 102)
      WRITE (*, 103)
      WRITE (*, 104)
      WRITE (*, 105)
      WRITE (*, 106)
      WRITE (*, 107)
      WRITE (*, 108)
      WRITE (*, 109)
      WRITE (*, 110)
C
  101 FORMAT (/////////)
  102 FORMAT (15X,'********************************************************')
  103 FORMAT (15X,'*                                                      *')
  104 FORMAT (15X,'*         PROGRAM: MCB2D      version 2.0              *')
  105 FORMAT (15X,'*  For simulating coupled 2-D groundwater flow         *')
  106 FORMAT (15X,'*  and mass transport in saturated zones               *')
  107 FORMAT (15X,'*                                                      *')
  108 FORMAT (15X,'*              By  Prof. Ne-Zheng Sun                   *')
  109 format (15x,'*                                                      *')
  110 FORMAT (15X,'********************************************************')
C
      WRITE (*, 111)
  111 FORMAT (/,5X, 'Please wait.....',
     &          /,5X, 'The program is reading control integers')
C
C Reading control integers
C
      CALL JUMP(1, 2)
      READ (1, 2) CHA
```

```
    2   FORMAT (70A1)
        READ (1, 4) NP, NE, NB, NZON, IHC, IST, IPR
    4   FORMAT (7I10)
        IF ( IPR .LT. 1) THEN
           WRITE (6, 2) CHA
           WRITE (6, 4) NP, NE, NB, NZON, IHC, IST, IPR
        ENDIF
        IF (IHC .EQ. 2 ) ISO=0
C
        CALL JUMP(1, 2)
        READ (1, 2) CHA
        READ (1, 4)  NW, NTW, NHB, NCB, NTB, NS, NTS
        IF ( IPR .LT. 1) THEN
           WRITE (6, 2) CHA
           WRITE (6, 4) NW, NTW, NHB, NCB, NTB, NS, NTS
        ENDIF
C
        CALL JUMP(1, 2)
        READ (1, 2) CHA
        READ (1, 6) NOBW, NOBT, ISO
    6   FORMAT (3I10)
        IF ( IPR .LT. 1) THEN
           WRITE (6, 2) CHA
           WRITE (6, 6) NOBW, NOBT, ISO
        ENDIF
C
C
C Allocating storage spaces
C
        NP1=NP+1
        IG1=1
        JG1=1
C
C X, Y, TR, BR, TH, PS, IJM, MA, MCPN, NPZON, NSZON
C
        IG2=IG1+NP
        IG3=IG2+NP
        IG4=IG3+NP
        IG5=IG4+NP
        IG6=IG5+NP
        IB1=IG6+NP
        JG2=JG1+3*NE
        JG3=JG2+NP1
        JG4=JG3+NB*NP
        JG5=JG4+NE
        JB1=JG5+NE
C
C Reading and processing geometric and geologic information
C
        CALL DATAG (A(IG1),A(IG2),A(IG3),A(IG4),A(IG5),A(IG6),N(JG1),
       *            N(JG2),N(JG3),N(JG4),N(JG5),NP,NP1,NE,NB,IH,IPR)
C
C Continuously allocating storage spaces
C
C
C H00, C00, HB1, CB1, STEPB, NHBP, NCBP
C
        IB2=IB1+NP
        IB3=IB2+NP
        IB4=IB3+NHB*NTB
        IB5=IB4+NCB*NTB
```

```
      IS1=IB5+NTB
      JB2=JB1+NHB
      JE1=JB2+NCB
C
C WRAT, WCON, WLX, WLY, LEW, STEPW, SRAT, SCON, STEPS,
C DWR, DWC, DSR, DSC
C
      IS2=IS1+NW*NTW
      IS3=IS2+NW*NTW
      IS4=IS3+NW
      IS5=IS4+NW
      IS6=IS5+NW
      IS7=IS6+NTW
      IS8=IS7+NS*NTS
      IS9=IS8+NS*NTS
      IS10=IS9+NTS
      IS11=IS10+NP
      IS12=IS11+NP
      IS13=IS12+NP
      IP1=IS13+NP
C
C ZTK, ZSE, ZPR, ZAL, ZAT, ZAD, DTK, DSE, DPR, DAL, DAT, DAD,
C PSE, PPR, PAD
C
      IP2=IP1+NZON
      IP3=IP2+NZON
      IP4=IP3+NZON
      IP5=IP4+NZON
      IP6=IP5+NZON
      IP7=IP6+NZON
      IP8=IP7+NE
      IP9=IP8+NE
      IP10=IP9+NE
      IP11=IP10+NE
      IP12=IP11+NE
      IP13=IP12+NE
      IP14=IP13+NP
      IP15=IP14+NP
      IH1=IP15+NP
C
C H0, H1, H2, C0, C1, C2, HF00, HF0, CF00, CF0
C
      IH2=IH1+NP
      IH3=IH2+NP
      IH4=IH3+NP
      IH5=IH4+NP
      IH6=IH5+NP
      IH7=IH6+NP
      IH8=IH7+NP
      IH9=IH8+NP
      IH10=IH9+NP
      IE1=IH10+NP
C
C DXX, DXY, DYY, VX, VY, CFA, CFB, IE, IEE
C
      IE2=IE1+NE
      IE3=IE2+NE
      IE4=IE3+NE
      IE5=IE4+NE
      IE6=IE5+NE
      IE7=IE6+IH
```

```
         IO1=IE7+IH
         JE2=JE1+NP
         JD1=JE2+NP
C
C HOB, COB, OBWX, OBWY, LEOB, OBT
C
         IO2=IO1+NOBW*NOBT
         IO3=IO2+NOBW*NOBT
         IO4=IO3+NOBW
         IO5=IO4+NOBW
         IO6=IO5+NOBW
         ID1=IO6+NOBT
C
C Calculating the required total storage space
C
         NTOTAL=ID1+JD1
         IF (IPR .LT. 1) THEN
            WRITE  (6,20) NTOTAL
   20    FORMAT (//,1X,'THE REQUIRED TOTAL STORAGE SPACE IS:',I15,//)
         ENDIF
         IF(ID1 .GT. MA .OR. JD1 .GT. MN) THEN
         WRITE (*, 30)
   30    FORMAT (/,5X, 'The decleared dimensions of arrays A and N are exc
        &eeded')
         STOP
         END IF
C
         WRITE (*, 40)
   40    FORMAT (/,5X, 'The program is running')
         WRITE (6, 50)
   50    FORMAT (/,15X,'*****  Principal control integers *******')
         WRITE (6, 60) IHC, IST, IPR
   60    FORMAT (/,10X,'IHC=',I2,10X,'IST=',I2,10X,'IPR=',I2)
C
C Solving the forward problem
C
         CALL MCB2D (A(IG1),A(IG2),A(IG3),A(IG4),A(IG5),A(IG6),N(JG1),
        1           N(JG2),N(JG3),N(JG4),N(JG5),
        2           A(IB1),A(IB2),A(IB3),A(IB4),A(IB5),
        3           N(JB1),N(JB2),
        4           A(IS1),A(IS2),A(IS3),A(IS4),A(IS5),A(IS6),A(IS7),
        5           A(IS8),A(IS9),A(IS10),A(IS11),A(IS12),A(IS13),
        6           A(IP1),A(IP2),A(IP3),A(IP4),A(IP5),A(IP6),A(IP7),
        7           A(IP8),A(IP9),A(IP10),A(IP11),A(IP12),A(IP13),A(IP14),
        8           A(IP15),A(IH1),A(IH2),A(IH3),A(IH4),A(IH5),
        1           A(IH6),A(IH7),A(IH8),A(IH9),A(IH10),
        2           A(IE1),A(IE2),A(IE3),A(IE4),A(IE5),A(IE6),
        3           A(IE7),N(JE1),N(JE2),
        4           A(IO1),A(IO2),A(IO3),A(IO4),A(IO5),A(IO6),
        5           NP,NP1,NE,NB,NZON,IH,IHC,IST,ISO,IPR,
        6           NW,NTW,NHB,NCB,NTB,NS,NTS,NOBW,NOBT)
C
C
C
         CLOSE(1)
         CLOSE(2)
         CLOSE(3)
         CLOSE(4)
         CLOSE(5)
         CLOSE(6)
         CLOSE(7)
```

```
        CLOSE(8)
        CLOSE(9)
        CLOSE(10)
        CLOSE(11)
        CLOSE(12)
C
        STOP
        END
C
        SUBROUTINE JUMP (J,I)
        DO K=1,I
        READ (J, *)
        END DO
        RETURN
        END
C
C
C**********************************************************************
C
C     Subroutine MCB2D solves coupled flow and mass transport problems
C
C**********************************************************************
C
        SUBROUTINE MCB2D (X,Y,TR,BR,TH,PS,IJM,MA,MCPN,NPZON,NSZON,
     1                    H00,C00,HB1,CB1,STEPB,NHBP,NCBP,
     2                    WRAT,WCON,WLX,WLY,LEW,STEPW,SRAT,SCON,STEPS,
     3                    DWR,DWC,DSR,DSC,
     4                    ZTK,ZSE,ZPR,ZAL,ZAT,ZAD,
     5                    DTK,DSE,DPR,DAL,DAT,DAD,PSE,PPR,PAD,
     6                    H0,H1,H2,C0,C1,C2,HF00,HF0,CF00,CF0,
     7                    DXX,DXY,DYY,VX,VY,CFA,CFB,IE,IEE,
     8                    HOB,COB,OBWX,OBWY,LEOB,OBT,
     1                    NP,NP1,NE,NB,NZON,IH,IHC,IST,ISO,IPR,
     2                    NW,NTW,NHB,NCB,NTB,NS,NTS,NOBW,NOBT)
C
        IMPLICIT REAL*8 (A-H, O-Z)
        DIMENSION  X(NP),Y(NP),TR(NP),BR(NP),TH(NP),PS(NP),IJM(3,NE),
     1             MA(NP1),MCPN(NB,NP),NPZON(NE),NSZON(NE),
     2             H00(NP),C00(NP),HB1(NHB,NTB),CB1(NCB,NTB),STEPB(NTB),
     3             NHBP(NHB),NCBP(NCB),
     4             WRAT(NW,NTW),WCON(NW,NTW),WLX(NW),WLY(NW),LEW(NW),
     5             STEPW(NW),SRAT(NS,NTS),SCON(NS,NTS),STEPS(NTS),
     6             DWR(NP),DWC(NP),DSR(NP),DSC(NP),
     7             ZTK(NZON),ZSE(NZON),ZPR(NZON),ZAL(NZON),ZAT(NZON),
     8             ZAD(NZON),DTK(NE),DSE(NE),DPR(NE),DAL(NE),DAT(NE),
     1             DAD(NE),PSE(NP),PPR(NP),PAD(NP),
     2             H0(NP),H1(NP),H2(NP),C0(NP),C1(NP),C2(NP),
     3             HF00(NP),HF0(NP),CF00(NP),CF0(NP),CFA(IH),CFB(IH),
     4             VX(NE),VY(NE),DXX(NE),DXY(NE),DYY(NE),IE(NP),IEE(NP),
     5             HOB(NOBW,NOBT),COB(NOBW,NOBT),OBWX(NOBW),OBWY(NOBW),
     6             LEOB(NOBW),OBT(NOBT),DTF(2),DTL(2),ALF(2),EPS(2)
C
C Reading initial and boundary conditions, sink/source terms,
C parameters, observations and computation control parameters.
C
        WRITE (*, 101)
 101    FORMAT (/,5X, 'The program is reading input data')
C
        CALL DATAI (H00,C00,NP,IHC)
        CALL DATAB (HB1,CB1,NHBP,NCBP,STEPB,NHB,NCB,NTB,IHC)
        CALL DATAP (ZTK,ZSE,ZPR,ZAL,ZAT,ZAD,DTK,DSE,DPR,DAL,DAT,
```

```
       *                    DAD,NPZON,RDC,DFU,NE,NZON,IHC)
            CALL DATAW  (WRAT,WCON,WLX,WLY,LEW,STEPW,NW,NTW,IHC)
            CALL DATAS  (SRAT,SCON,STEPS,NS,NTS,IHC)
            CALL DATAO  (OBWX,OBWY,LEOB,OBT,NOBW,NOBT)
            CALL DATAC  (DTI,DTF,DTL,ALF,EPS,PT,IHC)
C
            WRITE (*, 102)  NOBT
  102       FORMAT (/, 5X, 'The total number of time periods is:', i5,/)
            IF (IST .EQ. 1) THEN
            WRITE (*, 103)
  103       FORMAT (/, 5X, 'Finding the steady state ....',/)
            END IF
C
C Setting initial conditions
C
            STEPW(NTW)=OBT(NOBT)+1.
            STEPS(NTS)=OBT(NOBT)+1.
            TSUM=0.
            DT00=DTI
            DT0=DTI
            DT=DTI
C
            DO N=1, NP
               HF00(N)=0.
               HF0(N)=0.
               H0(N)=H00(N)
               IF (IHC .EQ. 2) THEN
                  CF00(N)=0.
                  CF0(N)=0.
                  C0(N)=C00(N)
               ENDIF
            END DO
C
C Calculating the mass transport in transient flow field
C
            L=1
            IF (IST .EQ. 1) GOTO 45
  10        TSUM=TSUM + DT
            CALL FDB  (TSUM,HB1,CB1,H0,C0,STEPB,NHBP,NCBP,
       *              NP,NHB,NCB,NTB,IHC)
            CALL FDW  (TSUM,X,Y,IJM,WRAT,WCON,WLX,WLY,LEW,DWR,DWC,
       *              STEPW,NP,NE,NW,NTW,IHC)
            CALL FDS  (TSUM,X,Y,IJM,SRAT,SCON,DSR,DSC,STEPS,
       *              NSZON,NP,NE,NS,NTS,IHC)
C
C
C  Solving the flow problem
C
            CALL FVD  (X,Y,IJM,DTK,DPR,DAL,DAT,DAD,DFU,DXX,DXY,DYY,VX,VY,
       1              H0,NP,NE,1)
            CALL COEFF (X,Y,IJM,MA,MCPN,VX,VY,DXX,DXY,DYY,TR,BR,TH,H0,PS,
       1    DTK,DSE,DPR,DAD,PSE,PPR,PAD,CFA,CFB,PT,NP,NP1,NE,NB,ISO,IH,1)
            CALL SNIM  (MA,MCPN,IE,IEE,CFA,CFB,NHBP,H0,H1,H2,HF00,HF0,
       1               DWR,DWC,DSR,DSC,TH,PS,PPR,PAD,EPS,ALF,RDC,
       2               DT00,DT0,DT,NP,NP1,NB,NHB,IH,1)
C
C  Solving the mass transport problem
C
            IF (IHC .EQ. 2) THEN
C
            CALL FVD  (X,Y,IJM,DTK,DPR,DAL,DAT,DAD,DFU,DXX,DXY,DYY,VX,VY,
```

```
     1            H0,NP,NE,2)
      CALL COEFF (X,Y,IJM,MA,MCPN,VX,VY,DXX,DXY,DYY,TR,BR,TH,H0,PS,
     1   DTK,DSE,DPR,DAD,PSE,PPR,PAD,CFA,CFB,PT,NP,NP1,NE,NB,ISO,IH,2)
      CALL SNIM  (MA,MCPN,IE,IEE,CFA,CFB,NCBP,C0,C1,C2,CF00,CF0,
     1            DWR,DWC,DSR,DSC,TH,PS,PPR,PAD,EPS,ALF,RDC,
     2            DT00,DT0,DT,NP,NP1,NB,NCB,IH,2)
C
      DO N=1, NP
        IF ( C0(N) .LT. 0. ) THEN
           NA=MA(N+1)
           NC=MA(N)
           ND=NA-NC
           SS=0.
           DD=ND
           DO K=2, ND
              MC=MCPN(K,N)
              SS=SS+C0(MC)
           END DO
           C0(N)=SS/DD
           IF (C0(N) .LT. 0) C0(N)=0.
        END IF
      END DO
C
C
      ENDIF
C
C writing head and concentration distributions
C
      IF ( ABS(OBT(L)-TSUM) .LT. 1.E-5) THEN
C
         IF (IPR .LT. 3) THEN
         WRITE (6, 20) OBT(L)
  20     FORMAT (/,1X, ' THE OBSERVATION TIME IS:', F10.2,/)
         IF (IHC .EQ. 1) THEN
           WRITE (6, 22)
  22       FORMAT (/,6X,'NODE',11X, 'HEAD',/)
           DO N=1,NP
           WRITE (6,24) N, H0(N)
  24       FORMAT (I10, F15.2)
           END DO
         ELSE
           WRITE (6, 26)
  26       FORMAT (/,6X,'NODE',11X, 'HEAD',2X,'CONCENTRATION',/)
           DO N=1, NP
           WRITE (6, 28) N, H0(N), C0(N)
  28       FORMAT (I10, 2F15.2)
           END DO
         END IF
         END IF
C
C
C
      WRITE (11, 20) OBT(L)
      IF (IHC .EQ. 1) THEN
         WRITE (11, 30)
  30     FORMAT (/,9X,'X-COOR', 9X, 'Y-COOR', 16X, 'HEAD',/)
         DO N=1,NP
         WRITE (11,32) X(N), Y(N), H0(N)
  32     FORMAT (2F15.2, F20.3)
         END DO
      ELSE
```

```
          WRITE (11, 34)
 34       FORMAT (/,9X,'X-COOR',9X, 'Y-COOR', 16X, 'HEAD',
    * 7X,'CONCENTRATION',/)
          DO N=1, NP
            WRITE (11, 36) X(N), Y(N), H0(N), C0(N)
 36       FORMAT (2F15.2, 2F20.3)
          END DO
          END IF
C
C writing solutions corresponding to observation wells and times
C
          CALL FOB (X,Y,IJM,H0,C0,HOB,COB,OBWX,OBWY,LEOB,
    *             L,NP,NE,NOBW,NOBT,IHC)
          IF (IPR .LT. 4) THEN
C
            WRITE (6, 20) OBT(L)
          IF (IHC .EQ. 1) THEN
          WRITE (6, 22)
          DO N=1,NOBW
          WRITE (6,24) N, HOB(N,L)
          END DO
        ELSE
          WRITE (6, 38)
 38       FORMAT (/,4X,'NUMBER',11X, 'HEAD',2X,'CONCENTRATION',/)
          DO N=1, NOBW
          WRITE (6, 28) N, HOB(N,L), COB(N,L)
          END DO
          END IF
C
        END IF
C
C turning to the next time step
C
      WRITE (*, 110) L
110   FORMAT(/,3X,'The simulation of time period',i5,2x,'is completed')
C
        L=L+1
C
      END IF
C
      IF (L .GT. NOBT)  GOTO 90
      IF (DT .LT. DTL(1)) DT=DT*DTF(1)
      IF (DT+TSUM .GT. OBT(L)) DT=OBT(L)-TSUM
C
      GOTO 10
C
C generating the steady flow field
C
 45   CALL FDB (TSUM,HB1,CB1,H0,C0,STEPB,NHBP,NCBP,
    *           NP,NHB,NCB,NTB,IHC)
      CALL FDW (TSUM,X,Y,IJM,WRAT,WCON,WLX,WLY,LEW,DWR,DWC,
    *           STEPW,NP,NE,NW,NTW,IHC)
      CALL FDS (TSUM,X,Y,IJM,SRAT,SCON,DSR,DSC,STEPS,
    *           NSZON,NP,NE,NS,NTS,IHC)
C
      CALL FVD (X,Y,IJM,DTK,DPR,DAL,DAT,DAD,DFU,DXX,DXY,DYY,VX,VY,
    1           H0,NP,NE,1)
      CALL COEFF (X,Y,IJM,MA,MCPN,VX,VY,DXX,DXY,DYY,TR,BR,TH,H0,PS,
    1 DTK,DSE,DPR,DAD,PSE,PPR,PAD,CFA,CFB,PT,NP,NP1,NE,NB,ISO,IH,1)
C
 50   TSUM = TSUM + DT
```

```
      CALL SNIM   (MA,MCPN,IE,IEE,CFA,CFB,NHBP,H0,H1,H2,HF00,HF0,
     1              DWR,DWC,DSR,DSC,TH,PS,PPR,PAD,EPS,ALF,RDC,
     2              DT00,DT0,DT,NP,NP1,NB,NHB,IH,1)
C
C
C  Testing the steady condition
C
      DO N=1, NP
         DD=HF0(N)
         IF ( ABS(DD)/DT0 .GT. EPS(1) ) THEN
            IF (DT .LT. 10.) DT=DT*DTF(1)
            GOTO 50
         ENDIF
      END DO
C
      IF (IPR .LT. 4) THEN
         WRITE (6, 52)
   52 FORMAT (/,1X,'THE STEADY STATE HEAD DISTRIBUTION',/,6X,'NODE',
     *          11X, 'HEAD',/)
         DO  N=1, NP
            WRITE (6, 54) N, H0(N)
   54 FORMAT (I10, F15.2)
         END DO
      END IF
C
         WRITE (11, 56)
   56 FORMAT (/,1X,'THE STEADY STATE HEAD DISTRIBUTION',/)
         WRITE (11, 30)
         DO  N=1, NP
          WRITE (11, 32) X(N), Y(N), H0(N)
         END DO
C
C calculating mass transport in the steady flow field
C
      IF (IHC .EQ. 2) THEN
C
         CALL FVD (X,Y,IJM,DTK,DPR,DAL,DAT,DAD,DFU,DXX,DXY,DYY,VX,VY,
     1              H0,NP,NE,2)
         CALL COEFF (X,Y,IJM,MA,MCPN,VX,VY,DXX,DXY,DYY,TR,BR,TH,H0,PS,
     1      DTK,DSE,DPR,DAD,PSE,PPR,PAD,CFA,CFB,PT,NP,NP1,NE,NB,ISO,IH,2)
C
         L=1
         TSUM=0.
         DT00=DTI
         DT0=DTI
         DT=DTI
   60    TSUM=TSUM + DT
         CALL SNIM (MA,MCPN,IE,IEE,CFA,CFB,NCBP,C0,C1,C2,CF00,CF0,
     1              DWR,DWC,DSR,DSC,TH,PS,PPR,PAD,EPS,ALF,RDC,
     2              DT00,DT0,DT,NP,NP1,NB,NCB,IH,2)
C
      DO N=1, NP
         IF ( C0(N) .LT. 0. ) THEN
            NA=MA(N+1)
            NC=MA(N)
            ND=NA-NC
            SS=0.
            DD=ND
            DO K=2, ND
               MC=MCPN(K,N)
```

```
                  SS=SS+C0(MC)
              END DO
              C0(N)=SS/DD
              IF (C0(N) .LT. 0) C0(N)=0.
          END IF
      END DO
C
C
C

      IF ( ABS(OBT(L)-TSUM) .LT. 1.E-5) THEN
C
          IF (IPR .LT. 3) THEN
              WRITE (6, 20) OBT(L)
              WRITE (6, 62)
   62         FORMAT (/,6X, 'NODE', 2X, 'CONCENTRATION',/)
              DO N=1, NP
              WRITE (6, 24) N, C0(N)
              END DO
          END IF
C
              WRITE (11, 20) OBT(L)
              WRITE (11, 64)
   64         FORMAT (/,9X, 'X-COOR',9X,'Y-COOR', 7X, 'CONCENTRATION',/)
              DO N=1, NP
              WRITE (11, 32) X(N), Y(N), C0(N)
              END DO
C
C writing solutions corresponding to observation wells and times
C
          CALL FOB (X,Y,IJM,H0,C0,HOB,COB,OBWX,OBWY,LEOB,
     *              L,NP,NE,NOBW,NOBT,IHC)
          IF (IPR .LT. 4) THEN
              WRITE (6, 20) OBT(L)
              WRITE (6, 62)
              DO N=1, NOBW
              WRITE (6, 24) N, COB(N,L)
              END DO
          ENDIF
C
C turning to the next time period
C
      WRITE (*, 110) L
C
              L=L+1
C
          ENDIF
C
          IF (L .LE. NOBT) THEN
              IF (DT .LT. DTL(2)) DT=DT*DTF(2)
              IF (DT+TSUM .GT. OBT(L)) DT=OBT(L)-TSUM
              GOTO 60
          ENDIF
C
      ENDIF
C
C Form data file DRAWT
C
   90 DO N=1, NOBW
      WRITE (12, 92) N
   92 FORMAT (/, 2X, 'THE TIME-HEAD-CONCENTRATION FOR OBS. WELL', I3)
      WRITE (12, 94)
```

```
   94    FORMAT (/,11X, 'TIME', 16X, 'HEAD', 7X, 'CONCENTRATION',/)
         DO L=1, NOBT
         WRITE (12, 96) OBT(L), HOB(N,L), COB(N,L)
   96    FORMAT (F15.3, 2F20.3)
         END DO
         END DO
C
C
         RETURN
         END
C
C
C**********************************************************************
C
C
C           Subroutine DATAG reads and processes geometric and
C           geologic information
C
C**********************************************************************
C
         SUBROUTINE DATAG (X,Y,TR,BR,TH,PS,IJM,MA,MCPN,NPZON,NSZON,
        1                NP,NP1,NE,NB,IH,IPR)
         IMPLICIT REAL*8 (A-H, O-Z)
         DIMENSION   X(NP),Y(NP),TR(NP),BR(NP),TH(NP),PS(NP),
        1            IJM(3,NE),MA(NP1),MCPN(NB,NP),NPZON(NE),
        2            NSZON(NE),ME(3),CHB(70)
C
C   reading nodal coordinates
C
         CALL JUMP (2, 2)
         READ (2, 1) CHB
   1     FORMAT (70A1)
C
         DO   N=1, NP
           READ (2, 2) X(N), Y(N), TR(N), BR(N)
           TH(N)=TR(N)-BR(N)
   2     FORMAT (10X, 2F15.2, 2F10.2)
         END DO
         IF (IPR .LT. 1) THEN
           WRITE (6, 4) CHB
   4     FORMAT (//, 70A1, /)
           DO   N=1, NP
             WRITE(6, 6) N, X(N), Y(N), TR(N), BR(N)
   6     FORMAT (I10, 2F15.2, 2F10.2)
           END DO
         ENDIF
C
C   reading node numbers of each element and zonation patterns
C
         CALL JUMP (2, 2)
         READ (2, 1) CHB
C
         DO   M=1, NE
           READ (2, 12) IJM(1,M),IJM(2,M),IJM(3,M),NPZON(M),NSZON(M)
   12    FORMAT(10X, 5I10)
         END DO
         IF (IPR .LT. 1) THEN
           WRITE (6, 14) CHB
   14    FORMAT (/, 70A1,/)
           DO   M=1, NE
             WRITE (6, 16) M,IJM(1,M),IJM(2,M),IJM(3,M),NPZON(M),NSZON(M)
   16    FORMAT (6I10)
```

```
            END DO
         ENDIF
C
C   calculating arrays MCPN, MA and integer IH
C
         DO  I=1, NP
           MCPN(1,I)=I
           DO  J=2, NB
           MCPN(J,I)=0
           END DO
         END DO
         DO  I=1,NP1
           MA(I)=1
         END DO
         DO  N=1,NE
           DO  J=1,3
             ME(J)=IJM(J,N)
           END DO
           DO  L=1,3
             DO  K=2,3
               IBR=0
               M1=ME(1)
               NG=MA(M1+1)
               DO  I=1, NG
                 IF(ME(K).EQ.MCPN(I+1,M1))  IBR=1
               END DO
                 IF(IBR.EQ.0) THEN
                 MA(M1+1)=MA(M1+1)+1
                 NH=MA(M1+1)
                 MCPN(NH,M1)=ME(K)
                 ENDIF
               END DO
             J=ME(1)
             ME(1)=ME(2)
             ME(2)=ME(3)
             ME(3)=J
             END DO
         END DO
C
C writing arrays MCPN and MA
C
         IF (IPR .LT. 2) THEN
           WRITE  (6, 40)
  40     FORMAT (1X, 'MCPN'/)
           DO  J=1,NP
             WRITE (6, 42) (MCPN(I,J), I=1,NB)
  42     FORMAT (1X, 9I8)
           END DO
           WRITE  (6, 44) MA
  44     FORMAT (1X,'MA'/,(1X,10I5))
         ENDIF
C
C modifying array MA
C
         MA(1)=1
         DO  N=1,NP
           MA(N+1)=MA(N)+MA(N+1)
         END DO
         IH=MA(NP+1)
         IF (IPR .LT. 2) THEN
           WRITE  (6, 46) MA
```

```
   46   FORMAT (1X,'THE MODIFIED MA IS:'/,(1X,10I5))
        WRITE  (6, 48) IH
   48   FORMAT (1X,'THE LENTH OF COEFFICIENT MATRIX IS',I6)
        ENDIF
C
C calculating the area of each exclusive subdomain
C
        DO  I=1,NP
          PS(I)=0.
        END DO
        DO  N=1,NE
          DO J=1, 3
            ME(J)=IJM(J,N)
          END DO
        MI=ME(1)
        MJ=ME(2)
        MK=ME(3)
        BI=Y(MJ)-Y(MK)
        BJ=Y(MK)-Y(MI)
        CI=X(MK)-X(MJ)
        CJ=X(MI)-X(MK)
        TRS=ABS(CJ*BI-BJ*CI)/6.
        PS(MI)=PS(MI)+TRS
        PS(MJ)=PS(MJ)+TRS
        PS(MK)=PS(MK)+TRS
        END DO
        IF (IPR .LT. 2) THEN
          WRITE  (6, 56)
   56   FORMAT (1X,'THE AREA OF EXCLUSIVE SUBDOMAIN OF EACH NODE:',/)
          DO  N=1, NP
            WRITE (6, 58) N, PS(N)
   58   FORMAT (I10, F15.2)
          END DO
        ENDIF
  100   RETURN
        END
C
C***********************************************************************
C
C          Subroutine DATAI reads initial conditions
C
C***********************************************************************
C
        SUBROUTINE DATAI (H00,C00,NP,IHC)
        IMPLICIT REAL*8 (A-H, O-Z)
        DIMENSION  H00(NP),C00(NP)
C
C Reading initial conditions
C
        CALL JUMP (3, 3)
C
        DO  I=1, NP
          IF (IHC .EQ. 1) THEN
            READ (3, 10) H00(I)
          ELSE
            READ (3, 20) H00(I), C00(I)
   10   FORMAT (10X, F20.5)
   20   FORMAT (10X, 2F20.5)
          END IF
        END DO
C
```

```
       RETURN
       END
C
C************************************************************************
C
C          Subroutine DATAB reads boundary conditions
C
C************************************************************************
C
       SUBROUTINE DATAB (HB1,CB1,NHBP,NCBP,STEPB,NHB,NCB,NTB,IHC)
       IMPLICIT REAL*8 (A-H, O-Z)
       DIMENSION  HB1(NHB,NTB),CB1(NCB,NTB),NHBP(NHB),
      1           NCBP(NCB),STEPB(NTB)
C
C Reading boundary conditions of the flow problem
C
       IF (NHB .GT. 0) THEN
C
       CALL JUMP (4, 2)
C
       DO K=1, NTB
         CALL JUMP (4, 3)
         READ (4, 4) STEPB(K)
    4  FORMAT (10X, F15.2)
         CALL JUMP (4,1)
         DO  I=1, NHB
           READ (4, 10) NHBP(I), HB1(I,K)
   10  FORMAT (10x, I10, F20.5)
         END DO
       END DO
       ENDIF
C
C Reading boundary conditions of the transport problem
C
       IF (IHC .EQ. 2) THEN
C
       IF (NCB .GT. 0) THEN
C
       CALL JUMP (4, 2)
C
       DO K=1, NTB
         CALL JUMP (4, 3)
         READ (4, 4) STEPB(K)
         CALL JUMP (4,1)
         DO I=1, NCB
           READ (4, 20) NCBP(I), CB1(I,K)
   20  FORMAT (10x, I10, F20.5)
         END DO
       END DO
       ENDIF
       ENDIF
C
       RETURN
       END
C
C************************************************************************
C
C          Subroutine DATAP reads model parameters
C
C************************************************************************
C
```

```
C
      SUBROUTINE DATAP (ZTK,ZSE,ZPR,ZAL,ZAT,ZAD,DTK,DSE,DPR,DAL,DAT,
     *                  DAD,NPZON,RDC,DFU,NE,NZON,IHC)
      IMPLICIT REAL*8 (A-H, O-Z)
      DIMENSION  ZTK(NZON),ZSE(NZON),ZPR(NZON),ZAL(NZON),ZAT(NZON),
     *           ZAD(NZON),DTK(NE),DSE(NE),DPR(NE),DAL(NE),DAT(NE),
     *           DAD(NE),NPZON(NE)
C
C Reading parameters associated with each zone
C
      CALL JUMP (5, 3)
C
      DO  I=1,NZON
      READ (5, 20) ZTK(I),ZSE(I),ZPR(I),ZAL(I),ZAT(I),ZAD(I)
  20  FORMAT (5x, F15.5, 5F10.5)
      END DO
C
C Reading radioactive decay coefficient
C
      IF (IHC .EQ. 2) THEN
        CALL JUMP (5, 3)
        READ (5, 30) RDC, DFU
  30  FORMAT (2F15.6)
      END IF
C
C Generating distributed parameters
C
      DO N=1, NE
         DTK(N)=ZTK(NPZON(N))
         DSE(N)=ZSE(NPZON(N))
         DPR(N)=ZPR(NPZON(N))
         DAL(N)=ZAL(NPZON(N))
         DAT(N)=ZAT(NPZON(N))
         DAD(N)=ZAD(NPZON(N))
      END DO
C
      RETURN
      END
C
C************************************************************************
C
C           Subroutine DATAW reads extraction and injection data
C
C************************************************************************
C
      SUBROUTINE DATAW (WRAT,WCON,WLX,WLY,LEW,STEPW,NW,NTW,IHC)
      IMPLICIT REAL*8 (A-H, O-Z)
      DIMENSION  WRAT(NW,NTW),WCON(NW,NTW),WLX(NW),WLY(NW),
     *           LEW(NW),STEPW(NTW)
C
C Reading extraction and injection data
C
      IF (NW .GT. 0) THEN
C
      CALL JUMP (7, 3)
C
C
      DO K=1, NTW
        READ (7, 4) STEPW(K)
      END DO
   4  FORMAT (10X, F15.2)
```

```
C
      CALL JUMP (7, 3)
      DO  I=1, NW
        READ (7, 10) LEW(I), WLX(I), WLY(I)
      END DO
  10  FORMAT (10X, I15, 2F20.5)
C
      CALL JUMP (7, 2)
      DO I=1, NW
        CALL JUMP (7, 3)
        DO K=1, NTW
          IF (IHC .EQ. 1) THEN
            READ (7, 20) WRAT(I,K)
          ELSE
            READ (7, 30) WRAT(I,K), WCON(I,K)
          END IF
  20    FORMAT (10X, F20.5)
  30    FORMAT (10X, 2F20.5)
        END DO
      END DO
C
      ENDIF
C
      RETURN
      END
C
C***************************************************************************
C
C
C      Subroutine DATAS reads precipitation and return water data
C
C***************************************************************************
C
      SUBROUTINE DATAS (SRAT,SCON,STEPS,NS,NTS,IHC)
      IMPLICIT REAL*8 (A-H, O-Z)
      DIMENSION   SRAT(NS,NTS),SCON(NS,NTS),STEPS(NTS)
C
C Reading sink/source data
C
      IF (NS .GT. 0) THEN
C
      CALL JUMP (8, 3)
C
C
      DO K=1, NTS
        READ (8, 4) STEPS(K)
      END DO
  4   FORMAT (10X, F15.2 )
C
      CALL JUMP (8, 2)
      DO  I=1, NS
        CALL JUMP(8, 3)
        DO K=1, NTS
          IF (IHC .EQ. 1) THEN
            READ (8, 20) SRAT(I,K)
          ELSE
            READ (8, 30) SRAT(I,K), SCON(I,K)
          END IF
  20    FORMAT (10X, F20.5)
  30    FORMAT (10X, 2F20.5)
        END DO
      END DO
```

```
C
      ENDIF
C
      RETURN
      END
C
C
C**********************************************************************
C
C            Subroutine DATAO reads observation data
C
C**********************************************************************
C
C
      SUBROUTINE DATAO (OBWX,OBWY,LEOB,OBT,NOBW,NOBT)
      IMPLICIT REAL*8 (A-H, O-Z)
      DIMENSION   OBWX(NOBW),OBWY(NOBW),LEOB(NOBW),OBT(NOBT)
C
C Reading locations of observation wells
C
      CALL JUMP (9, 3)
C
      DO I=1, NOBW
      READ (9, 10) LEOB(I), OBWX(I), OBWY(I)
  10  FORMAT (10X, I15, 2F20.5)
      END DO
C
C Reading observation times
C
      CALL JUMP (9, 3)
C
      DO    I=1, NOBT
      READ (9, 20) OBT(I)
  20  FORMAT (10X, F20.5)
      END DO
C
      RETURN
      END
C
C
C**********************************************************************
C
C         Subroutine DATAC reads computation control numbers
C
C**********************************************************************
C
C
      SUBROUTINE DATAC (DTI,DTF,DTL,ALF,EPS,PT,IHC)
      IMPLICIT REAL*8 (A-H, O-Z)
      DIMENSION   DTF(2),DTL(2),ALF(2),EPS(2)
C
C Reading computation control numbers for the flow problem
C
      CALL JUMP (10, 3)
      READ (10, 10) DTI, DTF(1), DTL(1), ALF(1), EPS(1)
  10  FORMAT (5F10.5)
C
C Reading computation control numbers for the transport problem
C
      IF (IHC .EQ. 2) THEN
        CALL JUMP(10, 3)
```

```
        READ (10, 10)  PT, DTF(2), DTL(2), ALF(2), EPS(2)
        END IF
C
        RETURN
        END
C
C*************************************************************************
C
C       Subroutine FDB forms boundary conditions for each time step
C
C*************************************************************************
C
        SUBROUTINE FDB (TSUM,HB1,CB1,H0,C0,STEPB,NHBP,NCBP,
       *               NP,NHB,NCB,NTB,IHC)
        IMPLICIT REAL*8 (A-H, O-Z)
        DIMENSION  HB1(NHB,NTB),CB1(NCB,NTB),H0(NP),C0(NP),
       1           STEPB(NTB),NHBP(NHB),NCBP(NCB)
C
C
        IF (NHB .GT. 0) THEN
C
        DO K=1, NTB
          IF (STEPB(K) .GT. TSUM) THEN
            DO J=1, NHB
              A=HB1(J, K-1)
              T1=STEPB(K-1)
              B=HB1(J, K)
              T2=STEPB(K)
              CALL LIP(TSUM, A, B, T1, T2, C)
              H0(NHBP(J))=C
            END DO
C
            IF (IHC .EQ. 2 .AND. NCB .GT. 0) THEN
              DO J=1, NCB
                A=CB1(J, K-1)
                T1=STEPB(K-1)
                B=CB1(J, K)
                T2=STEPB(K)
                CALL LIP(TSUM, A, B, T1, T2, C)
                C0(NCBP(J))=C
              END DO
            END IF
            RETURN
C
          END IF
        END DO
C
        END IF
C
        RETURN
        END
C
C
C
        SUBROUTINE LIP (T, A, B, T1, T2, C)
        IMPLICIT REAL*8 (A-H, O-Z)
C
        C=A+(B-A)*(T-T1)/(T2-T1)
        RETURN
        END
C
```

```
C***********************************************************************
C
C      Subroutine FDW forms data of wells for each time step
C
C***********************************************************************
C
       SUBROUTINE FDW (TSUM,X,Y,IJM,WRAT,WCON,WLX,WLY,LEW,DWR,DWC,
      *              STEPW,NP,NE,NW,NTW,IHC)
       IMPLICIT REAL*8 (A-H, O-Z)
       DIMENSION  X(NP),Y(NP),IJM(3,NE),WRAT(NW,NTW),WCON(NW,NTW),
      1            WLX(NW),WLY(NW),LEW(NW),DWR(NP),DWC(NP),STEPW(NTW)
C
       DO I=1, NP
         DWR(I)=0.
         DWC(I)=0.
       END DO
C
       IF (NW .GT. 0) THEN
C
       DO K=1, NTW
       IF (STEPW(K) .GT. TSUM) THEN
         DO J=1, NW
           MI=IJM(1, LEW(J))
           MJ=IJM(2, LEW(J))
           MK=IJM(3, LEW(J))
           AI=X(MJ)*Y(MK)-X(MK)*Y(MJ)
           AJ=X(MK)*Y(MI)-X(MI)*Y(MK)
           AK=X(MI)*Y(MJ)-X(MJ)*Y(MI)
           BI=Y(MJ)-Y(MK)
           BJ=Y(MK)-Y(MI)
           BK=Y(MI)-Y(MJ)
           CI=X(MK)-X(MJ)
           CJ=X(MI)-X(MK)
           CK=X(MJ)-X(MI)
           TRS=ABS(CJ*BI-BJ*CI)
           WI=(AI+BI*WLX(J)+CI*WLY(J))/TRS
           WJ=(AJ+BJ*WLX(J)+CJ*WLY(J))/TRS
           WK=(AK+BK*WLX(J)+CK*WLY(J))/TRS
           WI=ABS(WI)
           WJ=ABS(WJ)
           WK=ABS(WK)
           WW=WI+WJ+WK
           WI=WI/WW
           WJ=WJ/WW
           WK=WK/WW
C
           R=WRAT(J, K)
           RI=R*WI
           RJ=R*WJ
           RK=R*WK
           DWR(MI)=DWR(MI) + RI
           DWR(MJ)=DWR(MJ) + RJ
           DWR(MK)=DWR(MK) + RK
C
           IF (IHC .EQ. 2 ) THEN
             C=WCON(J, K)
             DWC(MI)=DWC(MI) + C*RI
             DWC(MJ)=DWC(MJ) + C*RJ
             DWC(MK)=DWC(MK) + C*RK
           END IF
         END DO
```

```
C
       DO I=1, NP
         IF ( ABS(DWR(I)) .GT. 0.0001)    DWC(I)=DWC(I)/DWR(I)
       END DO
       RETURN
C
       END IF
C
       END DO
       END IF
C
       RETURN
       END
C
C***********************************************************************
C
C      Subroutine FDS forms source/sink data for each time step
C
C***********************************************************************
C
       SUBROUTINE FDS (TSUM,X,Y,IJM,SRAT,SCON,DSR,DSC,STEPS,
      *                NSZON,NP,NE,NS,NTS,IHC)
       IMPLICIT REAL*8 (A-H, O-Z)
       DIMENSION  X(NP),Y(NP),IJM(3,NE),SRAT(NS,NTS),SCON(NS,NTS),
      *           DSR(NP),DSC(NP),NSZON(NE),STEPS(NTS)
C
       DO I=1, NP
         DSR(I)=0.
         DSC(I)=0.
       END DO
C
       IF (NS .GT. 0) THEN
C
       DO K=1, NTS
       IF (STEPS(K) .GT. TSUM) THEN
         DO J=1, NE
           MI=IJM(1, J)
           MJ=IJM(2, J)
           MK=IJM(3, J)
           BI=Y(MJ)-Y(MK)
           BJ=Y(MK)-Y(MI)
           CI=X(MK)-X(MJ)
           CJ=X(MI)-X(MK)
           TRS=ABS(CJ*BI-BJ*CI)
C
           R=SRAT(NSZON(J), K)
           R=R*TRS/6.
           DSR(MI)=DSR(MI) + R
           DSR(MJ)=DSR(MJ) + R
           DSR(MK)=DSR(MK) + R
C
           IF (IHC .EQ. 2 ) THEN
             C=SCON(NSZON(J), K)
             DSC(MI)=DSC(MI) + C*R
             DSC(MJ)=DSC(MJ) + C*R
             DSC(MK)=DSC(MK) + C*R
           END IF
         END DO
C
         DO I=1, NP
           IF ( ABS(DSR(I)) .GT. 0.00001) DSC(I)=DSC(I)/DSR(I)
```

```
            END DO
C
        RETURN
C
      END IF
C
      END DO
C
      END IF
C
      RETURN
      END
C
C*********************************************************************
C
C     Subroutine FVD generates velocities and dispersion coefficients.
C
C*********************************************************************
C
      SUBROUTINE FVD (X,Y,IJM,TK,PR,AL,AT,AD,DFU,DXX,DXY,DYY,VX,VY,
     *                H0,NP,NE,IHC)
      IMPLICIT REAL*8 (A-H, O-Z)
      DIMENSION  X(NP),Y(NP),IJM(3,NE),TK(NE),PR(NE),
     1           AL(NE),AT(NE),AD(NE),DXX(NE),DXY(NE),DYY(NE),
     2           VX(NE),VY(NE),H0(NP)
C
C For flow in an anisotropic aquifer
C
      IF (IHC .EQ. 1) THEN
         DO N=1, NE
         DXX(N)=AL(N)
         DYY(N)=AT(N)
         DXY(N)=AD(N)
         END DO
      ELSE
      DO N=1, NE
         MI=IJM(1,N)
         MJ=IJM(2,N)
         MK=IJM(3,N)
         BI=Y(MJ)-Y(MK)
         BJ=Y(MK)-Y(MI)
         BK=Y(MI)-Y(MJ)
         CI=X(MK)-X(MJ)
         CJ=X(MI)-X(MK)
         CK=X(MJ)-X(MI)
         TRS=ABS(CJ*BI-BJ*CI)
C
C Calculating average values of parameters in each element
C
         A=-TK(N)/PR(N)/TRS
         RET=1. + (1.-PR(N))*AD(N)/PR(N)
C
C Calculating the velocity distribution
C
         VX(N)=A*(BI*H0(MI)+BJ*H0(MJ)+BK*H0(MK))/RET
         VY(N)=A*(CI*H0(MI)+CJ*H0(MJ)+CK*H0(MK))/RET
         V=SQRT(VX(N)*VX(N)+VY(N)*VY(N))
         IF (V .LT. 1.E-10) V=1.E-10
C
C Calculating dispersion coefficients
```

```
C
            DXX(N)=(AL(N)*VX(N)*VX(N)+AT(N)*VY(N)*VY(N))/V + DFU
            DYY(N)=(AT(N)*VX(N)*VX(N)+AL(N)*VY(N)*VY(N))/V + DFU
            DXY(N)=(AL(N)-AT(N))*(VX(N)*VY(N))/V
         END DO
C
         ENDIF
C
C        WRITE (6, 20)
C  20 FORMAT(//,3X,'ELEMEN',8X,'VX',8X,'VY',7X,'DXX',7X,'DXY',7X,'DYY')
C        DO M=1, NE
C        WRITE (6, 25) M, VX(M), VY(M), DXX(M), DXY(M), DYY(M)
C  25    FORMAT (I10, 5E10.5)
C        END DO
C
         RETURN
         END
C
C
C************************************************************************
C
C        Subroutine COEFF forms coefficient matrices of MCB equations
C
C************************************************************************
C
C
         SUBROUTINE COEFF (X,Y,IJM,MA,MCPN,VX,VY,DXX,DXY,DYY,TR,BR,TH,H0,
      1     PS,TK,SE,PR,AD,PSE,PPR,PAD,CFA,CFB,PT,NP,NP1,NE,NB,ISO,IH,IHC)
         IMPLICIT REAL*8 (A-H, O-Z)
         DIMENSION  X(NP),Y(NP),IJM(3,NE),MA(NP1),MCPN(NB,NP),
      1              VX(NE),VY(NE),DXX(NE),DXY(NE),DYY(NE),TR(NP),BR(NP),
      2              TH(NP),TK(NE),SE(NE),PR(NE),AD(NE),PSE(NP),PPR(NP),
      3              PAD(NP),PS(NP),H0(NP),CFA(IH),CFB(IH),ME(3)
         DO  K=1, IH
            CFA(K)=0.
            CFB(K)=0.
         END DO
         DO I=1, NP
            PSE(I)=0.
            PPR(I)=0.
            PAD(I)=0.
         END DO
C
C
C
         DO  I=1, NP
            IF (H0(I) .GT. TR(I)) THEN
               TH(I)=TR(I)-BR(I)
            ELSE
               TH(I)=H0(I)-BR(I)
            ENDIF
         END DO
C
C
C
         DO N=1, NE
            MI=IJM(1,N)
            MJ=IJM(2,N)
            MK=IJM(3,N)
               BI=Y(MJ)-Y(MK)
               BJ=Y(MK)-Y(MI)
```

```
                CI=X(MK)-X(MJ)
                CJ=X(MI)-X(MK)
                TRS=ABS(CJ*BI-BJ*CI)
C
C Calculating PPR and PAD
C
         TRS1=TRS/6.
         PSE(MI)=PSE(MI)+ TRS1*SE(N)
         PSE(MJ)=PSE(MJ)+ TRS1*SE(N)
         PSE(MK)=PSE(MK)+ TRS1*SE(N)
         PPR(MI)=PPR(MI)+ TRS1*PR(N)
         PPR(MJ)=PPR(MJ)+ TRS1*PR(N)
         PPR(MK)=PPR(MK)+ TRS1*PR(N)
         PAD(MI)=PAD(MI)+ TRS1*AD(N)
         PAD(MJ)=PAD(MJ)+ TRS1*AD(N)
         PAD(MK)=PAD(MK)+ TRS1*AD(N)
C
            IF (IHC .EQ. 1) THEN
                AVH0=(H0(MI)+H0(MJ)+H0(MK))/3.
                AVTR=(TR(MI)+TR(MJ)+TR(MK))/3.
                AVBR=(BR(MI)+BR(MJ)+BR(MK))/3.
                  IF (AVH0 .GT. AVTR) THEN
                    AVTH=AVTR-AVBR
                    SS=SE(N)
                  ELSE
                    AVTH=AVH0-AVBR
                    SS=PR(N)
                  ENDIF
                    TT=TK(N)*AVTH
                    IF (ISO .EQ. 0) THEN
                      D11=TT
                      D12=0.
                      D22=TT
                    ELSE
                      D11=DXX(N)*AVTH
                      D12=DXY(N)*AVTH
                      D22=DYY(N)*AVTH
                    ENDIF
            ELSE
                SS=1.
                D11=DXX(N)
                D12=DXY(N)
                D22=DYY(N)
            ENDIF
C
C
C
         DO  J=1, 3
            ME(J)=IJM(J,N)
         END DO
         DO L=1, 3
            MI=ME(1)
            MJ=ME(2)
            MK=ME(3)
            XM=(X(MI)+X(MJ)+X(MK))/3
            YM=(Y(MI)+Y(MJ)+Y(MK))/3
            BI=Y(MJ)-Y(MK)
            BJ=Y(MK)-Y(MI)
            BK=Y(MI)-Y(MJ)
            CI=X(MK)-X(MJ)
            CJ=X(MI)-X(MK)
```

```
              CK=X(MJ)-X(MI)
              BJI=Y(MJ)-YM
              BJJ=YM-Y(MI)
              BJM=Y(MI)-Y(MJ)
              CJI=XM-X(MJ)
              CJJ=X(MI)-XM
              CJM=X(MJ)-X(MI)
              BKI=YM-Y(MK)
              BKM=Y(MK)-Y(MI)
              BKK=Y(MI)-YM
              CKI=X(MK)-XM
              CKM=X(MI)-X(MK)
              CKK=XM-X(MI)
              TRS=ABS(CJ*BI-BJ*CI)
C
C introducing upstream weights
C
              IF (IHC .EQ. 1) THEN
C
                  WI=1./3.
                  WJ=1./3.
                  WK=1./3.
              ELSE
          CALL WEIT(BI,BJ,BK,CI,CJ,CK,DXX(N),VX(N),VY(N),WI,WJ,WK,PT)
              ENDIF
              BBJI=BJI+WI*BJM
              BBJJ=BJJ+WJ*BJM
              BBJK=WK*BJM
              CCJI=CJI+WI*CJM
              CCJJ=CJJ+WJ*CJM
              CCJK=WK*CJM
              BBKI=BKI+WI*BKM
              BBKJ=WJ*BKM
              BBKK=BKK+WK*BKM
              CCKI=CKI+WI*CKM
              CCKJ=WJ*CKM
              CCKK=CKK+WK*CKM
C
C generating coefficient matrices of discretization equations
C
              NI=MA(MI)
              CFA(NI+1)=CFA(NI+1)
     1            +((D11*BBJI+D12*CCJI)*(BI-BJ)
     2            + (D12*BBJI+D22*CCJI)*(CI-CJ)
     3            + (D11*BBKI+D12*CCKI)*(BI-BK)
     4            + (D12*BBKI+D22*CCKI)*(CI-CK))/(2.*TRS)
              IF (IHC .EQ. 2) THEN
                  VV=(VX(N)*(BBJI+BBKI)+VY(N)*(CCJI+CCKI))/4.
                  CFA(NI+1)=CFA(NI+1)+VV
              ENDIF
              CFB(NI+1)=CFB(NI+1)+TRS*SS/6.
              NS=MA(MI+1)-MA(MI)
              DO I=1,NS
                  IF(MCPN(I,MI) .EQ. MJ) THEN
                      CFA(NI+I)=CFA(NI+I)
     1                    +((D11*BBJJ+D12*CCJJ)*(BI-BJ)
     2                    + (D12*BBJJ+D22*CCJJ)*(CI-CJ)
     3                    + (D11*BBKJ+D12*CCKJ)*(BI-BK)
     4                    + (D12*BBKJ+D22*CCKJ)*(CI-CK))/(2.*TRS)
                      IF (IHC .EQ. 2) THEN
                          VV=(VX(N)*(BBJJ+BBKJ)+VY(N)*(CCJJ+CCKJ))/4.
```

```
                        CFA(NI+I)=CFA(NI+I)+VV
                     ENDIF
                     CFB(NI+I)=CFB(NI+I)
                  ENDIF
            END DO
            DO  I=1,NS
               IF(MCPN(I,MI) .EQ. MK) THEN
                  CFA(NI+I)=CFA(NI+I)
     1               +((D11*BBJK+D12*CCJK)*(BI-BJ)
     2               + (D12*BBJK+D22*CCJK)*(CI-CJ)
     3               + (D11*BBKK+D12*CCKK)*(BI-BK)
     4               + (D12*BBKK+D22*CCKK)*(CI-CK))/(2.*TRS)
                  IF (IHC .EQ. 2) THEN
                     VV=(VX(N)*(BBJK+BBKK)+VY(N)*(CCJK+CCKK))/4.
                     CFA(NI+I)=CFA(NI+I)+VV
                  ENDIF
                  CFB(NI+I)=CFB(NI+I)
               ENDIF
            END DO
            J=ME(1)
            ME(1)=ME(2)
            ME(2)=ME(3)
            ME(3)=J
         END DO
      END DO
C
      DO I=1, NP
        PSE(I)=PSE(I)/PS(I)
        PPR(I)=PPR(I)/PS(I)
        PAD(I)=PAD(I)/PS(I)
      END DO
C
      RETURN
      END
C
C************************************************************************
C
C     Subroutine  WEIT  generates adaptive upstream weights.
C
C************************************************************************
C
      SUBROUTINE WEIT (BI,BJ,BK,CI,CJ,CK,DXX,VX,VY,WI,WJ,WK,PT)
      IMPLICIT REAL*8 (A-H, O-Z)
      VIJ=CK*VX-BK*VY
      VIK=BJ*VY-CJ*VX
      VKJ=BI*VY-CI*VX
      IF (DXX .GT. 1.E-20) THEN
         WI=( VIJ+VIK)/DXX
         WJ=(-VIJ-VKJ)/DXX
         WK=(-VIK+VKJ)/DXX
      ELSE
         WI=0.
         WJ=0.
         WK=0.
      ENDIF
      WI=1./3.+PT*WI
      WJ=1./3.+PT*WJ
      WK=1./3.+PT*WK
C
      RETURN
      END
```

```
C
C*********************************************************************
C
C         Subroutine  SNIM  solves discretization equations.
C
C*********************************************************************
C
      SUBROUTINE SNIM (MA,MCPN,IE,IEE,CFA,CFB,NHBP,H0,H1,H2,DF00,DF0,
     1                 DWR,DWC,DSR,DSC,TH,PS,PPR,PAD,EPS,ALF,RDC,
     2                 DT00,DT0,DT1,NP,NP1,NB,NHB,IH,IHC)
      IMPLICIT REAL*8 (A-H, O-Z)
      DIMENSION  MA(NP1),MCPN(NB,NP),IE(NP),IEE(NP),CFA(IH),
     1           CFB(IH),NHBP(NHB),H0(NP),H1(NP),H2(NP),
     2           DF00(NP),DF0(NP),DWR(NP),DWC(NP),DSR(NP),
     3           DSC(NP),TH(NP),PS(NP),PPR(NP),PAD(NP),EPS(2)
C
C Identifying given head or given concentration boundary nodes
C
      DO I=1, NP
         IE(I)=0
         IF (NHB .GT. 0) THEN
         DO  J=1, NHB
         IF (I .EQ. NHBP(J)) THEN
            H1(I)=H0(I)
            IE(I)=2
         ENDIF
         END DO
         ENDIF
      END DO
C
C extrapolation
C
      DO N=1, NP
         IF(IE(N).LT.2) THEN
            IF(ABS(DF00(N)).GE.1.E-5) THEN
               DD=(DT00/DT0)*(DF0(N)/DF00(N))
            ELSE
               DD=1.
            ENDIF
            IF(DD. GT. 5.) DD=5.
            IF(DD. LT. 0.) DD=0.
            H1(N)=H0(N)-(DT1*DD*DF0(N))/DT0
         ENDIF
      END DO
C
C forming iteration equations
C
   40 DO N=1, NP
         IF(IE(N) .EQ. 0) THEN
            NA=MA(N+1)
            NC=MA(N)
            TRS=CFA(NC+1)+CFB(NC+1)/DT1
            IF (IHC .EQ. 2) TRS=TRS+RDC*PS(N)
            DD=CFB(NC+1)/DT1*H0(N)
            ND=NA-NC
            DO K=2, ND
               MC=MCPN(K,N)
               DD=DD+CFB(NC+K)/DT1*(H0(MC)-H1(MC))-CFA(NC+K)*H1(MC)
            END DO
C
C incorporating sink/source terms
```

```
C
                IF (IHC .EQ. 1) THEN
                    DD=DD+DWR(N)
                    DD=DD+DSR(N)
                ELSE
                    RET=1.+(1.-PPR(N))*PAD(N)/PPR(N)
                    IF (DWR(N) .GT. 0.) THEN
                    DD=DD+DWR(N)*(DWC(N)-H0(N))/TH(N)/PPR(N)/RET
                    END IF
                    DD=DD+DSR(N)*(DSC(N)-H0(N))/TH(N)/PPR(N)/RET
                ENDIF
C
C The selecting node iteration procedure
C
                H2(N)=(1.-ALF)*H1(N)+ALF*DD/TRS
                IF ( ABS(H2(N)-H1(N)) .LT. EPS(IHC)) IE(N)=1
                H1(N)=H2(N)
C
            ENDIF
        END DO
C
C
C
        DO  N=1, NP
            IEE(N)=1
        END DO
        DO N=1,NP
            IF(IE(N) .EQ. 0) THEN
                IEE(N)=0
                NA=MA(N+1)
                NC=MA(N)
                ND=NA-NC
                DO K=2,ND
                    MC=MCPN(K,N)
                    IF(IE(MC). LT. 2) IEE(MC)=0
                END DO
            ENDIF
        END DO
        DO  N=1, NP
            IF(IE(N) .LT. 2) IE(N)=IEE(N)
        END DO
C
        DO N=1 ,NP
            IF(IE(N) .EQ. 0) GO TO 40
        END DO
C
C The solution is obtained for this time step.
C
        DT00=DT0
        DT0=DT1
        DO N=1,NP
            IF(IE(N).LT.2) THEN
                DF00(N)=DF0(N)
                DF0(N)=H0(N)-H1(N)
                H0(N)=H1(N)
            ENDIF
        END DO
C
        RETURN
        END
C
```

```
C
C**********************************************************************
C
C      Subroutine FOB calculates solutions at observation locations
C
C**********************************************************************
C
C
       SUBROUTINE FOB (X,Y,IJM,H0,C0,HOB,COB,OBWX,OBWY,LEOB,
      *                L,NP,NE,NOBW,NOBT,IHC)
       IMPLICIT REAL*8 (A-H, O-Z)
       DIMENSION  H0(NP),C0(NP),HOB(NOBW,NOBT),COB(NOBW,NOBT),
      1           X(NP),Y(NP),IJM(3,NE),OBWX(NOBW),OBWY(NOBW),
      2           LEOB(NOBW)
C
       DO J=1,NOBW
           MI=IJM(1, LEOB(J))
           MJ=IJM(2, LEOB(J))
           MK=IJM(3, LEOB(J))
           AI=X(MJ)*Y(MK)-X(MK)*Y(MJ)
           AJ=X(MK)*Y(MI)-X(MI)*Y(MK)
           AK=X(MI)*Y(MJ)-X(MJ)*Y(MI)
           BI=Y(MJ)-Y(MK)
           BJ=Y(MK)-Y(MI)
           BK=Y(MI)-Y(MJ)
           CI=X(MK)-X(MJ)
           CJ=X(MI)-X(MK)
           CK=X(MJ)-X(MI)
           TRS=ABS(CJ*BI-BJ*CI)
           WI=(AI+BI*OBWX(J)+CI*OBWY(J))/TRS
           WJ=(AJ+BJ*OBWX(J)+CJ*OBWY(J))/TRS
           WK=(AK+BK*OBWX(J)+CK*OBWY(J))/TRS
           WI=ABS(WI)
           WJ=ABS(WJ)
           WK=ABS(WK)
           WW=WI+WJ+WK
           WI=WI/WW
           WJ=WJ/WW
           WK=WK/WW
C
           HOB(J,L)=H0(MI)*WI+H0(MJ)*WJ+H0(MK)*WK
           IF (IHC .EQ. 2) THEN
           COB(J,L)=C0(MI)*WI+C0(MJ)*WJ+C0(MK)*WK
           END IF
C
       END DO
C
       RETURN
       END
```

Auxiliary Program: FINPUT.FOR

```
C
C*******************************************************************
C
C               Auxiliary Program:  Finput.for
C
C*******************************************************************
C
      DIMENSION CHA(70)
C
      OPEN(1,FILE='DATAN')
      OPEN(2,FILE='DATAG')
      OPEN(3,FILE='DATAI')
      OPEN(4,FILE='DATAB')
      OPEN(5,FILE='DATAP')
      OPEN(6,FILE='RESUL')
      OPEN(7,FILE='DATAW')
      OPEN(8,FILE='DATAS')
      OPEN(9,FILE='DATAO')
      OPEN(10,FILE='DATAC')
C
C Reading and writting control integers
C
      READ (1, 2) CHA
      READ (1, 2) CHA
      READ (1, 2) CHA
    2 FORMAT (70A1)
      READ (1,4)  NP, NE, NB, NZON, IHC, IST, IPR
    4 FORMAT (7I10)
C
      WRITE (6, 2) CHA
      WRITE (6, 4) NP,NE,NB,NZON,IHC,IST,IPR
C
C
      READ (1, 2) CHA
      READ (1, 2) CHA
      READ (1, 2) CHA
      READ (1, 4)  NW, NTW, NHB, NCB, NTB, NS, NTS
C
      WRITE (6, 2) CHA
      WRITE (6, 4) NW,NTW,NHB,NCB,NTB,NS,NTS
C
      READ (1, 2) CHA
      READ (1, 2) CHA
      READ (1, 2) CHA
      READ (1, 4)  NOBW, NOBT, ISO
C
      WRITE (6, 2) CHA
      WRITE (6, 4) NOBW, NOBT, ISO
C
C Generating the format for input data file "DATAG"
C
C
   10    FORMAT (I10)
         WRITE (2, 12)
   12 FORMAT (/,' NODE COORDINATES, ELEVATIONS OF TOP AND BED ROCKS')
         WRITE (2, 14)
   14 FORMAT (6X,'NODE',5X, 'X COORD.00', 5X, 'Y COORD.00', 5X,
```

```
     *           'TR.00', 5X, 'BR.00')
          DO N=1, NP
          WRITE (2, 10) N
          END DO
C
C
C
          WRITE (2, 16)
   16 FORMAT (/,' NODE NUMEBERS OF EACH ELEMENT AND ZONATION PATTERNS')
          WRITE (2, 18)
   18 FORMAT (3X,'ELEMENT', 9X, 'I', 9X, 'J', 9X, 'K',5X, 'NPZON',
     *           5X, 'NSZON')
          DO N=1, NE
          WRITE (2, 10) N
          END DO
C
      WRITE (2, 20)
   20 FORMAT (1X, '//')
C
C Generating the format for input data file 'DATAI'
C
C
      WRITE (3, 30)
   30 FORMAT (/, ' INITIAL CONDITIONS OF HEAD AND CONCENTRATION')
          WRITE (3, 34)
   34     FORMAT (6X, 'NODE', 11X, 'H00.0', 15X, 'C00.0')
          DO I=1, NP
           WRITE (3, 10) I
           END DO
C
      WRITE (3, 20)
C
C Generating the format for input data file 'DATAB'
C
      WRITE (4, 36)
   36 FORMAT (/, ' **** GIVEN HEAD BOUNDARY CONDITION **** ' )
          DO K=1, NTB
           WRITE (4, 38) K
   38     FORMAT (/, ' THE NUMBER OF TIME STEP IS:', I10)
           WRITE (4, 40)
   40     FORMAT (10X, 'MEASURE TIME.00')
           IF (K .EQ. 1) THEN
           WRITE (4, 42)
   42     FORMAT (21X, '0.00')
           ELSE
           WRITE (4, 44)
   44     FORMAT (1X)
           END IF
           WRITE (4, 46)
   46     FORMAT (4X, 'NUMBER', 6X, 'NODE', 10X, 'HEAD.00')
           DO I=1, NHB
           WRITE (4, 10) I
           END DO
          END DO
C
      IF (IHC .EQ. 2) THEN
C
      WRITE (4, 47)
   47 FORMAT (/, ' **** GIVEN CONCENTRATION BOUNDARY CONDITION **** ')
          DO K=1, NTB
           WRITE (4, 38) K
```

```fortran
         WRITE (4, 40)
         IF (K .EQ. 1) THEN
         WRITE (4, 42)
         ELSE
         WRITE (4, 44)
         END IF
         WRITE (4, 48)
   48    FORMAT (4X, 'NUMBER', 6X, 'NODE', 8X, 'CONCEN.00')
         DO  I=1, NCB
         WRITE (4, 10) I
         END DO
      END DO
C
      END IF
C
      WRITE (4, 20)
C
C Generating the format for input data file 'DATAP'
C
         WRITE (5, 50)
   50    FORMAT (/, ' **** GIVEN VALUES OF PARAMETERS **** ' )
         WRITE (5, 52)
   52    FORMAT (1X, 'ZONE', 6X, 'ZTK.', 6X, 'ZSE.', 6X, 'ZPR.',
        *        6X, 'ZAL.', 6X, 'ZAT.', 6X, 'ZAD.')
         DO I=1, NZON
         WRITE (5, 54) I
   54    FORMAT (I5)
         END DO
C
         WRITE (5, 56)
   56    FORMAT (/, ' RADIOACTIVE AND MOLECULAR DIFFUSION COEFFICIENTS ')
         WRITE (5, 58)
   58    FORMAT (6X, 'RDC.00000', 6X, 'DFU.00000',/)
C
      WRITE (5, 20)
C
C Generating the format for input data file 'DATAW'
C
         WRITE (7, 60)
   60    FORMAT (/, ' TIME STEPS OF EXTRACTION AND INJECTION ')
         WRITE (7, 62)
   62    FORMAT (6X, 'STEP', 8X, 'TIME.00')
         DO K=1,NTW
         WRITE (7, 10) K
         END DO
C
         WRITE (7, 66)
   66    FORMAT (/, ' LOCATIONS OF PUMPING AND RECHARGE WELLS')
         WRITE (7, 68)
   68    FORMAT(4X,'NUMBER',5X,'IN ELEMENT',8X,'X COORD.0',11X,'Y COORD.0')
         DO J=1,NW
         WRITE (7, 10) J
         END DO
C
         WRITE (7, 70)
   70    FORMAT (/,' PUMPING AND RECHARGE RATES, CONC. OF RECHARGE WATER')
         DO J=1, NW
         WRITE (7, 72) J
   72    FORMAT (/, ' WELL NUMBER IS ', I10)
         WRITE (7, 74)
   74    FORMAT (6X, 'STEP', 11X, 'RATE.00', 11X, 'CONC.00')
```

```
      DO K=1,NTW
      WRITE (7, 10) K
      END DO
      END DO
C
      WRITE (7, 20)
C
C Generating the format for input data file 'DATAS'
C
      WRITE (8, 80)
  80  FORMAT (/, ' TIME STEPS FOR INPUTTING SINK/SOURCE RECORDS')
      WRITE (8, 62)
      DO K=1,NTS
      WRITE (8, 10) K
      END DO
C
      WRITE (8, 82)
  82  FORMAT (/, ' RATES AND CONC. OF SINK/SOURCE TERMS')
      DO J=1, NS
      WRITE (8, 84) J
  84  FORMAT (/, ' THE NUMBER OF SINK/SOURCE ZONE', I10)
      WRITE (8, 74)
      DO K=1,NTS
      WRITE (8, 10) K
      END DO
      END DO
C
      WRITE (8, 20)
C
C  Generating the format for input data file 'DATAO'
C
      WRITE (9, 90)
  90  FORMAT (/, ' LOCATIONS OF OBSERVATION WELLS ' )
      WRITE (9, 92)
  92  FORMAT(4X,'NUMBER',5X,'IN ELEMENT',7X,'X COORD.0',7X,'Y COORD.0')
      DO I=1,NOBW
      WRITE (9, 10) I
      END DO
C
C
C
      WRITE (9, 94)
  94  FORMAT (/, ' OBSERVATION TIMES ' )
      WRITE (9, 96)
  96  FORMAT (4X, 'NUMBER', 11X, 'TIME.00')
      DO K=1,NOBT
      WRITE (9, 10) K
      END DO
C
C
      WRITE (9, 20)
C
C Generating the format for input data file 'DATAC'
C
      WRITE (10, 100)
 100  FORMAT (/, ' COMPUTATIONAL PARAMETERS FOR FLOW PROBLEM')
      WRITE (10, 102)
 102  FORMAT (2X, 'DTI.0000', 4X, 'DTF.00', 4X, 'DTL.00', 4X,
     *           'ALF.00', 2X, 'ESP.0000',/)
      WRITE (10, 104)
 104  FORMAT (/, ' COMPUTATIONAL PARAMETERS FOR TRANSPORT PROBLEM')
```

```
      WRITE (10, 106)
 106  FORMAT (3X, 'PT.0000', 4X, 'DTF.00', 4X, 'DTL.00', 4X,
     *          'ALF.00', 2X, 'ESP.0000',/)
C
      WRITE (10, 20)
C
C
C
      CLOSE(1)
      CLOSE(2)
      CLOSE(3)
      CLOSE(4)
      CLOSE(5)
      CLOSE(6)
      CLOSE(7)
      CLOSE(8)
      CLOSE(9)
      CLOSE(10)
C
      STOP
      END
```

References

Ahlfeld, D.P., J.M. Mulvey, G.F. Pinder, and E.F. Wood, Contaminated groundwater remediation design using simulation, optimization, and sensitivity theory, 1, Model development, *Water Resour. Res.*, *24*(3), 431–441, 1988.

Anderson, M.P., Using models to simulate the movement of contaminants through groundwater flow systems, *Crit. Rev. Environ. Control*, *9*, 97–156, 1979.

Andricevic, R., and P.K. Kitanidis, Optimization of the pumping schedule in aquifer remediation under uncertainty, *Water Resour. Res.*, *26*(5), 875–886, 1990.

Babu, D.K., and G.F. Pinder, A finite element-finite difference alternating direction algorithm for three-dimensional ground-water transport, in *Proceedings of the Fifth International Conference on Finite Element in Water Resources*, Burlington, VT, USA, pp. 165–174, 1984.

Bard, Y., *Nonlinear Parameter Estimation*, Academic, San Diego, Calif., 1974.

Basak, P., and V.V. Murty, Groundwater quality improvement through nonlinear diffusion, *J. Hydrol.*, *vol. 53*, 151–159, 1981.

Bear, J., and Y. Bachmat, A general theory on hydrodynamic dispersion in porous media, *I.A.S.H. Symp.* Artificial Recharge and Management of Aquifers, Haifa, Israel, IASH, P. N. 72, 7–16, 1967.

Bear, J., *Dynamics of Fluids in Porous Media*, Elsevier, New York, 1972.

Bear, J., *Hydraulics of Groundwater*, McGraw-Hill, New York, 1979.

Bear, J., and A. Veruijt, *Modeling Groundwater Flow and Pollution*, D. Reidel, Hingham, Mass., 1987.

Bear, J., C.-F. Tsang, and G. de Marsily, eds., *Flow and Contaminant Transport in Fractured Rock*, Academic, New York, 1993.

Bentley, L.R., and G.F. Pinder, Eulerian-Lagrangian solution of the vertically averaged groundwater transport equation, *Water Resour. Res.*, *28*(11), 3011–3020, 1992.

Bibby, R., Mass transport of solute in dual-porosity media, *Water Resour. Res.*, *17*(4), 1075–1081, 1981.

Boggs, J.M., S.C. Young, L.M. Beard, L.W. Gelhar, K.R. Rehfeldt, and E.E. Adams, Field study of dispersion in a heterogeneous aquifer, 1, Overview and site description, *Water Resour. Res.*, *28*(12), 3281–3291, 1992.

Bredehoeft, J.D., and R.A. Young, Conjunctive use of groundwater and surface water for irrigated agriculture: risk aversion, *Water Resour. Res.*, *19*(5), 1111–1121, 1983.

Brooks, A.N., and T.J.R. Hughes, Streamline Upwind/Petrov—Galerkin formulations for convection dominated flow with particular emphasis on the impressible Navier-Stokes equations, *Compt. Meths. Appl. Mech. Engrg.*, *vol. 32*, 199–259, 1982.

Cady, R., and S.P. Neuman, Advection-dispersion with adaptive Eulerian-Lagrangian finite elements, in *Advances in Transport Phenomena in Porous Media*, edited by J. Bear and M.Y. Corapcioglu, *NATO/ASI Series E, No. 128*, Martinus Nijhoff Publishers, Dordrecht, 921–952, 1987.

Carrera, J., and S.P. Neuman, Estimation of aquifer parameters under transient and steady state conditions, 1, Maximum likelihood method incoporating prior information, *Water Resour. Res.*, *22*(2), 199–210, 1986a.

Carrera, J., and S.P. Neuman, Estimation of aquifer parameters under transient and steady state conditions, 2, Uniqueness, stability, and solution algorithms, *Water Resour. Res.*, *22*(2), 211–227, 1986b.

Carrera, J., and S.P. Neuman, Estimation of aquifer parameters under transient and steady state conditions, 3, Application to synthetic and field data, *Water Resour. Res.*, *22*(2), 228–242, 1986c.

Carrera, J., State of the art of the inverse problem applied to the flow and solute transport equations, in *Groundwater Flow and Quality Modeling, NATO ASI Ser.*, edited by E. Curtodio, A. Gurgui, and J.P. Lobo Ferreira, D. Reidel, Hingham, Mass., 1987.

Celia, M.A., and I. Herrera, Solution of general ordinary differential equations by unified theory approach, *Num. Methods of Partial Differential Equations*, *3*(2), 117–129, 1987.

Celia, M.A., J.S. Kindred, and I. Herrera, Contaminant transport and biodegradation, 1, A numerical model for reactive transport in porous media, *Water Resour. Res.*, *25*(6), 1141–1148, 1989.

Celia, M.A., T.F. Russell, I. Herrera, and R.E. Ewing, An Eulerian-Lagrangian localized adjoint method for the advection-dispersion equation, *Adv. Water Resour.* *13*(4), 187–206, 1990.

Chankong, V., and Y.Y. Haimes, *Multiobjective Decision Making: Theory and Methodology*, North-Holland, New York, Amsterdam, Oxford, 1983.

Charbeneau, R.J., Groundwater contaminant transport with absorption and ion exchange chemistry: method of characteristics for the case without dispersion, *Water Resour. Res.*, *17*(3), 705–713, 1981.

Chavent, G., M. Dupuy, and P. Lemonnier, History matching by use of optimal theory, *Soc. Pet. Eng. J.*, *15*(1), 74–86, 1975.

Chavent, G., Identification of distributed parameter system: About the output least square method, its implementation, and identifiability, in *Proceedings of the Fifth IFAC Symposium, vol. 1*, Identification and System Parameter Estimation, edited by R. Igermann, 85–97, Pergamon, New York, 1979.

Chavent, G., Identifiability of parameters in the output least square formulation, in *Identifiability of Parametric Models*, edited by E. Walter, Pergamon, New York, 1987.

Chen, C.-S., Solution for radionuclide transport from an injection well into a single fracture in a porous formation, *Water Resour. Res.*, *22*(4), 508–518, 1986.

Chen, C.-S., Analytical solution for radial dispersion with Cauchy boundary at injection well, *Water Resour. Res.*, *23*(7), 1217–1224, 1987.

Chen, C.-S., Solutions approximating solute in a leaky aquifer receiving wastewater injection. *Water Resour. Res.*, *25*(1), 61–72, 1989.

Cheng, R.T., V. Casulli, and S.N. Milford, Eulerian-Lagrangian solution of the convection-dispersion equation in natural coordinates, *Water Resour. Res.*, *20*(7), 944–952, 1984.

Cherry, J.A., G.E. Grisak, and W.E. Chister, Hydrogeologic studies at a subsurface radioactive waste management site in West-central Canada, in *Underground Waste Management and Artificial Recharge, American Association of Petroleum Geologists*, 436, 1973.

Cherry, J.A., R.W. Gillham, and J.F. Pickens, Contaminant hydrogeology, 1, Physical processes, *Geosci. Can., vol. 2*, 1975.

Chiang, C.Y., M.F. Wheeler, and P.B. Bedient, A modified method of characteristics technique and mixed finite element method for simulation of groundwater solute transport, *Water Resour. Res., 25*(7), 1541–1550, 1989.

Ciccioli, P., W.P. Cooper, P.M. Hammer, and J.M. Hayes, Organnic solute-mineral surface interactions: A new method for the determination of groundwater velocities, *Water Resour. Res., 16*(1), 217–223, 1980.

Cordes, C., and W. Kinzelbach, Continuous groundwater velocity fields and path lines in linear, bilinear, and trilinear finite elements, *Water Resour. Res., 28*(11), 2903–2912, 1992.

Cooley, R.L., Incorporation of prior information on parameters into nonlinear regression groundwater flow models, 1, Theory, *Water Resour. Res., 18*(4), 965–976, 1982.

Cushman, J.H., and D. Kirkham, Solute travel times to well in single or multiple layered aquifers, *J. Hydrol., 37*, 169–184, 1978.

Cushman, J.H., Multiphase transport equations, 1, General equation for macroscopic statistical, local, space-time homogeneity, *Transp. Theor. Stat. Phys., 12*, 35–71, 1983.

Dagan, G., Solute transport in heterogeneous porous formations, *J. Fluid Mech., 145*, 151–177, 1984.

Dagan, G., Stochastic modeling of groundwater flow by unconditional and conditional probabilities: The inverse problem, *Water Resour. Res., 21*(1), 65–72, 1985.

Dagan, G., Statistical theory of groundwater flow and transport: pore to laboratory, laboratory to formation, and formation to reginal scale, *Water Resour. Res., 22*(9), 120s–134s, 1986.

Dagan, G., and Y. Rubin, Stochastic identification of recharge, transmissivity, and storativity in aquifer transient flow: A quasi-steady approach, *Water Resour. Res., 24*(10), 1698–1710, 1988.

Dagan, G., *Flow and Transport in Porous Formations*, Springer-Verlag, Berlin, 1989.

Dagan, G., Transport in heterogeneous formations: Spatial moments, ergodicity, and effective dispersion, *Water Resour. Res., 26*, 1281–1290, 1990.

Dagan, G., and S.P. Neuman, Nonasymptotic behavior of common Eulerian approximation for transport in random velocity fields, *Water Resour. Res., 27*(12), 3249–3256, 1991.

Delhomme, J.P., Spatial variability and uncertainty in groundwater flow parameters: A geostatistical approach, *Water Resour. Res., 15*(2), 269–280, 1979.

Desai, C.S., and D.N. Contractor, Finite element analysis of flow, diffusion, and salt water instrusion in porous media, in *Formulations and Computational Algorithms in Finite Element Analysis*, U.S.–Germany Symposium, 1977.

Devary, J.L., and P.G. Doctor, Pore velocity estimation uncertainties, *Water Resour. Res., 18*(4), 1157–1164, 1982.

Domenico, P.A., and G.A. Robbins, A dispersion scale effect in model calibrations and field tracer experiments, *J. Hydrol., vol. 70*, 123–132, 1984.

Donea, J., A Taylor-Galerkin method for convective transport problems, *Int. J. for Num. Meth. in Eng.*, vol. 20, 101–129, 1984.

Douglass, Jr. J., and T.F. Russell, Numerical methods for convection dominated diffusion problems based on combining the method of characteristics with finite element or finite difference procedures, SIAM. *J. on Numerical Analysis*, 19(5), 871–885, 1982.

Elimelech, M., J. Gregory, X. Jia, and R. A. Williams, *Particle Deposition and Aggregation: Measurement, Modeling, and Simulation*, Butterworth-Heinemann, Oxford, 1995.

Emigdio, Z., and W. Flores, Stochastic management of a stream aquifer system, in *Advances in Groundwater Hydrology*, edited by Z.A. Saleem, *AGU*, 1976.

Emsellem, Y., and G. de Marsily, An automatic solution for the inverse problem, *Water Resour. Res.*, 7(5), 1264–1283, 1971.

Ewing, R.E., and M.F. Wheeler, Computational aspects of mixed finite element methods, in *Numerical Methods for Scientific Computing*, edited by R.S. Stepleman, North-Holland, 1983.

Ewing, R.E., T. Lin, and R. Falk, Inverse and ill-posed problems in reservoir simulation, in *Inverse and Ill-Posed Problems*, edited by H.W. Engl and C.W. Groetsch, Academic, San Diego, Calif., 1987.

Falta, R.W., K. Pruess, I. Javandel, and P.A. Witherspoon, Numerical modeling of steam injection for the removal of nonaqueous phase liquids from the subsurface, 1, Numerical formulation, *Water Resour. Res.*, 28(2), 433–450, 1992.

Farmer, C.L., and R.A. Norman, The implementation of moving point methods for convection-diffusion equations, in *Numerical Methods for Fluid Dynamics*, edited by K.W. Mordon and M.J. Baines, Oxford University Press, Oxford, 1985.

Faust, C.R., and J.W. Mercer, Groundwater modeling: recent developments, *Groundwater*, 18(6), 569–577, 1980.

Freeze, R.A., A stochastic-conceptual analysis of one-dimensional groundwater flow in nonuniform homogeneous media, *Water Resour. Res.*, 11(5), 725–741, 1975.

Freeze, R.A., and J.A. Cherry, *Groundwater*, Prentice-Hall, Englewood Cliffs, New Jersey, 1979.

Fried, J.J., *Groundwater Pollution*, Elsevier Scientific, Amsterdam, 1975.

Friedman, R., C. Ansell, S. Diamond, and Y.Y. Haimes, The use of models for water resources management, planning, and policy, *Water Resour. Res.*, 20(7), 793–802, 1984.

Galeati, G., and G. Gambolati, On boundary conditions and point sources in the finite element integration of the transport equation, *Water Resour. Res.*, 25(5), 847–856, 1989.

Galeati, G, G. Gambolati, and S.P. Neuman, Coupled and partially coupled Eulerian-Lagrangian Model of Freshwater-Seawater Mixing, *Water Resour. Res.*, 28(1), 149–166, 1992.

Garabedian, S.P., D.R. LeBlanc, L.W. Gelhar, and M.A. Celia, Large-scale natural gradient tracer test in sand and gravel, Cape Cod, Massachusetts, 2, Analysis of spatial moments for a nonreactive tracer, *Water Resour. Res.*, 27(5), 911–924, 1991.

Garder, Jr., A.O., D.W. Peaceman, and Jr., A.L. Pozzi, Numerical calculation of multi-dimensional miscible displacement by the method of characteristics, *Society of Petroleum Engineers J.*, 4(1), 26–36, March, 1964.

Gelhar, L.W., and J.L. Wilson, Groundwater quality modeling, *Groundwater*, 12(6), 399–408, 1974.

Gelhar, L.W., Stochastic analysis of flow in aquifer, in *Advances in Groundwater Hydrology*, edited by Z.A. Saleem, *AGU*, 1976.

Gelhar, L.W., and C.L. Axness, Three-dimensional stochastic analysis of macro-dispersion in aquifers, *Water Resour. Res.*, *19*(1), 161–180, 1983.

Gelhar, L.W., Stochastic subsurface hydrology from theory to applications, *Water Resour. Res.*, *22*(9), 135s–145s, 1986.

Gelhar, L.W., C. Welty, and K.R. Rehfeldt, A critical review of data on field-scale dispersion in aquifers, *Water Resour. Res.*, *28*(7), 1955–1974, 1992.

Germain, D., and E.O. Frind, Modeling contaminant migration in fracture networks: Effects of matrix diffusion, in *Contaminant Transport in Groundwater*, edited by H.E. Kobus and W. Kinzelbach, 267–274, A.A. Balkema, Rotterdam, Netherlands, 1989.

Goode, D.J., Particle velocity interpolation in block-centered finite difference groundwater flow models, *Water Resour. Res.*, *26*(5), 925–940, 1990.

Gorelick, S.M., and I. Remson, Optimal dynamic management of groundwater pollutant sources, *Water Resour. Res.*, *18*(1), 71–76, 1982.

Gorelick, S.M., A review of distributed parameter groundwater management modeling methods, *Water Resour. Res.*, *19*(2), 305–319, 1983.

Gorelick, S.M., C.I. Voss, P.E. Gill, W. Murray, M.A. Saunders, and M.H. Wright, Aquifer reclamation design: the use of contaminant transport simulation combined with nonlinear programming, *Water Resour. Res.*, *20*(4), 415–427, 1984.

Gray, W.G., and G.F. Pinder, An analysis of the numerical solution of the transport equation, *Water Resour. Res.*, *12*(3), 547–555, 1976.

Griffin, R.A., K. Cartwright, N.F. Shimp, J.D. Steele, R.R. Ruch, W.A. White, G.M. Hughes, and R.H. Gilkeson, Attenuation of pollutants in municipal landfill leachate by clay minerals: Part 1, Column leaching and field verification, *Illinois State Geological Survey Environmental Geology Notes, No. 78*, Urbana, 1976.

Grisak, G.E., and J.F. Pickens, Solute transport through fractured media, 1, The effect of matrix diffusion, *Water Resour. Res.*, *16*(4), 719–730, 1980.

Grove, D.B., An analysis of the flow field of a discharging recharging pair of wells, *U.S. Geol. Surv. Rep.*, USGS, 474, 99, 1971.

Gupta, S.K., K.K. Tanji, and J.N. Luthin, A three-dimensional finite element groundwater model, California Water Research Center, Contribution, 152, 1975.

Gupta, S.K., and K.K. Tanji, Computer program for solution of large spare, unsymmetric system of linear equations, *Int. J. for Num. Meth. in Eng.*, vol. *11*, 1251–1259, 1977.

Gupta, A.D., and P.N.D.D. Yapa, Saltwater encroachment in an aquifer: A case study, *Water Resour. Res.*, *18*(3), 546–556, 1982.

Haidari, M., Application of linear system's theory and linear programming to groundwater management in Kansas, *Water Resour. Bull.*, *18*(6), 1003–1012, 1982.

Haimes, Y.Y., W.A. Hall, and H.T. Freedman, *Multiobjective Optimization in Water Resources Systems: The Surrogate Worth Trade-off Method*, Elsevier, Amsterdam, 1975.

Haimes, Y.Y., *Hierachical Analysis of Water Resources System: Modeling and Optimization of Large-Scale Systems*, McGraw Hill, New York, 1977.

Haimes, Y.Y., and K. Tarvainen, Hierachical-multiobjective frame-work for large-scale systems, in *Multicriteria Analysis in Practice*, edited by P. Nijkamp and J. Spronk, Gower, London, 1980.

Han, Z.-S., Optimal management of water resources in the area of Shimenzhai, Liaoning, China (in chinese), research paper, Graduated Student School of Wuhan Geology University, Beijing, 1985.

Healy, R.W., and T.F. Russell, A finite-volume Eulerian-Lagrangian localized adjoint method for solution of the advection-dispersion equation, *Water Resour. Res.*, 29(7), 2399–2413, 1993.

Heinrich, J.C., P.S. Huyakorn, O.C. Zienkiewicz, and A.R. Mitchell, An "Upwind" finite element scheme for two-dimensiona, convective-transport equations, *Int. J. for Num. Meth. in Eng.*, vol. 11, 131–143, 1977.

Heinrich, J.C., and O.C. Zienkiewicz, Quadratic finite element schemes for two-dimensional convective-transport problems, *Int. J. for Num. Meth. in Eng.*, vol. 11, 1831–1844, 1977.

Herrera, I., Unified approach to numerical methods, 1, Green's formula for operators in discontinuous fields, *Num. Methods for Partial Differential Equations*, 1(1), 25–44, 1985a.

Herrera, I., Unified approach to numerical methods, 2, Finite elements, boundary elements, and their couping, *Num. Methods for Partial Differential Equations*, 1(3), 159–186, 1985b.

Herweijer, J.C., G.A. Van Luijn, and C.A.J. Appelo, Calibration of a mass transport model using environmental tritium, *J. Hydrol.*, vol. 78, 1–17, 1985.

Hinstrup, P., A. Kej, and U. Kroszynski, A high accuracy two-dimensional transport-dispersion model for environmental applications, *proc. XVIII Int. Assoc. Hydraul. Res. Congress*, Baden, Germany, 13, 129–137, 1977.

Hoeksema, R.J., and P.K. Kitanidis, Analysis of the spatial structure of properties of selected aquifers, *Water Resour. Res.*, 21(4), 563–572, 1985.

Hoopes, J.A., and D.R.F. Harleman, Dispersion in radial flow from a recharge well, *J. Geophys. Res.*, vol. 72, 3595, 1967.

Hsieh, P.A., A new formula for the analytical solution of the radial dispersion problem, *Water Resour. Res.*, 22(11), 1597–1605, 1986.

Hsu, N.-S., and W.W.-G. Yeh, Optimum experimental design for parameter identification in groundwater hydrology, *Water Resour. Res.*, 25(2), 1025–1040, 1989.

Huang, K.-L., An alternative direction Galerkin method combined with characteristic technique for modeling of saturated-unsaturated solute transport, in *Development in Water Science*, vol. 35, 115–120, 1988.

Hutton, T.A., and E.N. Lightfoot, Dispersion of trace solutes in flowing groundwater, *Water Resour. Res.*, 20(9), 1253–1259, 1984.

Huyakorn, P.S., Solution of steady state, convective transport equation using an upwind finite element scheme, *Appl. Math. Modelling*, 1, 187–195, 1977.

Huyakorn, P.S., and K. Nilkuha, Solution of a transient transport equation using an upstream finite element scheme, *Appl. Math. Modelling*, 3, 7–17, 1979.

Huyakorn, P.S., B.H. Lester, and C.R. Faust, Finite element techniques for modeling groundwater flow in fractured aquifers, *Water Resour. Res.*, 19(4), 1019–1035, 1983.

Huyakorn, P.S., and G.F. Pinder, *Computational Methods in Subsurface Flow*, Academic Press, New York, 1983.

Huyakorn, P.S., B.G. Jones, and P.F. Anderson, Finite element algorithm for simulating three-dimensional groundwater flow and solute transport in multilayer systems, *Water Resour. Res.*, 22(3), 361–374, 1986.

Huyakorn, P.S., P.F. Andersen, J.W. Mercer, and H.O. White, Jr., Saltwater intrusion in aquifers: Development and testing of a three-dimensional finite element model, *Water Resour. Res.*, 23(2), 293–312, 1987.

Hwang, J.C., C.-J. Chen, M. Sheikhoslami, and K. Panigrahi, Finite analytic numerical solution for two-dimensional groundwater solute transport, *Water Resour. Res.*, 21(9), 1354–1360, 1985.

Hydrogeology Department of Shaanxi Province and Shandong University, Modeling of groundwater quality of Xian, Shaanxi Province, China, Research Report (in chinese), 1985.

Jennings, A.A., D.J. Kirkner, and T.L. Theis, Multicomponent equilibrium chemistry in groundwater quality models, *Water Resour. Res.*, 18(4), 1089–1096, 1982.

Jensen, K.H., K. Bitsch, and P.L. Bjerg, Large-scale dispersion experiments in a sandy aquifer in Denmark: Observed tracer movements and numerical analyses, *Water Resour. Res.*, 29(3), 673–696, 1993.

Johns, R.A., and P.V. Roberts, A solute transport model for channelized flow in a fracture, *Water Resour. Res.*, 27(8), 1797–1808, 1991.

Johnston, H.M., and R.W. Gillham, A review of selected radio-nuclide distribution coefficients of geologic materials, Atomic Energy of Canada Limited Report, TR-90, 1980.

Kapoor, V., and L.W. Gelhar, Transport in three-dimensionally heterogeneous aquifers, 1, Dynamics of concentration fluctuations, *Water Resour. Res.*, 30(6), 1775–1788, 1994.

Kapoor, V., and L.W. Gelhar, Transport in three-dimensionally heterogeneous aquifers, 2, Predictions and observations of concentration fluctuations, *Water Resour. Res.*, 30(6), 1789–1801, 1994.

Kaunas J.R., and Y.Y. Haimes, Risk management of groundwater comtamination in a multiobjective framework, *Water Resour. Res.*, 21(11), 1721–1730, 1985.

Kindred J.S., and M.A. Celia, Contaminant transport and biodegradation, 2, Conceptual model and test simulations, *Water Resour. Res.*, 25(6), 1149–1159, 1989.

Killey, R.W.D., and G.L. Moltyaner, Twin Lake tracer tests: 1, setting, methodology, and hydaulic conductivity distribution, *Water Resour. Res.*, 24(10), 1585–1612, 1988.

King, L.D., Description of soil characteristics for partially saturated flow, *Soil Sci. Soc. Amer. Proc.*, 29, 359–362, 1965.

Kirkham, D., and M.O. Stotres, Casing depths and solute travel times to well, *Water Resour. Res.*, 14(2), 237–244, 1978.

Kitanidis, P.K., and E.G. Vomvoris, A geostatistical approach to the inverse problem in groundwater modeling (steady state) and one-dimensional simulations, *Water Resour. Res.*, 19(3), 677–690, 1983.

Kitamura, S., and S. Nakagiri, Identifiability of spatially varying and constant parameters in distributed systems of parabolic type, *SIAM J. Control Optimization*, 15(5), 785–802, 1977.

Klotz, D., K.P. Seiler, H. Moser, and F. Neumaier, Dispersivity and velocity relationship from laboratory and field experiments, *J. of Hydrol.*, 45(3/4), 169–184, 1980.

Knopman, D.S., and C.I. Voss, Behavior of sensitivities in the one-dimensional advection-dispersion equation: Implications for parameter estimation and sampling design, *Water Resour. Res.*, 23(1), 253–272, 1987.

Konikow, L.F., and J.D. Bredehoeft, Modeling flow and chemical quality changes in an irrigated stream-aquifer system, *Water Resour. Res.*, 10(3), 546–562, 1974.

Konikow, L.F., Modeling chloride movement in the alluvial aquifer at the Rocky Mountain Arsenal Colorado: U.S. Geol. Survey Water Supply Paper, 2044, 1977.

Konikow, L.F., and J.D. Bredehoeft, *Computer Model of Two-Dimensional Solute Transport and Dispersion in Groundwater*, U.S. Geol. Survey Techniques of Water Resources Investigations, Book 7, C2, 1978.

Kool, J.B., J.C. Parker, M. Th. Van Genuchten, Parameter estimation for unsaturated flow and transport models—A review, *J. Hydrol., vol. 91*, 255–293, 1987.

Kool, J.B., and J.C. Parker, Analysis of the inverse problem for transient unsaturated flow, *Water Resour. Res.*, 24(6), 817–830, 1988.

Lapidus, L., and G.F. Pinder, *Numerical Solution of Partial Differential Equations in Science and Engineering*, John Wiley & Sons Inc., New York, 1982.

Leij, F.J., and J.H. Dane, Analytical solution of the one-dimensional advection equation and two- or three-dimensional dispersion equation, *Water Resour. Res.*, 26(7), 1475–1482, 1990.

Leij, F.J., T.H. Shaggs, and M.T. van Genuchten, Analytical solution for solute transport in three-dimensional semi-infinite porous media, *Water Resour. Res.*, 27(10), 2719–2733, 1991.

Leonard, B.P., A survey of finite differences of opinion on numerical modeling of the incomprehensible defective confusion equation, in *AMD vol. 34, Finite Element Methods for Convection Dominated Flows*, edited by T.J.R. Hughes, ASME, New York, 1979.

Li, S., and D. McLaughlin, A nonstationary spectral method for solving stochastic groundwater problems: Unconditional analysis, *Water Resour. Res.*, 27(7), 1589–1605, 1991.

Li, S.-G., R. Feng, and D. Mclaughlin, A space-time accurate method for solving solute transport problems, *Water Resour. Res.*, 28(9), 2297–2306, 1992.

Lindstrom, F.T., and L. Boersma, Analytical solutions for convective-dispersive transport in confined aquifers with different initial and boundary conditions, *Water Resour. Res.*, 25(2), 241–256, 1989.

Lowell, R.P., Contaminant transport in a single fracture: Periodic boundary and flow conditions, *Water Resour. Res.*, 25(5), 774–780, 1989.

Luenberger, D.G., *Linear and Non-Linear Programming*, Addision-Wesley, 1984.

Lynch, D.R., and K. O'Neill, Elastic grid deformation for moving boundary problem in two space dimensions, in *Finite Element in Water Resources, vol. 3*, edited by S.Y. Wang, C.V. Alonso, C.A. Brebbia, W.G. Gray, C.F. Pinder, University of Mississippi, 1980.

Maddock, T., III, The operation of stream-aquifer system under stochastic demands, *Water Resour. Res.*, 10(1), 1–10, 1974.

Maddock, T., III, and Y.Y. Haimes, A tax system for groundwater management, *Water Resour. Res.*, 11(1), 7–14, 1975.

Marsily, G. de., *Quantitative Hydrogeology*, Academic, San Diego, Calif., 1986.

Massmann, J., and R.A. Freeze, Groundwater contamination from waste management sites: The interaction between risk-based engineering design and regulatory policy, 1, Methodology, *Water Resour. Res.*, 23(2), 351–367, 1987a.

Massmann, J., and R.A. Freeze, Groundwater contamination from waste management sites: The interaction between risk-based engineering design and regulatory policy, 2, Result, *Water Resour. Res.*, 23(2), 368–380, 1987b.

Mclaughlin, D., and E.F. Wood, A distributed parameter approach for evaluating the accuracy of groundwater model predictions, 1, Theory, *Water Resour. Res.*, 24(7), 1037–1047, 1988.

Mclaughlin, D., and E.F. Wood, A distributed parameter approach for evaluating the accuracy of groundwater model predictions, 2, Application to groundwater flow, *Water Resour. Res.*, 24(7), 1048–1060, 1988.

Mercado, A., and E. Halevy, Determining the average porosity and permeability of a stratified aquifer with the aid of radio-active tracers, *Water Resour. Res.*, 2(2), 525–532, 1966.

Mercado, A., Nitrate and chloride pollution of aquifers—a regional study with the aid of a single cell model, *Water Resour. Res.*, 12(4), 731–747, 1976.

Mercer, J.W., S.P. Larson, and C.R. Faust, Simulation of saltwater interface motion, *Ground Water*, 18(4), 374–385, 1980.

Mercer, J.W., and C.R. Faust, Groundwater Modeling: Applications, *Ground Water*, 18(5), 486–497, 1980.

Mironenko, E.V., and Ya. A. Pachepsky, Analytical solution for chemical transport with non-equilibrium mass transfer, adsorption and biological transformation, *J. Hydrol.*, vol. 70, 167–175, 1984.

Mishra, S., and S.C. Parker, Parameter estimation for coupled unsaturated flow and transport, *Water Resour. Res.*, 25(3), 670–682, 1989.

Moltyaner, G.L., and R.W.D. Killey, Twin Lake tracer tests: 2, Logitudinal dispersion, *Water Resour. Res.*, 24(10), 1613–1627, 1988a.

Moltyaner, G.L., and R.W.D. Killey, Twin Lake tracer tests: 3, Transverse dispersion, *Water Resour. Res.*, 24(10), 1628–1637, 1988b.

Moltyaner, G.L., Advection in geologic media, *Water Resour. Res.*, 29(10), 3407–3415, 1993.

Moltyaner, G.L., M.H. Klukas, C.A. Wills, and R.W.D. Killey, Numerical simulations of Twin Lake natural-gradient tracer tests: A comparison of methods, *Water Resour. Res.*, 29(10), 3433–3452, 1993.

Muzukami, A., and T.J.R. Hughes, A Petrov-Galerkin finite element method for convection dominated flows: an accurate upwinding technique for satisfying the maximum principle, *Compt. Meths. Appl. Mech. Engrg.*, 50(2), 1985.

Moench, A.F., and A. Ogata, A numerical inversion of the Laplace Transform solution to radial dispersion in a porous medium, *Water Resour. Res.*, 17(1), 250–252, 1981.

Morel-Seytoux, H.J., G. Peters, R. Yaung, and T. Illangasekare, Groundwater modeling for management, paper presented at the international symposium on Water Resources Systems, University of Roorkee, India, 1980.

Naff, R.L., An Eulerian scheme for the second-order approximation of subsurface transport moments, *Water Resour. Res.*, 30(5), 1439–1455, 1994.

Narasimhan, T.N., Multidimensional numerical simulation of fluid flow in fractured porous media, *Water Resour. Res.*, 18(4), 1235–1247, 1982.

Nelson, R.W., In-place determination of permeability distribution for heterogeneous porous media through analysis of energy dissipation, *Soc. Pet. Eng. J.*, 8(1), 33–42, 1968.

Nelson, R.W., Evaluating the environmental consequences of groundwater contamination, I–IV, *Water Resour. Res.*, 14(3), 409–450, 1978.

Nelson, R.W., E.A. Jacobson, and W. Conbere, An approach to uncertainty assessment for fluid flow and contaminant transport modeling in heterogeneous groundwater systems, in *Advances in Transport Phenomena in Porous Media*, edited by J.

Bear and M.Y. Corapcioglu, NATO/ASI Series E, No. 128, Martinus Nijhoff Publishers, Dordrecht, 701–726, 1987.

Neuman, S.P., Calibration of distributed parameter groundwater flow models viewed as a multiple objective decision process under uncertainty, *Water Resour. Res.*, 9(4), 1006–1021, 1973.

Neuman, S.P., and S. Yakowitz, A statistical approach to the inverse problem of aquifer hydrology: 1, Theory, *Water Resour. Res.*, 15(4), 845–860, 1979.

Neuman, S.P., A statistical approach to the inverse problem of aquifer hydrology: 3, Improved solution method and added perspective, *Water Resour. Res.*, 16(2), 331–346, 1980.

Neuman, S.P., An Eulerian-Lagrangian numerical scheme for the dispersion-convection equation using conjugate space-time grids, *J. of Computational Physics*, 41(2), 270–294, 1981.

Neuman, S.P., and S. Sorek, Eulerian-Lagrangian methods for advection-dispersion, In: *Proc. of 4th Intern. Conference on Finite Element in Water Resources*, Hannover, Germany, Berlin, 1982.

Neuman, S.P., Adaptive Eulerian-Lagrangian finite element method for advection-dispersion, *Int. J. for Num. Methods in Eng.*, vol. 20, 321–337, 1984.

Neuman, S.P., C.L. Winter, and C.M. Neuman, Stochastic theory of field-scale Fickian dispersion in anisotropic porous media, *Water Resour. Res.*, 23(3), 453–466, 1987.

Neuman, S.P., and Y.-K. Zhang, A quasi-linear theory of non-Fickian subsurface dispersion, 1, Theoretical analysis with application to isotropic media, *Water Resour. Res.*, 26(5), 887–902, 1990.

Noorished, J., and M. Mchran, An upstream finite element method for solution of transient transport equation in fractured porous media, *Water Resour. Res.*, 18(3), 588–596, 1982.

Noorished, J., C.F. Tsang, P. Perrochet, and A. Musy, A perspective on the numerical solution of convection-dominated transport problems: a price to pay for the easy way out, *Water Resour. Res.*, 28(2), 551–562, 1992.

Novakowski, K.S., An evaluation of boundary conditions for one-dimensional solute transport, 1, Mathematical development, *Water Resour. Res.*, 28(9), 2399–2410, 1992.

O'Neill, K., Highly efficient, oscillation-free solution of the transport equation over long times and spaces, *Water Resour. Res.*, 17(6), 1665–1675, 1981.

Pickens, J.F., and R.W. Gillham, Finite element analysis of solute transport under hysteretic unsaturated flow conditions, *Water Resour. Res.*, 16(6), 1071–1078, 1980.

Pickens, J.F., and G.E. Grisak, Scale-dependent dispersion in a stratified granular aquifer, *Water Resour. Res.*, 17(4), 1191–1221, 1981.

Pinder, G.F., and H.H. Cooper, A numerical technique for calculating the transient position of the saltwater front, *Water Resour. Res.*, 6(3), 875–882, 1970.

Pinder, G.F., and H. Page, Finite element simulation of saltwater intrusion on the South Fork of Long Island, in *Proc. of Intern. Conf. on Finite Element Methods in Water Resources*, Princeton University, Princeton, N.J., 1976.

Pinder, G.F., and G. Gray, *Finite Element Simulation in Surface and Subsurface Hydrology*, Academic, New York, 1977.

Pinder, G.F., and Shapiro, A new collocation method for the solution of the convection-dominated transport equation, *Water Resour. Res.*, 15(5), 1177–1182, 1979.

Prickett, T.A., T.G. Naymik, and C.G. Lonnguist, A 'Random-Walk' solute transport model for selected groundwater quality evaluations, *ISWS/BUL-65/81, Bulletin 65*,

State of Illinois, Illinois Department of Energy and Natural Resources, Champaign, 1981.

Rasmuson, A., Diffusion and adsorption in particles and two-dimensional dispersion in porous medium, *Water Resour. Res.*, *17*(2), 321–328, 1981.

Rasmuson, A., T.N. Narasimhan, and I. Neretnieks, Chemical transport in a fissured rock: Verification of a numerical model, *Water Resour. Res.*, *18*(5), 1479–1492, 1982.

Rasmussen, T.D., and D.D. Evans, Fluid flow and solute transport modeling in three-dimensional networks of variably saturated discrete fractures, *Rep. NUREG/CR-5239*, U.S. Nucl. Regul. Comm., Washington, D.C., 1989.

Raviart, P.A., Particle numerical models in fluid dynamics, in *Numerical Methods for Fluid Dynamics*, edited by K.W. Morton and M.J. Baines, Oxford University Press, Oxford, 1985.

Rehfeldt, K.H., and L.W. Gelhar, Stochastic analysis of dispersion in unsteady flow in heterogeneous aquifers, *Water Resour. Res.*, *28*(8), 2085–2099, 1992.

Remson, I., and S.M. Gorelick, Management models incorporating groundwater variables, in *Operations Research in Agriculture and Water Resources*, North-Holland, Amsterdam, 1980.

Robinson, R. A., and R. H. Stokes, *Electrolyte Solutions*, Butterworth Press, London, 1965.

Rubin, Y., and G. Dagan, Stochastic identification of transmissivity and effective recharge in steady groundwater flow, 1, Theory, *Water Resour. Res.*, *23*(7), 1185–1192, 1987a.

Rubin, Y., and G. Dagan, Stochastic identification of transmissivity and effective recharge in steady groundwater flow, 2, Case study, *Water Resour. Res.*, *23*(7), 1193–1200, 1987b.

Rubin, Y., Stochastic modeling of macrodispersion in heterogeneous porous media, *Water Resour. Res.*, *26*(1), 133–141, 1990.

Rubin, Y., Transport in heterogeneous porous media: Prediction and uncertainty, *Water Resour. Res.*, *27*(7), 1723–1738, 1991.

Russell T.F., and M.F. Wheeler, Finite element and finite difference methods for continuous flows in porous media, in *The Mathematics of Reservoir Simulation, Frontiers in Applied Mathematics, 1*, Society for Industrial and Applied Mathematics, Philadelphia, 1983.

Russell, T.F., Eulerian-Lagrangian localized adjoint methods for advection-dominated problems, in *Numerial Analysis 1989, Pitman Res. Notes Math. Ser., vol. 228*, edited by D.F. Griffiths and G.A. Watson, 206–228, Longman Scientific and Technical, Harlow, England, 1990.

Russo, D., Stochastic modeling of solute flux in a heterogeneous partially saturated porous formation, *Water Resour. Res.*, *29*(6), 1731–1744, 1993.

Sagar, B., Galerkin finite element procedure for analyzing flow through random media, *Water Resour. Res.*, *14*(6), 1035–1044, 1978.

Sauty, J-P., An analysis hydrodispersive transfer in aquifers, *Water Resour. Res.*, *16*(1), 145–158, 1980.

Segol, G., G.F. Pinder, and W.G. Gray, A Galerkin-Finite Element technique for calculating the transient position of the saltwater front, *Water Resour. Res.*, *11*(2), 343–347, 1975.

Selim, H.M., and R.S. Mansell, Analytical solution of the equation for transport of reactive solute, *Water Resour. Res.*, *12*(3), 528–532, 1976.

Silvey, S.D., *Optimal Design—An Introduction to the Theory for Parameter Estimation*, Chapman and Hall, New York, 1980.

Smith, L., and F.W. Schwartz, Mass transport, 1, A stochastic analysis of macroscopic dispersion, *Water Resour. Res.*, *16*, 303–313, 1980.

Sposito, G., V.K. Gupta, and R.N. Bhattacharya, Foundational theories of solute transport in porous media: A critical review, *Adv. Water Resour.*, *2*, 56–68, 1979.

Sposito, G., W.A. Jury, and V.K. Gupta, Fundamental problems in the stochastic convection-dispersion model of solute transport in aquifers and field soils, *Water Resour. Res.*, *22*(1), 77–88, 1986.

Strecker, E.W., and W. Chu, Parameter identification of a groundwater contaminant transport model, *Ground Water*, *24*(1), 56–62, 1986.

Sudicky, E.A., and E.O. Frind, Contaminant transport in fractured porous media; analytical solutions for a system of parallel fractures, *Water Resour. Res.*, *18*(6), 1634–1642, 1982.

Sudicky, E.A., J.A. Cherry, and E.O. Frind, Hydrogeological studies of a sandy aquifer at an abandoned landfill, 4, A natural gradient dispersion test, *J. Hydrol.*, *63*, 81–108, 1983.

Sudicky, E.A., and E.O. Frind, Contaminant transport in fractured porous media: analytical solution for a two-member decay chain in a single fracture, *Water Resour. Res.*, *20*(7), 1021–1029, 1984.

Sudicky, E.A., A natural gradient experiment on solute transport in a sand aquifer: Spatial variability of hydraulic conductivity and its role in the dispersion process, *Water Resour. Res.*, *22*(13), 2069–2082, 1986.

Sudicky, E.A., The Laplace transform Galerkin technique: A time-continuous finite element theory and application to mass transport in groundwater, *Water Resour. Res.*, *25*(8), 1833–1846, 1989.

Sudicky, E.A., and Melaren, The Laplace transform Galerkin technique for large-scale simulation of mass transport in discretely fractured porous formations, *Water Resour. Res.*, *28*(2), 499–514, 1992.

Sun, N.-Z., A three-dimensional numerical technique for evaluating the mine water (in chinese), *Hydrogeology and Engineering Geology*, *13*(4), Publishing House of Geology, Beijing, 1979.

Sun, N.-Z., Mathematical Models and Numerical Methods of Groundwater Flow (in chinese), Publishing House of Geology, Beijing, 1981.

Sun, N.-Z., and W.W.-G. Yeh, A proposed upstream weight numerical method for simulating pollutant transport in groundwater, *Water Resour. Res.*, *19*(6), 1489–1500, 1983.

Sun, N.-Z., W.W.-G. Yeh, and C.C. Wang, An upstream weight finite element method for solving three-dimensional convection-dispersion equations, in *Proc. 5th Intern. Conf. on Finite Element Methods in Water Resources*, Burlington, Vermont, U.S.A., 1984.

Sun, N.-Z., and W.K. Liang, An advection control method for the solution of advection-dispersion equations, in *Development in Water Science, vol. 36*, 51–56, 1988.

Sun, N.-Z., Applications of numerical methods to simulate the movement of contaminants in groundwater, in *Environmental Health Perspectives, vol. 83*, 97–115, 1989.

Sun, N.-Z., G.-M. Qi, J. Li, M.-Q. Liu, and C. Sun, A new approach for predicting mine water inflow in the coal field with fractures and faults (in chinese), *J. of Coal Society, No. 2*, 1–10, 1989.

Sun, N.-Z., and W.W.-G. Yeh, Coupled inverse problem in groundwater modeling, 1, Sensitivity analysis and parameter identification, *Water Resour. Res.*, *26*(10), 2507–2525, 1990a.

Sun, N.-Z., and W.W.-G. Yeh, Coupled inverse problem in groundwater modeling, 2, Identifiability and experimental design, *Water Resour. Res.*, *26*(10), 2527–2540, 1990b.

Sun, N.-Z., and W.W.-G. Yeh, A stochastic inverse solution for transient groundwater flow: parameter identification and reliability analysis, *Water Resour. Res.*, *28*(12), 3269–3280, 1992.

Sun, N.-Z., *Inverse Problems in Groundwater Modeling*, Kluwer Publishers, The Netherlands, 1994.

Taigbenu, A., and J.A. Liggett, An integral solution for the diffusion-advection equation, *Water Resour. Res.*, *22*(8), 1237–1246, 1986.

Tang, D.H., and D.K. Babu, Analytical solution of a velocity dependent dispersion problem, *Water Resour. Res.*, *15*(6), 1471–1478, 1979.

Tang, D.H., and G.F. Pinder, Analysis of mass transport with uncertain physical parameters, *Water Resour. Res.*, *15*(5), 1147–1155, 1979.

Tang, D.H., E.O. Frind, and E.A. Sudicky, Contaminant transport in fractured porous media: analytical solution for a single fracture, *Water Resour. Res.*, *17*(3), 555–564, 1981.

Tang, Y., and M.M., Aral, Contaminant transport in layered porous media, 1, General solution, *Water Resour. Res.*, *28*(5), 1389–1398, 1992.

Thomson, N.R., J.F. Sykes, and W.C. Lennox, A Lagrangian porous media mass transport model, *Water Resour. Res.*, *20*(3), 389–399, 1984.

Tsang, Y.W., and C.F. Tsang, Channel model of flow through fractured media, *Water Resour. Res.*, *23*(3), 467–479, 1987.

Tsang, C.F., *Coupled Processes Associated with Nuclear Waste Repositories*, Academic Press Inc., USA, 1987.

Umari, A., R. Willis, and P.L.-F. Liu, Identification of aquifer dispersivities in two-dimensional transient groundwater contaminant transport: an optimization approach, *Water Resour. Res.*, *15*(4), 815–831, 1979.

Valocchi, A.J., Describing the transport of ion-exchanging contaminants using an effective Kd approach, *Water Resour. Res.*, *20*(4), 499–503, 1984.

Van der Heijde, P., Y. Bachmat, J. Bredehoeft, B. Andrews, D. Holtz, and S. Sebastian, Groundwater Management: The Use of Numerical Models, *Geophys. Monogr. Ser.*, vol. 5, AGU, Washington, DC, 1985.

Van Genuchten, M. Th., A closed-form equation for predicting the hydraulic conductivity of unsaturated soils, *Soil Sci. Am. J.*, *44*, 892–989, 1980.

Van Genuchten, M. Th., Analytical solutions for chemical transport with simultaneous adsorption, zero-order production and first-order decay, *J. Hydrol.*, vol. 49, 213–233, 1981.

Vomvoris, E.G., and L.W. Gelhar, Stochastic analysis of the concentration variability in a three-dimensional heterogeneous aquifer, *Water Resour. Res.*, *26*(10), 2591–2602, 1990.

Wagner, B.J., and S.M. Gorelick, A statistical methodology for estimating transport parameters: Theory and applications to one-dimensional advective-dispersive systems, *Water Resour. Res.*, *22*(8), 1303–1315, 1986.

Wagner, B.J., and S.M. Gorelick, Optimal groundwater quality management under parameter uncertainty, *Water Resour. Res.*, *23*(7), 1162–1174, 1987.

Wagner, B.J., and S.M. Gorelick, Reliable aquifer remediation in the presence of spatially variable hydraulic conductivity: From data to design, *Water Resour. Res.*, 25(10), 2211–2225, 1989.

Wang, C., N.-Z. Sun, and W.W.-G. Yeh, An upstream weight multiple cell balance finite element for solving three-dimensional convection-dispersion equations, *Water Resour. Res.*, 22(11), 1575–1589, 1986.

Willis, R., A planning model for management of groundwater quality, *Water Resour. Res.*, 15(6), 1035–1312, 1979.

Willis, R., A unified approach to regional groundwater management, in *Groundwater Hydraulics, Water Resour. Monogr. Ser.*, edited by J.S. Rosenhein and G.D. Bennett, AGU, Wasington, D.C., 1983.

Willis, R., and B.A. Finney, Optimal control of nonlinear groundwater hydraulics: Theoretical development and numerical experiments, *Water Resour. Res.*, 21(10), 1476–1482, 1985.

Woodbury, A.D., and L. Smith, Simultaneously inversion of hydrogeologic and thermal data, 2, Incorporation of thermal data, *Water Resour. Res.*, 24(3), 356–372, 1988.

Woodbury, A.D., L. Smith, and W.S. Dunbar, Simultaneously inversion of hydrogeologic and thermal data, 1, Theory and application using hydraulic heat data, *Water Resour. Res.*, 23(8), 1586–1606, 1987.

Xiang, Y., J.F. Sykes, and N.R. Thomson, A composite L_1 parameter estimator for model fitting in groundwater flow and solute transport simulation, *Water Resour. Res.*, 29(6), 1661–1674, 1993.

Xiao, S.-T., Z.-D. Huang, and C.-Z. Zhou, The infiltration problem with constant rate in partially saturated porous media, *Acta Math. Appl. Sinica*, 1, 108–126, 1984.

Yakowitz, S., and L. Duckstein, Instability in aquifer identification: Theory and case study, *Water Resour. Res.*, 16(6), 1045–1064, 1980.

Yates, S.R., Three-dimensional radial dispersion in a variable velocity flow field, *Water Resour. Res.*, 24(7), 1083–1090, 1988.

Yeh, G.T., Comparisons of successive iteration and direct methods to solve finite element equations of aquifer contaminant transport, *Water Resour. Res.*, 21(3), 272–280, 1985.

Yeh, G.T., A Lagrangian-Eulerian method with zoomable hidden fine-mesh approach to solving advection-dispersion equations, *Water Resour. Res.*, 26(6), 1133–1144, 1990.

Yeh, T.-C., L.W. Gelhar, and A.L. Gutjahr, Stochastic analysis of unsaturated flow in heterogeneous soil, 1, Statistically isotropic media, *Water Resour. Res.*, 21(2), 447–456, 1985a.

Yeh, T.-C., L.W. Gelhar, and A.L. Gutjahr, Stochastic analysis of unsaturated flow in heterogeneous soil, 2, Statistically anisotropic media with variable α, *Water Resour. Res.*, 21(2), 457–464, 1985b.

Yeh, W.W.-G., and Y.S. Yoon, Aquifer parameter identification with optimum dimension in parameterization, *Water Resour. Res.*, 17(3), 664–762, 1981.

Yeh, W.W.-G., and N.-Z. Sun, An extended identifiability in aquifer parameter identification and optimal pumping test design, *Water Resour. Res.*, 20(12), 1837–1847, 1984.

Yeh, W.W.-G., Review of parameter identification procedures in groundwater hydrology: The inverse problem, *Water Resour. Res.*, 22(2), 95–108, 1986.

Yeh, W.W.-G., L. Baker, W.-K. Liang, N.-Z. Sun, M.W. Davert, and M.-C. Jeng, Development of a multi-objective optimization model for water quality management planning in the upper Santa Ana basin, the final report to the Santa Ana Watershed Project Authority, prepared by the Civil & Environmental Engineering Department, University of California, Los Angeles, 1992.

Yeh, W. W.-G., and N.-Z. Sun, An extended identifiability approach for experimental design in groundwater modeling, in Proceedings of 10th International Conference on Computational Methods in Water Resources, edited by Peters A., G. Wittum, B. Herrling, U. Meissner, C. A. Brebbia, W. G. Gray and G. F. Pinder, vol. 2, 899–906, Kluwer Academic Publishers, 1994.

Yeh, W.W.-G., N.-Z. Sun, M.-C. Jeng, M.-J. Horng, and Y.-H. Sun, Development of an optimal data collection strategy for the EMWD groundwater basin management program, the final report to the Metropolitan Water District of Southern California (MWD) and Eastern Muncipial Water District (EMWD), prepared by the Civil & Environmental Engineering Department, University of California, Los Angeles, 1995.

Yoon, Y.S., and W.W.-G. Yeh, Parameter identification in an inhomogeneous medium with the finite element method, Soc. Pet. Eng. J., 16(4), 217–226, 1976.

Yang, J.-Z., A numerical method for the solution of convective-dispersive equations (in chinese), J. of Wuhan Hydrology and Electric Institute, No. 1, 1985.

Yang, J.-Z., Experimental and numerical studies of solute transport in two-dimensional saturated-unsaturated soil, J. of Hydrology, 97, 303–322, 1988.

Young, R.A., and T.D. Bredehoeft, Digital computer simulation for solving management problems of conjective groundwater and surface water systems, Water Resour. Res., 3(3), 533–556, 1972.

Zeitoun, D.G., and G.F. Pinder, An optimal control least squares method for solving coupled flow-transport systems, Water Resour. Res., 29(2), 217–227, 1993.

Zienkiewicz, O.C., Finite element methods in thermal problems, in Numerical Methods in Heat Transfer, edited by R.W. Lewis, K. Morgon, and O.C. Zienkiewicz, John Wiley & Sons Ltd., 1981.

Index